AF179968

153
Structure and Bonding

For further volumes:
http://www.springer.com/series/430

Aims and Scope

The series *Structure and Bonding* publishes critical reviews on topics of research concerned with chemical structure and bonding. The scope of the series spans the entire Periodic Table and addresses structure and bonding issues associated with all of the elements. It also focuses attention on new and developing areas of modern structural and theoretical chemistry such as nanostructures, molecular electronics, designed molecular solids, surfaces, metal clusters and supramolecular structures. Physical and spectroscopic techniques used to determine, examine and model structures fall within the purview of *Structure and Bonding* to the extent that the focus is on the scientific results obtained and not on specialist information concerning the techniques themselves. Issues associated with the development of bonding models and generalizations that illuminate the reactivity pathways and rates of chemical processes are also relevant

The individual volumes in the series are thematic. The goal of each volume is to give the reader, whether at a university or in industry, a comprehensive overview of an area where new insights are emerging that are of interest to a larger scientific audience. Thus each review within the volume critically surveys one aspect of that topic and places it within the context of the volume as a whole. The most significant developments of the last 5 to 10 years should be presented using selected examples to illustrate the principles discussed. A description of the physical basis of the experimental techniques that have been used to provide the primary data may also be appropriate, if it has not been covered in detail elsewhere. The coverage need not be exhaustive in data, but should rather be conceptual, concentrating on the new principles being developed that will allow the reader, who is not a specialist in the area covered, to understand the data presented. Discussion of possible future research directions in the area is welcomed.

Review articles for the individual volumes are invited by the volume editors.

In references *Structure and Bonding* is abbreviated *Struct Bond* and is cited as a journal.

D. Michael P. Mingos
Editor

Nitrosyl Complexes in Inorganic Chemistry, Biochemistry and Medicine I

With contributions by

H. Berke • Y. Jiang • H. Lewandowska • D.M.P. Mingos

Springer

Editor
D. Michael P. Mingos
Inorganic Chemistry Laboratory
University of Oxford
Oxford
United Kingdom

ISSN 0081-5993 ISSN 1616-8550 (electronic)
ISBN 978-3-662-51087-2 ISBN 978-3-642-41187-8 (eBook)
DOI 10.1007/978-3-642-41187-8
Springer Heidelberg New York Dordrecht London

Softcover reprint of the hardcover 1st edition 2014

Printed on acid-free paper

Springer is part of Springer Science+Business Media (www.springer.com)

Preface

Volumes 153 and 154 of *Structure and Bonding* have been devoted to nitrosyl complexes of the transition metals and their implications to catalysis, biochemistry and medicine. It is surprising that this topic has not been the subject of previous volumes of *Structure and Bonding* since their structural and chemical properties have attracted great interest from coordination and organometallic chemists. In the 1960s and 1970s the renaissance of coordination chemistry led to the detailed study of nitrosyl complexes and the emergence of new spectroscopic and structural techniques defined the metrics of nitric oxide when coordinated to transition metals and established that unlike CO and CN^- NO adopted alternative coordination geometries with transition metals. This ambivalence caused some interest and controversy in the inorganic community, but the research was considered to be of academic rather than practical importance. However, the discovery in the 1980s that NO played a very important role as a messenger molecule in biology provided the impetus for the widespread resurgence of interest in this molecule and its coordination chemistry. NO is produced in vivo by the nitric oxide synthase (NOS) family of enzymes and plays a key role in the nerve-signal transreduction, vasodilation, blood clotting and immune response by white blood cells. In these biological processes the coordination of nitric oxide to metal centres is crucial and therefore their detailed study is essential for an understanding of nitric oxide's functions at the molecular level. These volumes provide contemporary reviews of these important developments by leading experts in the field.

The first volume starts with an introductory chapter by myself on "Historical Introduction to Nitrosyl Complexes" recounts the discovery of nitric oxide and its complexes and serves as a general broad introduction to the two volumes. This is followed by a pair of chapters by Dr. Hanna Lewandowska on the "Coordination Chemistry of Nitrosyls and Its Biological Implications" and the "Spectroscopic Characterization of Nitrosyl Complexes." A comprehensive overview is presented of the biologically relevant coordination chemistry of nitrosyls and its biochemical consequences in the first chapter. Representative classes of metal nitrosyls are introduced along with the structural and bonding aspects that may have consequences for the biological function of these complexes. The biological targets and functions of nitrogen (II) oxide are also introduced. The second chapter reviews structural and spectroscopic data and provides descriptions of the spectroscopic characteristics of nitrosyl complexes. The results of IR, Raman, UV–Vis, EPR Mössbauer, magnetic circular dichroism, NRVS, X-ray absorption spectroscopy are reviewed and conclusions concerning the structure and reactivities of nitrosyls are summarised. The study of nitrosyl complexes has not only had implications for biology but also homogeneous catalysis and Professor Heinz Berke and Dr. Yangfeng Jiang have reviewed recent developments in this field in the chapter entitled "Nitrosyl Complexes in Homogeneous Catalysis." The ability of nitric oxide to coordinate in a flexible fashion has considerable implications for lowering the kinetic barriers of reactions of organic molecules at metal centres and Berke and Jiang give many examples of this charactristic.

Cellular actions are coordinated by sending signals to each other. This intercellular signalling is achieved by using neurotransmitters. Molecules which behave as neurotransmitters are compounds produced by neurons and stored in vesicles until stimulation of the neurons triggers their release. They bind to specific membrane receptors in a neighbouring cell to produce a physiological effect. 'Gasotransmitters' are a group of small gaseous molecules that exhibit a similar signalling function in the body but through a different mechanism. They function without receptors because they are freely permeable to cell membranes. The molecule must be produced within the body for a specific biological function. Three gasotransmitter molecules have been proposed – nitric oxide, carbon monoxide and hydrogen sulphide. They modulate cellular functions by influencing a range of intercellular signalling processes. The significance of this discovery was recognised by the award of the 1998 Nobel Prize for physiology to Murad, Furchgott and Ignarro for the discovery of the endogenous production of NO. In addition to the three accepted gasotransmitter molecules, recent reports suggest that the small gaseous sulphur dioxide molecule may also play a gasotransmitter role within the body, and other gases such as carbonyl sulphide and nitrous oxide have been suggested for investigation. Therefore the second volume highlights general electronic features of ambivalent molecules and the specific role of nitric oxide in biology and medicine.

The second volume also starts with an introductory chapter by myself on "Ambivalent Lewis Acid/Bases with Symmetry Signatures and Isolobal Analogies." This review emphasises that the nitric oxide belongs to an important class of ambivalent molecules which have the potential to act as messenger molecules in biology. The ambivalence of ligands may also have implications for understanding intermediates in nitrogen fixation processes. The subclass also encompasses ambiphilic molecules such as SO_2 and I_2. Professor Lijuan Li has contributed a chapter on "Synthesis of Nitrosyl Complexes" which reviews the synthesis of dinitrosyl complexes, particularly of iron, which are relevant to understanding their important biological role. Professors Peter Ford et al. have reviewed the photochemistry and reactivities of nitrosyl complexes in their chapter entitled "Mechanisms of NO Reactions Mediated by Biologically Relevant Metal Centres." They illustrate how understanding the basic coordination chemistry of nitric oxide is so important for understanding its biological role. They discuss the applications of both thermal and photochemical methodologies for investigating such reactions which provide quantitative data on fundamental reactions involving NO. Professor William Tolman and Deborah Salmon have reviewed "Synthetic Models of Copper Nitrosyl Species Proposed as Intermediates in Biological Denitrification" and thereby emphasise that iron is not the only metal which is important in defining the role of NO in biology. Professor Robert Scheidt and Nicolai Lehnert have contributed much to our understanding of the structures of porphyrin nitrosyl complexes over the last 40 years and the bonding of nitric oxide to transition metals and their chapter with Dr. Mathew Wolf on "Heme Nitrosyl Structures" summarises our current understanding of the geometric and electronic structures of ferrous and ferric heme-nitrosyls. In detail and in-depth correlations are made between these properties and the reactivities of these biologically important complexes. The second volume finishes with a very timely chapter on the "Medical Applications of Solid Nitrosyl Complexes" by Professors Russell Morris and Phoebe Allan. They review endogenous production and biological effects of nitric oxide before discussing the exogenous dosage of nitric oxide as a medical device. They summarise recent research work on chemical donors, e.g. polymers, porous materials, particularly zeolites and metal-organic frameworks, as delivery vessels for NO.

Those of us who are old enough to remember performing the "brown-ring test" in qualitative inorganic chemistry practical exams will realise that the subject has come a long way in the last 50 years. There is, however, still much to learn about the biological and catalytic implications of the fascinating NO molecule and I am sure that future generations will realise its potential through interdisciplinary studies.

Oxford, UK
June 2013

D. Michael P. Mingos

Contents

Struct Bond (2014) 153: 1–44
DOI: 10.1007/430_2013_116

Published online: 6 March 2014

Historical Introduction to Nitrosyl Complexes

D. Michael P. Mingos

Abstract This review provides a historical introduction to nitrosyl complexes. In the 1960s and 1970s, the renaissance of coordination chemistry led to the detailed study of nitrosyl complexes, and the emerging of new spectroscopic and structural techniques were used to define the metrics of nitric oxide when coordinated to transition metals and established that unlike CO and CN^-, NO adopted alternative geometries. The development of theoretical models to account for this geometric ambivalence was initially based on semiempirical molecular orbital calculations in the early 1970s. Subsequently more accurate and sophisticated calculations have deepened our understanding of the bonding in transition metal-nitrosyl complexes. The discovery in the 1980s of the key importance of these complexes as models for the biological functions and transformations of NO encouraged these studies. NO is produced in vivo by the nitric oxide synthase (NOS) family of enzymes and plays a key role in the nerve-signal transreduction, vasodilation, blood clotting, and immune response by white blood cells. In these biological processes, the coordination of nitric oxide to metal centres is crucial, and therefore, their detailed study is essential for an understanding of nitric oxide's functions at the molecular level.

Keywords Bent nitrosyls · Bridging nitrosyls · Crystallography · DFT calculations · Electronic structure · Intermediate nitrosyls · Linear nitrosyls · Nitric oxide · Nitrosonium · Nitrosyl · Nitroxyl · Non-innocent ligands · Notations for nitrosyl complexes · Spectroscopy

Contents

D.M.P. Mingos (✉)
Inorganic Chemistry Laboratory, Oxford University, South Parks Road, Oxford OX1 3QR, UK
e-mail: michael.mingos@seh.ox.ac.uk

Abbreviations

acac	Acetylacetonate
Ar	Aryl
Bn	Benzyl
Bpy	2,2'-Bipyridyl
Bu^t	*Tert*-butyl
cod	Cyclooctadiene
Cp	Cyclopentadienyl
cyclam	1,4,8,11-Tetraazocyclotetradecane
DFT	Density functional theory
DME	1,2-Dimethoxyethane
DMF	Dimethylformamide
DMSO	Dimethyl sulfoxide
dppe	Bis(diphenylphosphino)ethane
dppm	Bis(diphenylphosphino)methane
EAN	Effective atomic number
en	Ethylenediamine
Et	Ethyl
Me	Methyl
Mes	Mesityl, 2,4,6-trimethylphenyl
Nu	Nucleophile
Ph	Phenyl
Pr	Propyl
Pr^i	Isopropyl
py	Pyridine
THF	Tetrahydrofuran
VSEPR	Valence shell electron pair repulsion

1 Introduction

1.1 Discovery of Nitric Oxide

Nitric oxide (NO), which is a colourless gas, has been known since the thirteenth century when it was made from nitric acid [1, 2]. Its first recognisable synthetic route has been attributed to Johann Glauber (1604–1679), who observed its evolution when sulphuric acid was added to potassium nitrate. From its inception, its instantaneous conversion in the presence of oxygen to brown fumes of nitrogen dioxide (NO_2) was noted. Joseph Priestley (1733–1804) was the first chemist to characterise nitric oxide as a distinct chemical species and described its disproportionation to nitrous oxide (N_2O) and nitrogen dioxide (NO_2) when heated over iron powder. In the context of the present volume of *Structure and Bonding*, he reported for the first time that NO reacts with ferrous sulphate to give a black solution, which represented the first iron-nitric oxide complex. The black coloration formed the basis of the "brown ring test" which was used by generations of chemists as a qualitative test for nitrate ions and subsequently characterised as the complex cation $[Fe(NO)(H_2O)_5]^{2+}$. Henry Cavendish (1731–1810) and Sir Humphrey Davy (1778–1829) established the accepted chemical formula of NO by proving that it contained nitrogen and oxygen in equal proportions.

1.2 Nitric Oxide Complexes

As noted above, Priestley discovered the first transition metal complex of nitric oxide and in 1848 Lionel Playfair (1818–1898) reported red crystalline salts of $[Fe(CN)_5(NO)]^{2-}$, **1** (commonly described as nitroprusside) using the following routes based on either concentrated nitric acid or nitrite salts:

$$K_4[Fe(CN)_6] + 6\ HNO_3 \rightarrow H_2[Fe(CN)_5(NO)] + CO_2 + NH_4NO_3 + 4\ KNO_3$$

$$H_2[Fe(CN)_5NO] + Na_2CO_3 \rightarrow Na_2[Fe(CN)_5(NO)] + CO_2 + H_2O$$

$$[Fe(CN)_6]^{4-} + H_2O + NO_2^- \rightarrow [Fe(CN)_5(NO)]^{2-} + CN^- + 2\ OH^-$$

Subsequently the corresponding vanadium, chromium, manganese, and cobalt salts $[M(CN)_5(NO)]^{2-}$ anions were reported [3].

In solution sodium nitroprusside slowly releases cyanide ions, but not sufficiently rapidly, to prevent its historical use as an effective drug, which was administered intravenously, for malignant hypertension or for the rapid control of blood pressure during some surgical procedures [4–6]. The cyanide may be detoxified by thiosulphate ions, a process which is catalysed by the enzyme rhodanese. In the absence of sufficient thiosulphate, cyanide ions can quickly reach toxic levels. The half-life of nitroprusside is 1–2 min, but the metabolite thiocyanate has an excretion half-life of several days. The ability of nitroprusside to promote vasodilating effects in arterioles and venules is related to the role of nitric oxide as biochemical messenger. This application is no longer widely used in hospitals, but sodium nitroprusside is used in self-testing strips for detecting acetoacetic acid in urine samples by diabetics, because it develops a specific maroon colour when acetoacetic acid is present. Until the late 1980s, it was the accepted chemical and biochemical wisdom that NO was too toxic to have a significant biological role, but clever experimental work by Robert Furchgott, Louis Ignarro and Salvador Moncada established that NO is generated in vivo and plays an important role as a messenger molecule for controlling blood pressure, neurotransmission and the destruction of microbes [1]. It then became apparent that drugs such as nitroprusside were effective because they provided an alternative source of NO. Furthermore, it became apparent that transition metal-containing proteins and enzymes play an important role in transporting NO and moderating its toxicity. Thousands of papers have been published in the last two decades on the physiological role of nitric oxide, and this has been accompanied by a resurgence of interest in nitric oxide complexes of transition metals. This volume of *Structure and Bonding* provides a contemporary account of nitrosyl complexes, their synthesis, characterisation and chemical properties. The biological implications of these results are also developed in many of the subsequent chapters.

Since NO is toxic, biology has developed efficient detoxification pathways, which depend primarily on transition metal-containing enzymes and proteins. In mammals, the main degradation pathway of NO is based on oxyhaemoglobin and myoglobin, which results in the conversion of NO to nitrate (NO_3^-) and methaemoglobin and metmyoglobin. This reaction is also used by certain bacteria as a means of defence against nitrosative stress. In denitrifying bacteria and fungi, which are generally anaerobes, degradation of NO is achieved by reduction to N_2O, catalysed by NO reductases (NOR), which also contain heme and nonheme iron. Another potential pathway for NO degradation is disproportionation, $3NO \rightarrow NO_2 + N_2O$, which is catalysed by a number of transition metals, for example, copper and manganese [1].

Francois-Zacharie Roussin (1827–1894) discovered the first polynuclear cluster compound of nitric oxide as a result of his study of the reactions of iron(II) sulphate, ammonium sulphide and nitrous acid (HNO_2), and this black crystalline material was subsequently characterised as a salt of $[Fe_4S_3(NO)_7]^-$ (Roussin's black salt), **2**. It may be converted into a red salt on treatment with alkaline, which reacts with ethyl bromide to give Roussin's red ester, $[Fe_2(SEt)_2(NO)_4]$, **3** [7].

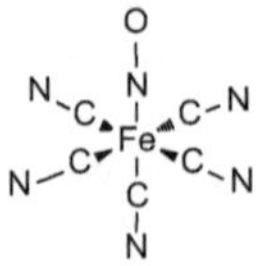

1

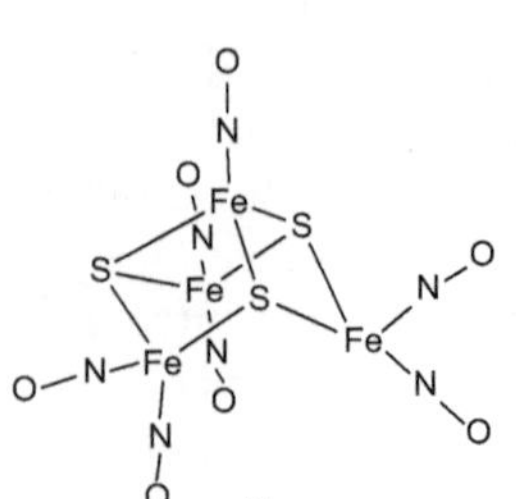

2

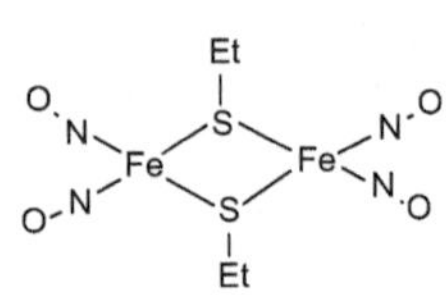

3

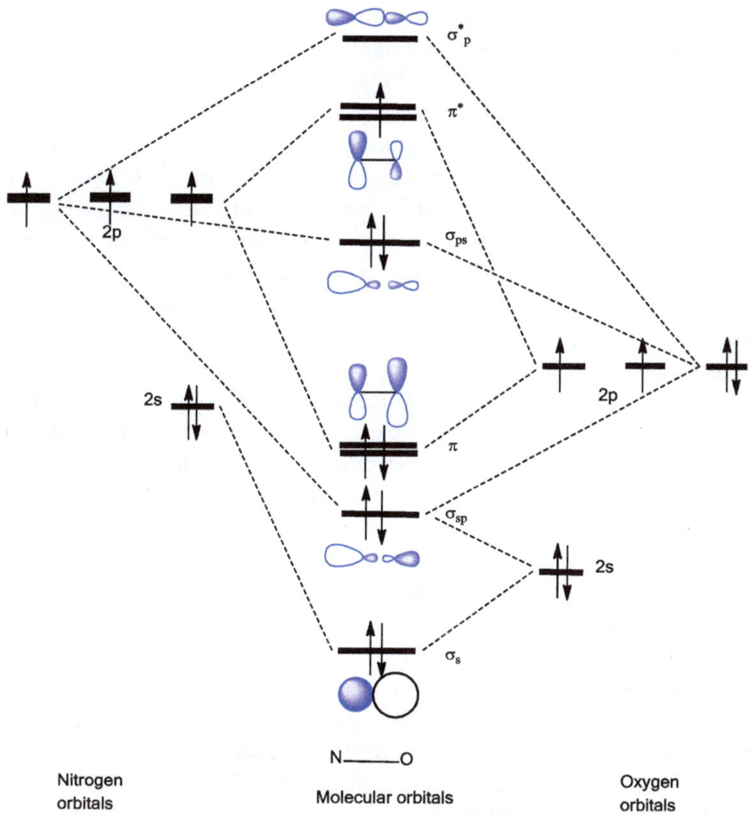

Fig. 1 Molecular orbital diagram for NO [11]

Free radicals are relatively rare in main group chemistry, although they do play an important role as reactive intermediates. There is, however, a concentration of them amongst the nitrogen oxides, and both nitric oxide and nitrogen dioxide are radicals in the gas phase. The free radical nature of nitric oxide and its relationship to neighbouring diatomic molecules and ions is conveniently rationalised by the generic molecular orbital diagram shown in Fig. 1. The odd electron in NO occupies the antibonding π^* orbital, and NO is associated with a formal bond order of 2.5. In an *aufbau* sense N_2 $((\pi^*)^0$ bond order 3), NO $((\pi^*)^1$ bond order 2.5) and O_2 $((\pi^*)^2$ bond order 2 and a triplet ground state) represent a trio of interrelated molecules. The related cation NO^+ is isoelectronic with CO and CN^-, which also form a wide range of complexes with transition metals. Since it was recognised in the 1980s that NO has an important role in biology, interest has focussed on metal-nitrosyl and metal dioxygen complexes and their interrelationships. The presence of the unpaired electron in NO may be used to rationalise the formation of weakly bonded dimers, N_2O_2, **4**, in the solid state and its ability to form a large range of transition metal complexes [7–10]. The dissociation enthalpy of the NO dimer is only 8.3 kJ mol^{-1}, and consequently it is only in the solid state at low temperatures that it is the predominant species. It is only a weakly perturbed version of the NO molecule, and the N–O bond length is essentially unchanged. The NO monomers show no tendency to form a strong N–N bond which would pair the electrons and lead to a diamagnetic ground state. The corresponding NO_2 dimer is somewhat more stable and has a dissociation enthalpy of 51 kJ mol^{-1}. This still falls below that of the corresponding dissociation enthalpy for N_2H_4, which is 156 kJ mol^{-1}. The weak bonding in the N_2O_2 dimer means that its electronic structure cannot be adequately described by conventional Hartree–Fock molecular orbital calculations, and it is necessary to incorporate electron correlation effects. The dimer is therefore only adequately, accurately represented by multi-determinant wave functions [10].

In contrast to N_2 and O_2, the π^* orbital of NO is not equally distributed and is localised 60% on nitrogen and 40% on oxygen, **5**, and this has interesting structural and chemical consequences. The geometry of N_2O_2 dimers and the occurrence of M–N–O rather than M–O–N geometries in complexes result from this asymmetry. The σ_{ps} molecular orbital of NO, shown in Fig. 1, is also more localised on nitrogen and may be viewed as a donor orbital by analogy with that for CO, although it is a much poorer donor because of the higher electronegativity of nitrogen. The π^* orbital of NO shown in Fig. 1 functions as degenerate pair of acceptor orbitals similar to those of CO, but the higher electronegativity of N vs C stabilises them and makes them stronger acceptors. Their maximum stabilisation is achieved when they overlap with a pair of donor orbitals with π symmetry and occupied by 3 electrons. Nitric oxide has an ionisation potential of 9.26 eV and an electron affinity of 0.024 eV which makes both NO^+ (isoelectronic with N_2) and NO^- (isoelectronic with O_2) chemically accessible [11]. NO undergoes typical radical reactions with the halogens (X_2) to give the diamagnetic angular XNO molecules. The related organonitroso compounds RNO have also been widely studied by organic chemists, **6**.

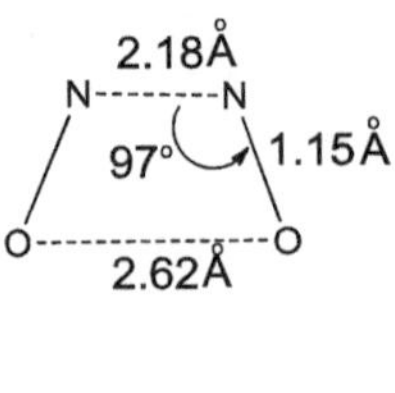

4

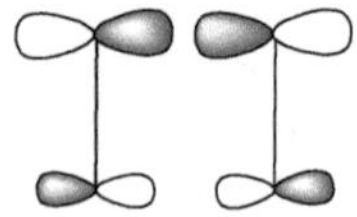

5

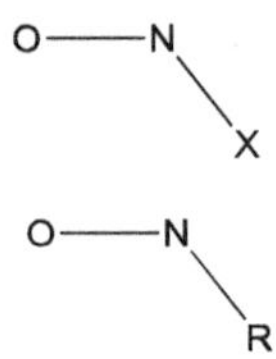

6

The nitrosonium ion has been isolated as a series of stable salts, e.g. [NO]BF_4, and is a useful synthetic oxidising agent, and indeed it has been used as an alternative to aqua regia for dissolving gold powder and a convenient source of NO for making complexes [12]. Hayton [13, 14] has used the salt to oxidise nickel and copper powders in the presence of nitromethane to form labile nitrosyl complexes of these metals, with interesting properties. For example, $[Cu(NO_2Me)_5(NO)]^{2+}$ has spectroscopic and structural properties which do not lead to a simple covalent Lewis acid/base description and indicate that the Cu ion is octahedral and has a d^9 configuration associated with Cu(II). The structure confirms a Jahn–Teller distorted elongated octahedral structure with the NO ligand coordinated in an equatorial position. The Cu–N distance is uncharacteristically long, and the Cu–N–O bond angle is 121°.

NO^+ is also a strong Lewis acid and forms adducts with donors which do not oxidise readily, e.g. Me_2SO. Formally the nitrosyl halides shown in **6** may be viewed as NO^+ adducts of the halide anions X^-. The independent chemistry of reduced nitric oxide (NO^-) is limited, and it probably has an extremely short independent life in biological media [1, 6]. NO^- is a Lewis base, which is isoelectronic with O_2 and formally forms adducts with Lewis acids such as H^+ giving angular HNO.

When Cotton and Wilkinson published their iconic textbook *Advanced Inorganic Chemistry* in 1962, nitric oxide complexes merited only 4 pages [15]. Their resumé emphasised that nitric oxide was very similar to carbon monoxide in its ability to form complexes with transition metals, but unlike CO it did not form adducts with simple acceptors or non-transition metals. It classified nitrosyl complexes in just two categories, those which formed complexes which may be regarded as derived from NO^+ and "other complexes." The former category represented those complexes which could be regarded as isoelectronic with CO and utilised a synergic bonding regime based on donation from the NO^+ lone pair to an empty metal orbital supplemented by back-bonding from filled metal d orbitals to the empty π^* orbitals of NO^+. Since the neutral NO has a single electron in the π^* orbitals, it is first necessary to formally transfer an electron to the metal atom, thereby reducing its formal oxidation state by 1.

There are very few homoleptic nitrosyl complexes, and the majority of nitrosyl complexes are associated with other ligands. If these additional ligands are π-acceptor ligands such as CO or cyclic aromatics, e.g. C_5H_5 or C_6H_6, the 18-electron rule is generally applicable. This property provides a simple account of the following series of isoelectronic and isostructural tetrahedral molecules: $Ni(CO)_4$, $Co(CO)_3(NO)$, $Fe(CO)_2(NO)_2$ and $Mn(CO)(NO)_3$ and $Cr(NO)_4$, **7**. If CO and NO are defined as 2e and 3e donors, respectively, then all the molecules in 7 conform to the 18-electron rule. In this series, the formal oxidation state of the metal decreases dramatically from Ni(0) to Cr(-4) in order to maintain a charge balance with the increasing number of NO^+ ligands. Nitrosyl complexes with spectator ligands which are not such good π-acceptors, e.g. CN^-, Cl^- and NH_3, do not obey the 18-electron rule as consistently [3, 7, 16]. For example, in the octahedral cyano-, chloro- and amine complexes $[M(NO)(CN)_5]^{2-}$ (M = V, Cr, Mn, Fe and Co) and $[M(NO)Cl_5]^{2-}$ (M = Mo, Os, Ru, Re, Tc) $[M(NO)(NH_3)_5]^{3+}$ (M = Cr, Ru, Os, Pt) [3, 16], the total electron counts vary between 15 and 20. Cotton and Wilkinson noted that in such complexes, NO was a better π-acceptor than CO and that in NO^+ complexes the NO frequency lay in the range 1,940–1,600 cm^{-1}, cf NO 1,878 and NO^+ 2,200 cm^{-1}. The limited number of X-ray structural determinations on nitrosyl complexes at that time suggested that NO^+ complexes had linear or very nearly linear M–N–O geometries, but they added the rider that $[Co(NO)(S_2CNMe_2)_2]$ which had a low NO frequency of 1,626 cm^{-1}

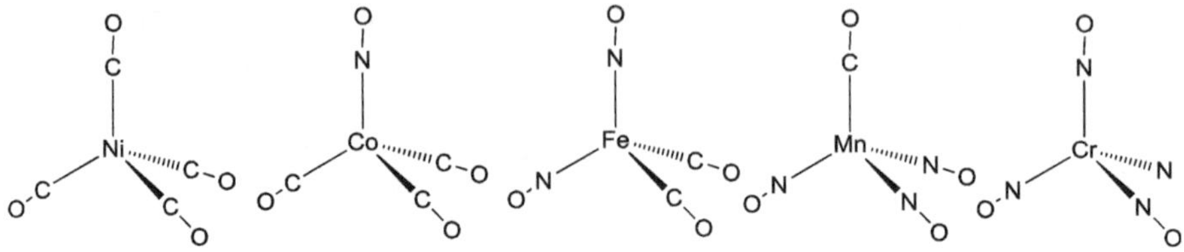

	Co	Fe	Mn	Cr
M-N Å	1.671	1.688	1.717	1.763
N-O Å	1.180	1.171	1.167	1.171

7

had a distinctly non-linear geometry and they speculated (wrongly) that this may result from a contribution of a π-bonded NO geometry. The second group of nitrosyl complexes which they identified was described very loosely as "other types of NO complexes." In this category, they noted that in the complex between heme and NO, the unpaired spin was perhaps not transferred to the metal but resided on the NO ligand. They also speculated on the possibility that NO may also act as an acceptor and thereby lead to a class of NO^- complexes. Although structural data were not available, they concluded that $[Co(NO)(NH_3)_5]^{2+}$ and $[Co(NO)(CN)_5]^{3-}$ (both of which were red in colour) may represent examples of NO^- complexes of Co(III), which would account for the very low NO stretching frequencies (1,100–1,200 cm^{-1}) in these complexes. Interestingly they also predicted that there may be complexes containing both NO^+ and NO^- ligands, e.g. $Fe(NO)_4$ and $Ni(NO)_2(PPh_3)_2$.

The research field has blossomed spectacularly in the last 50 years stimulated by the discovery of the important role played by nitric oxide as a biological signalling molecule and the realisation that nitrosyl complexes may play an important facilitating role in transporting and deactivating nitric oxide. In summary, even in the 1960s, the three essential bonding modes of nitric oxide shown in Scheme 1 (in valence bond terms) and in Scheme 2 (in molecular orbital terms) had been defined. Their geometries mirror those observed previously for NO_2^+ (O–N–O 180°; N–O 1.15 Å), NO_2 (O–N–O 134°, 1.19 Å), NO_2^- (ON–O 115°, N–O 1.24 Å) and related to changes in hybridisation at the central nitrogen and increased repulsion from the developing lone pair according to the valence bond and VSEPR theories. Within the molecular orbital framework, the Walsh diagram methodology provided an alternative interpretation based on the relative occupation of π*(O–N–O) in 16-, 17- and 18-electron NO_2^+, NO_2 and NO_2^- [17]. By analogy nitric oxide complexes maybe described as nitrosonium (NO^+) and nitroxyl (NO^-) complexes with linear and bent geometries, respectively. The radical nitric oxide may be associated with a partial lone pair and an intermediate geometry, but the odd electron may also be localised on the metal. For electron-counting purposes, the nitrosonium formulation requires the initial transfer of an electron to the metal followed by donation to the metal ion of the lone pair of NO^+, and for the nitroxyl

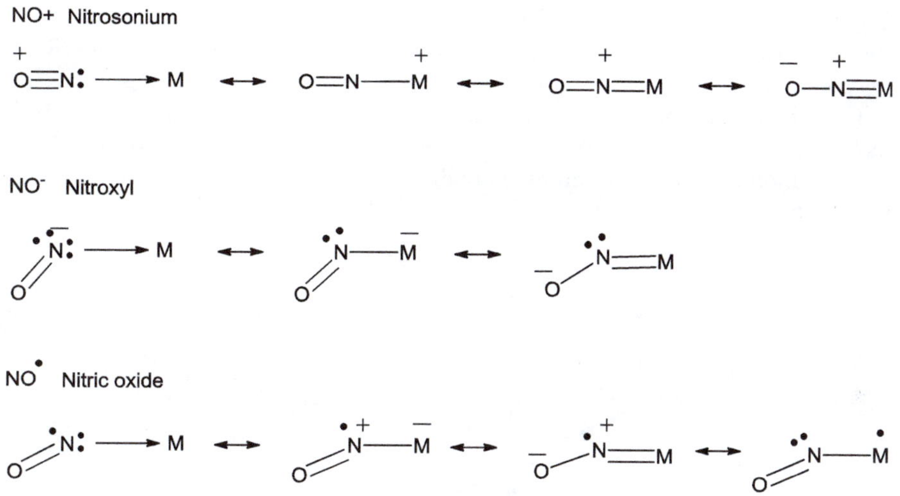

Scheme 1 Valence bond representations of M–NO bonding

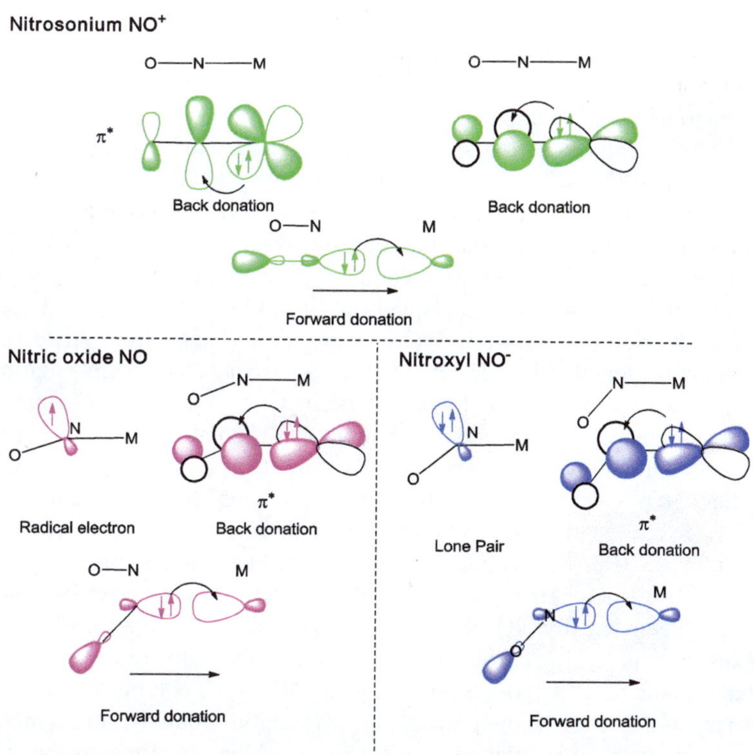

Scheme 2 Simplified MO pictures for M–NO binding in linear, intermediate and bent complexes

formulation, the initial transfer of an electron to NO followed by donation of one of the lone pairs of NO^-. In summary, *NO is a net 3-electron donor in linear complexes and 1-electron net donor in bent complexes.*

Both the valence bond and molecular orbital pictures in Schemes 1 and 2 permit back donation from filled orbitals on the metal to the empty π^* orbitals of NO and the formation of significant multiple bonding. The M–N bond is thereby strengthened at the expense of N–O bonding.

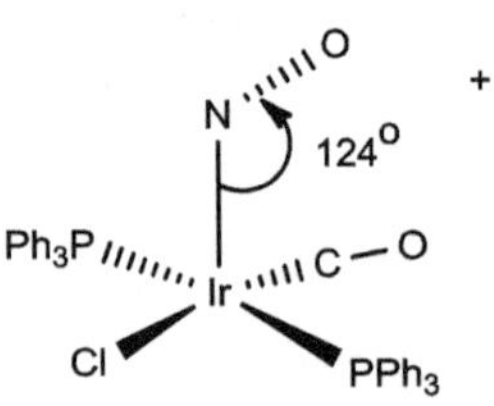

8

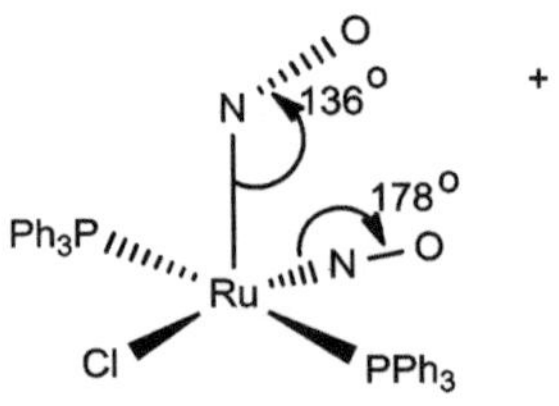

9

The first example of a structural determination of a transition metal complex with an unambiguously bent nitrosyl was $[Ir(NO)(CO)Cl(PPh_3)_2]^+$, **8**, which had an Ir–N–O bond angle of 124°, as determined by James Ibers and Derek Hodgson at Northwestern University in 1968 [18]. Cotton and Wilkinson's speculation that some transition metal complexes may contain both linear and bent nitrosyl complexes was realised in the structural determination of the isoelectronic ion $[Ru(NO)(NO)Cl(PPh_3)_2]^+$, **9**, by Eisenberg and Pierpont [19]. In this complex, the apical nitrosyl is clearly bent with Ru–N–O = 136°, and the basal nitrosyl is linear, Ru–N–O = 178°. The Ru–N distance to the equatorial linear nitrosyl (1.74(2) Å) is significantly shorter than that to the bent axial nitrosyl 1.85(2) Å, suggesting that the greater π-back donation for the equatorial ligand because of the double degeneracy of its π^* acceptor orbitals. The hybridisation change from sp to sp^2 also contributes to the lengthening. The N–O distances are not statistically, significantly different, 1.16(2) vs 1.17(2) Å. In $[Co(NO)(en)_2Cl]^+$ [20] which also has a bent nitrosyl ligand, the Co–N distance is considerably different from the average of the Co–N bonds to the ethylenediamine ligands, 1.820(11) vs 1.964(5) Å, confirming that the bent NO ligand retains significant multiple bond character because of back donation from a filled metal d orbital to the remaining empty π^* of the N–O ligand. Since NO^- is isoelectronic with O_2, which is known to form complexes with either bent or π-bonded geometries, such complexes were initially described as NO^- complexes. Electronically NO lies between N_2/CO and O_2 – the former generates complexes which are either linear or π-bonded and the latter forms either bent or π-bonded complexes. Although the majority of NO^+ complexes have linear geometries, it has been shown by X-ray diffraction studies using synchrotron radiation that metastable excited states formed photolytically may have either π-bonded or oxygen-bonded isomeric forms [21].

Scheme 1 provides not only resonance forms for complexes with NO^+ (nitrosonium) and NO^- (nitroxyl) ligands but also a neutral NO (nitric oxide) ligand. The most widely studied complexes in this category are those related to ferrous heme nitrosyls which have an obvious relationship to the biological function of these complexes. Addition of nitric oxide to iron(II) porphyrin ligands leads to either five or six coordinate nitrosyl complexes. The resultant complexes have a single unpaired electron ($S = 1/2$), and spectroscopic and structural data have been utilised to determine whether they have a low-spin d^6 Fe(II) configuration with the unpaired electron localised predominantly on the nitric oxide or whether the unpaired electron is more localised on the metal. In the five coordinate structures, the iron lies 0.2–0.3 Å out of the plane and towards the axial nitrosyl. The Fe–N–O bond angle is approximately 143°, the Fe–N distance is typically 1.73 Å and the N–O 1.17 Å. The corresponding 6-coordinate structures have similar bond lengths and angles 1.75 and 1.17 Å and an M–N–O bond angle of 137°. It is interesting to contrast these with the related ferric porphyrin complexes which have linear M–N–O geometries (Fe–N–O 174–177°, Fe–N(O) 1.64–1.67 Å, N–O 1.11–1.14 Å), and formally such complexes may be described as NO^+ complexes of Fe(II). These structures are discussed in greater detail in the chapter by Lehnert, Scheidt and Wolf [22–24]. There are other iron(II) nitrosyl complexes with macrocyclic ligands which have more complex electronic structures with intermediate spin states for iron, and antiferromagnetic coupling between the unpaired spins on the metal and the nitrosyl ligand has been proposed.

Although there has been a spectacular growth in the chemistries of transition metal-nitrosyl complexes, the corresponding compounds of lanthanides and actinides have been neglected. It was thought that the paucity of nitrosyl complexes of these elements was related to their high electropositive character and oxophilicity. In the last few years, the first examples of lanthanide and actinide nitrosyl complexes have recently been reported, and their structures are discussed in the next section. One may therefore anticipate a significant growth in this area.

2 Structural Aspects

2.1 *Structural Characterisation of Nitrosyl Complexes*

The bond length of free NO is 1.154 Å, which lies between that of a double (1.18 Å) and a triple (1.06 Å) bond and is consistent with its formal bond order of 2.5. Oxidation to NO^+ leads to an increase in formal bond order to 3 and the bond distance contracts to 1.06 Å. Reduction of NO to NO^- introduces an additional electron into π^*, a reduction in the formal bond order to 2 and an increase in bond length (1.26 Å) [25]. The bond length changes discussed above are reflected in ν(NO) which decreases from 2,377 (NO^+) to 1,875 (NO) to 1,470 cm^{-1} (NO^-). The nitrosyl halides, alkanes and arenes are "bent" molecules, **6**, the N–O distance varying from 1.13 to 1.22 Å and the X–N–O bond angle falling in the range from

Table 1 Analysis of structural determinations of metal-nitrosyl complexes [27]

Bond angle (°)	No. of structures	Description
180–170a	2,058	Linear(*l*)
170–160b	461	Linear(*l*)
160–150	76	Intermediate(*i*)
150–140	45	Intermediate(*i*)
140–130	44	Bent(*b*)
130–120	54	Bent(*b*)
120–110	15	Bent(*b*)
110–100	2	Bent(*b*)
100–90	1	π-bonded

101° to 134°; both dimensions depend on the electronegativities of the substituents. Substituent effects also influence ν(NO), which lies between 1,621 and 1,363 cm^{-1} [26].

The number of X-ray crystallographic determinations of nitrosyl complexes has increased exponentially since the 1980s when it was recognised that they may be important in understanding the biological role of nitric oxide. Table 1 summarises an analysis of the Cambridge Crystallographic Database [27]. The first point to make is that although more than 2,700 structural determinations have been determined, the vast majority of nitrosyl complexes (>90%) have bond angles between 160 and 180°. For the purpose of this review, we characterise all these complexes as *linear* (*l*) although mathematically this is imprecise. *Bent nitrosyls* (*b*) are defined as having M–N–O between 110 and 140°, and there are only 113 compounds having this geometry. There are approximately the same number of structures having bond angles between 130 and 150°, and they are described as *intermediate(i)*. The data indicate that there are 383 entries for dinitrosyls, $M(NO)_2$, but only 6 for $M(NO)_3$.

The great majority of complexes (1828) have N–O bond lengths between 1.10 and 1.20 Å, 717 structures with N–O between 1.20 and 1.30 Å and a smaller number (95) between 1.00 and 1.10 Å. Historically the presence of heavy metals in the proximity of the nitrosyl ligands limited the accuracy of X-ray determinations, but more recently low-temperature studies and more modern instrumentation have increased the accuracy of the determinations. The data suggest that the great majority have bond lengths not very different from that in the parent NO molecule and that the N–O bond length increases when the nitrosyl adopts a bent geometry, which is consistent with the simplified representations in Schemes 1 and 2.

In those complexes which have both carbonyl and nitrosyl ligands, the M–C distance is consistently shorter than the M–N distance by approximately 0.07 Å, i.e. comparable to the difference in radii, which suggests that both ligands owe their stabilities and geometries to synergic bonding effects. Photoelectron spectral studies have placed NO at the top of the following ordering of π-acceptor abilities [28, 29]:

$$NO > CO > MeCCMe > H_2CCH_2 > SO_2 > N_2 > PMe_3 > SiCl_3 > NH_3 > CN > Me > H > Cl > SiMe_3 > CCH$$

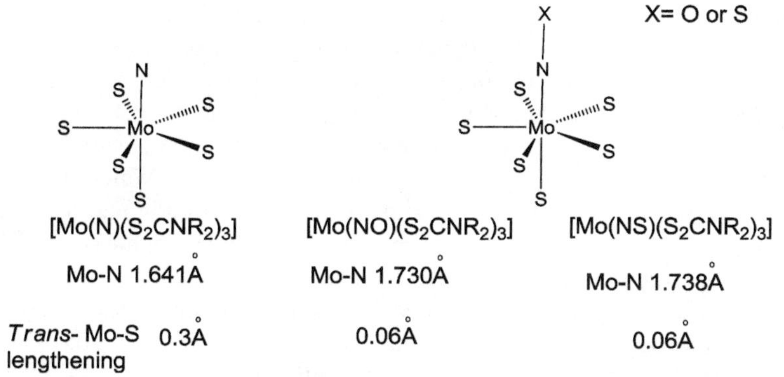

Scheme 3 Comparison of bond lengths in nitrido, nitrosyl and thionitrosyl molybdenum complexes

These bond length comparisons are consistent with the generally accepted view that the linear NO ligand is an outstanding π-acceptor, which leads to considerable shortening of the M–N(O) bond. In those low-spin d^6 18-electron octahedral complexes which have ammonia ligands *cis* to the nitrosyl ligand, the M–N(O) bond is 0.35–0.40 Å shorter than the M–N(H_3) bond length, and the ligand is very close to being linear. These complexes often show a small *trans* influence with the *trans* ligand less than 0.1 Å longer than the *cis* bonds. In some case, the metal is slightly displaced from the equatorial plane defined by the ligands. The *trans* influence and effect of the nitrosyl ligand and its catalytic implications have been discussed in some detail for organometallics by Berke and for biological systems by Lehnert [23, 30–33].

A comparison of the metal-nitrogen bond lengths in related nitrosyl and nitrido complexes (see Scheme 3) indicates that the M–N and M–NO bonds are shorter than that expected for a single Mo–N bond by at least 0.25 Å [34, 35]. Interestingly the Mo–N bond length in the nitrosyl is only about 0.09 Å longer than the related nitrido complex which has a formal triple bond, suggesting that the resonance form shown on the right-hand side makes a significant contribution for M–NO [34, 35].

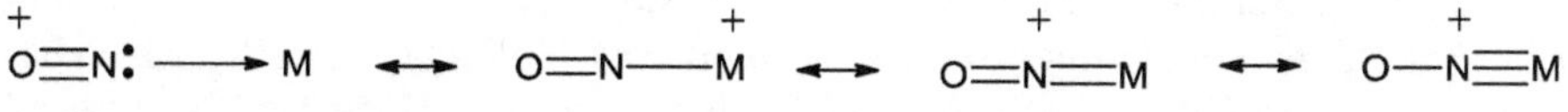

As early as 1966, Gray [36] had drawn attention to the similarity of the M–N bonding in nitrido and nitrosyl complexes, and Parkin [37] has recently reviewed this area and suggested that the covalent bond classification method introduced by Green, which assigns an X_3 valency to NO, provides a more accurate description of the metal-nitrosyl bond than the NO^+ dative bond description. The 60% localisation on nitrogen in π^*(NO) compared to 100% for nitrido reduces the overlap with the metal orbitals and reduces the multiple bond character. The seven coordinate structures illustrated in Scheme 3 indicate that the *trans* influence of the nitrido ligand is significantly larger than of the linear nitrosyl ligand and is associated with a significant movement of the metal from the equatorial plane and towards the nitrido ligand [38–40]. The bonding in thionitrosyl complexes is very similar to that described above for nitrosyls, and the similar M–N bond lengths shown above reinforce this view. There is some evidence to suggest that the thiocarbonyl and thionitrosyl ligands are better π-acceptors than the corresponding oxygen-containing ligands. In recent times, the range of complexes which have ligands analogous to NO and NS, e.g. PO, PS and AsS, has increased in number, and their structural parameters have been defined by accurate crystallographic determinations. The large *trans* influence of the nitrido ligand results in pairs of octahedral and square-pyramidal compounds with 18 and 16 valence electrons, respectively. Linear nitrosyl complexes show a similar, but less pronounced, tendency, and it has been suggested that this may have implications for their reactivities [23, 30–33].

In low-spin d^5octahedral complexes, e.g. $[Cr(NO)(NH_3)_5]^{2+}$, the depopulation of the "t_{2g}" orbitals does not change the bond lengths associated with the metal-nitrosyl moiety greatly [41]. The M–N–O bond angle is symmetry imposed at 180°, the M–N(O) bond is 0.9 Å shorter than the *cis* bonds to the amine ligands and the *trans* influence is 0.09 Å. In these complexes, the multiple bond character associated with d_{xz} and d_{yz} is retained, and the odd electron resides in the nonbonding d_{xy} orbital.

$[Ru(NO)(NO)Cl(PPh_3)_2]^+$ (see **9**) [19] is one of the few compounds which has linear and bent nitrosyls within one coordination sphere – the apical nitrosyl is clearly bent with Ru–N–O = 136° and the basal nitrosyl is linear, Ru–N–O = 178°. The Ru–N distance to the linear nitrosyl (1.74(2) Å) is significantly shorter than that to the bent axial nitrosyl (1.85(2) Å), confirming the greater π-back donation to the linear nitrosyl because of the double degeneracy of its π* acceptor orbitals. The N–O distances are not statistically, significantly different, 1.16(2) vs 1.17(2) Å. In $[Co(NO)(en)_2Cl]^+$ [20] which also has a bent nitrosyl ligand, the Co–N distance is considerably different from the average of the Co–N bonds to the ethylenediamine ligands, 1.820(11) vs 1.964(5) Å, suggesting that the bent NO ligand retains significant multiple bond character. Octahedral complexes containing a bent nitrosyl ligand show a large *trans* influence, and for nitrogen ligands, the *trans* ligand bond is at least 0.2 Å longer than the *cis* ligand bonds. The *trans* influence in bent nitrosyl complexes is larger than that in linear nitrosyl complexes, and in that respect, the bent nitrosyl resembles the nitrido ligand more closely [42–45]. It also forms related octahedral and square-pyramidal 18- and 16-electron complexes. $[Fe(NO)(py_4N)]Br_2$, which has one fewer electron than $[Co(NO)(en)_2 Cl]^+$, has an octahedral $[Fe(NO)(N_5)]$ ligand environment, and the Fe–N–O bond angle is 139° and the Fe–N(O) distance is 1.737 Å which is 0.28 Å shorter than the Fe–N (*cis*) ligands showing that significant multiple bond character is retained in this non-linear complex. The complex shows a smaller *trans* influence than the cobalt complex [46].

The origins of this *trans* effect were first interpreted within a framework of molecular orbital theory by Mingos in 1973 [43, 44]. This aspect will be discussed more fully in Sect. 3. The biological significance of the larger *trans* influence of NO compared to CO and NO was first recognised in 1976 by Perutz and co-workers [45] who used it to interpret differences in changes of the quartenary structures of haemoglobin adducts of these ligands. The chapter by Lehnert et al. discusses the biological implications of these *trans* influences in some detail [22, 23].

The remaining chapters discuss in greater detail recent developments in the structural, spectroscopic characterisation of nitrosyl complexes and the chemical and biochemical implications of the linear, intermediate and bent nitrosyl complexes. X-ray crystallography has become the routine method for structurally characterising crystalline samples of nitrosyl complexes. However, the accuracy of the structural determination may be reduced by disorder and thermal anisotropy effects. Bent nitrosyl complexes are particularly susceptible to crystallographic disorder (see **10** and **11**) although the more routine use of low temperature measurements has reduced the mean amplitudes of vibration of the atoms and assisted in the resolution of the components of the disorder more clearly [47–49]. The bent nitrosyl complexes frequently have the nitrogen atom located on the perpendicular to the spectator ligand plane, but in other cases, it may not lie on this line and this can lead to the non-axial disorder modes shown in **12** and **13** [22, 23]. If the nitrogen atom positions are close together, they are not clearly resolved and refined, and this may lead to inaccurate N–O bond lengths and M–N–O bond angles. This may be partially resolved by treating the disordered N–O units as constrained rigid groups whose centroids were sufficiently far apart to be refined within the least squares algorithms [49]. The corresponding disordered oxygen atoms are more separated and consequently more easily resolved (see **10-13**). In nitrosyl complexes where the spectator ligands are CO or N_2, crystallographic disorder involving the similar diatomic ligands may occur and limit the accuracy of the structural determinations.

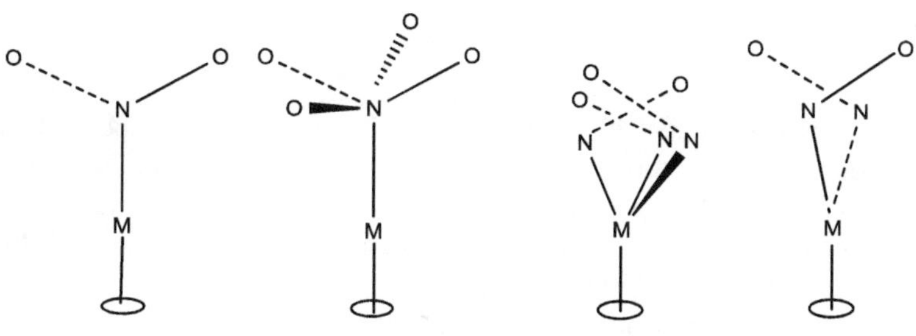

C_2 axis C_4 axis C_3 axis C_2 axis

10 11 12 13

The π-bonded nitrosyl ligand has been structurally characterised in metastable excited states of some complexes, e.g. [Ni(NO)(η-C_5Me_5)], by Coppens et al. [21, 50, 51], and they have the disorder shown in **14**. This research has taken on a special relevance with the appearance of a side-on nitrosyl in the structural determination of the native enzyme in bacterial *Alcaligenes faecalis CuNIR* [52, 53] but has not yet been duplicated in model copper complexes. This remains a synthetic challenge for future model studies [54]. Coppens [21] also structurally characterised metastable isomers where the nitrosyl is O bonded. The observation of these metastable isomers suggests that they may be important intermediates in the reactions of nitrosyl complexes and may be stabilised by specific hydrogen bonding effects in biological cavities.

In some of the earlier studies, it was not possible to distinguish disordered nitrosyls and nitro (NO_2) ligands, but the more routine use of low-temperature crystallographic studies have reduced these misleading determinations [16]. The asymmetric bonding mode shown in **15** has also been observed in some ferrous heme nitrosyl complexes and related complexes containing flat delocalised porphyrin-like ligands. A more detailed discussion of this coordination mode has been given by Scheidt and Lehnert [22, 23] who describe it as a tilt distortion which is particularly marked in the five coordinate complexes, where it is mirrored by symmetry in the equatorial Fe–N distances. In the five coordinate complexes, the Fe–N–O angles are approximately 143°, and so the nitrosyls are distinctly bent and the tilt angle α is approximately 7–8°. Similar tilt distortions have been noted in related carbonyl complexes which have linear geometries.

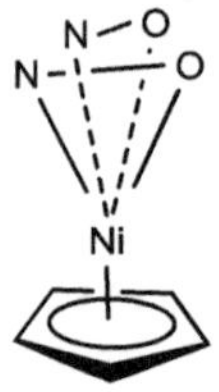

14

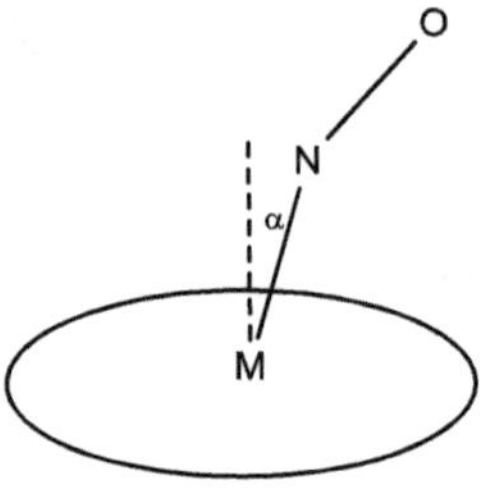

15

In complexes with several nitrosyl ligands, the linear vs bent distinction becomes more blurred, and the M–N–O bond angles vary over a considerable range. The dinitrosyl complexes of iron have been extensively studied recently because of their relevance to biological processes [55, 56]. If the nitrosyl ligands are related by a two- or threefold rotation axes, then the oxygen atoms may bend in the ONMNO plane towards or away from the other ligands as shown in Scheme 4a. The alternative distortions may be described as *repulso* and *attracto* distortions. The extent of distortion may be defined by comparing the N–M–N and O–N–O bond angles [57]. The former geometry is favoured for complexes of the later transition metals with phosphine ligands, whereas the latter is favoured for first-row transition metal complexes. For example, in $[Ir(NO)_2(PPh_3)_2]^+$ (*repulso*), the I–N–O bond angle is 164°, and the oxygen atoms bend towards the phosphine ligands, whereas in $[Fe(NO)_2(bipy)]$ (*attracto*) the oxygen atoms bend towards each other. The distortions probably have an electronic origin since the nitrosyl distortion appears to be related to the overall distortion of the complex from the idealised tetrahedral geometry. The N–Ir–N angle of 154° is considerably larger than the idealised tetrahedral angle [57]. Octahedral $[MoCl_2(NO)_2(PPh_3)_2]$ has a similar

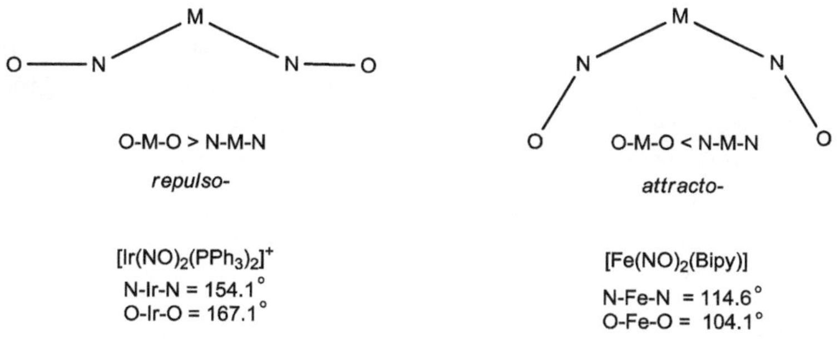

Scheme 4a *Repulso* and *attracto* distortions in polynitrosyls

intermediate geometry with Mo–N–O = 162°. These M–N–O distortions probably arise because more effective back donation results when M–N–O deviates from 180° [58]. The dinitrosyl complexes of iron (DNIC) are particularly relevant to the biological function of NO and, for example, may form as the result of the reactions of iron-sulphur clusters with NO [55, 59–61].

Octahedral and square-planar dicarbonyls and dinitrosyls have the possibility of *cis* or *trans* isomers [58, 62, 63]. Complexes which conform to the effective atomic number rule generally prefer a *cis* arrangement which maximises the back donation effects from the metal filled d orbitals to π*. However, for 17- and 16-electron octahedral complexes, the π-acceptor ligands prefer to occupy *trans* positions and thereby create an electron vacancy (or vacancies) in the nonbonding d_{xy} orbital. The carbene-stabilised nitrosyl cation $[Fe(NO)_3(IMes)]^+$ recently reported by Darensbourg and co-workers [64] represents an interesting and rare example of a tri-nitrosyl complex. The nitrosyls have essentially linear geometries Fe–N–O = 174°, and the idealised $Fe(NO)_3$ moiety has threefold symmetry and a small *attracto* geometry. There are some examples of polynitrosyls where the nitrosyls are not related by a symmetry axis and, for example, $[RuCl(NO)_2(PPh_3)_2]^+$ (**9**), whose structure was discussed in the previous section. Recently the structure of the anion $[Fe(NO)_4]^-$ has been reported and shown to have two linear and two very bent nitrosyl ligands [65]. It has been proposed that $M(NO)_3$ (M = Co, Rh and Ir) have C_{3v} pyramidal geometries [26].

Polynuclear metal-metal-bonded complexes may also be stabilised by nitric oxide in an analogous manner to carbon monoxide, and there are many examples of cluster compounds of the transition metals containing linear, doubly bridging and triply bridging nitrosyl ligands [16, 66]. There are also more unusual bridging modes, e.g. $[\{Re_3(CO)_{10}\}_2(\mu\text{-}NO)]$, **16**, the bridging NO bonded through both N and oxygen to two pairs of rhenium atoms from the two triangular Re_3 fragments [67]. There are also examples of clusters where the nitrosyl ligand bonds simultaneously through nitrogen and oxygen in a π-bonded manner to one of the metal atoms as shown in **17** and **18**.

There is one minor point of difference which is noteworthy. In terms of electron counting, carbonyl clusters have bridges which are formally associated with a metal-metal bond, whereas it is possible for nitrosyls to bridge two metal atoms without a formal metal-metal bond. For example, $[Ru_3(NO)_2(CO)_{10}]$ (**19**) has a triangular structure analogous to that observed in $[Os_3(OMe)_2(CO)_{10}]$ (**20**), which has a conventional 3-electron donor OMe bridges. In both the bridged Ru–Ru and Os–Os are significantly longer and suggest the absence of a formal metal–metal bond [68].

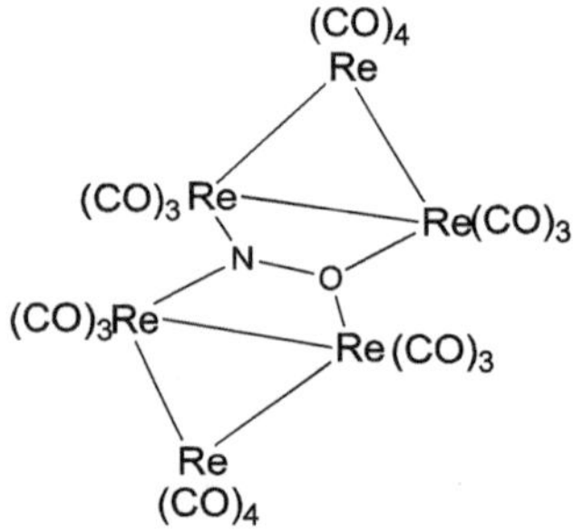

16

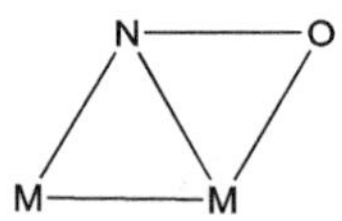

η^2, μ^2–NO

17

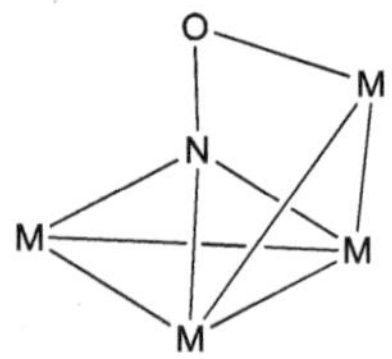

η^2, μ^4–NO

18

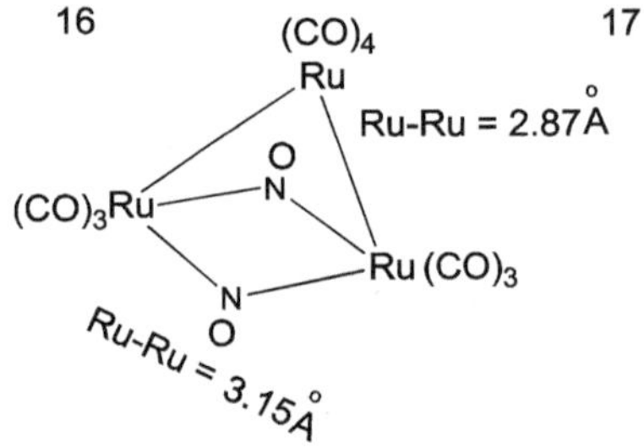

19

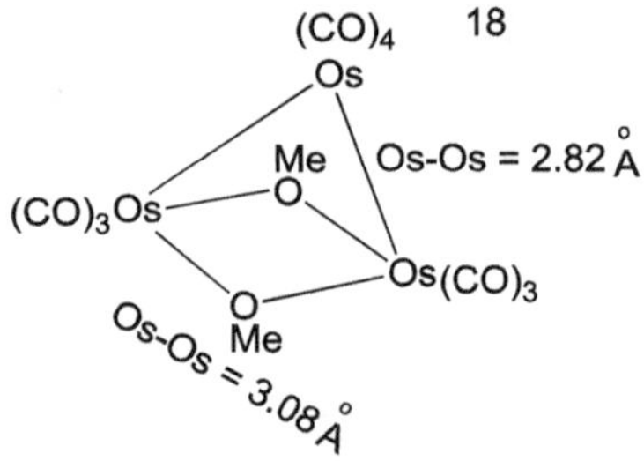

20

21 Provides a comparison of the relative bond lengths within one dimeric metal–metal-bonded molecule having a terminal and a bridging NO ligand. The shorter terminal Cr–N bond length suggests that it is associated with significantly greater multiple bond character than the bridging ligand [69]

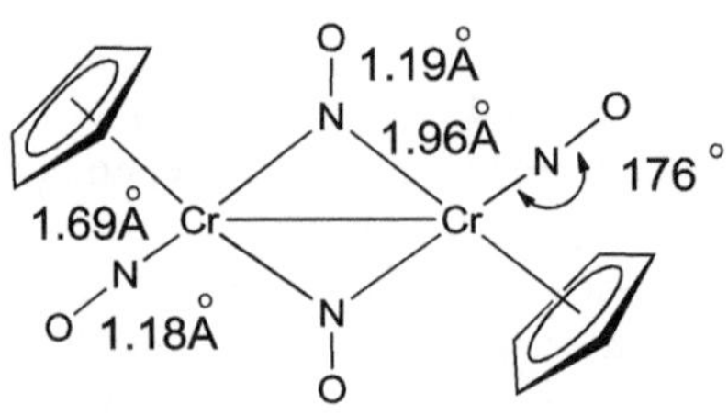

21

Recently the first examples of nitrosyl complexes of the lanthanides and actinides have been reported by Evans and co-workers, and they are illustrated in

22 and **23**. The yttrium dimer strongly resembles the previously reported compounds containing the N_2^{2-} anion sandwiched between the two Y^{3+} cations, and therefore by analogy, the nitrosyl compound is formulated as an NO^{2-} complex. This proposal has been confirmed by EPR measurements, and the bonding is thought to be mainly electrostatic. The long N–O bond is consistent with the high negative charge on the NO dianion and a formal bond order of only 1.5 according to Fig. 1. The bonding in the uranium compound is much more covalent, and the U–N bond is thought to have considerable multiple bond character. Evans et al. have proposed that the bonding in the singlet ground state is analogous to that in imido and nitrido uranium compounds (i.e. similar to that described in Scheme 3 above). However, there is very low-lying triplet excited state which has less multiple bond character and leads to temperature-independent paramagnetic properties [70, 71].

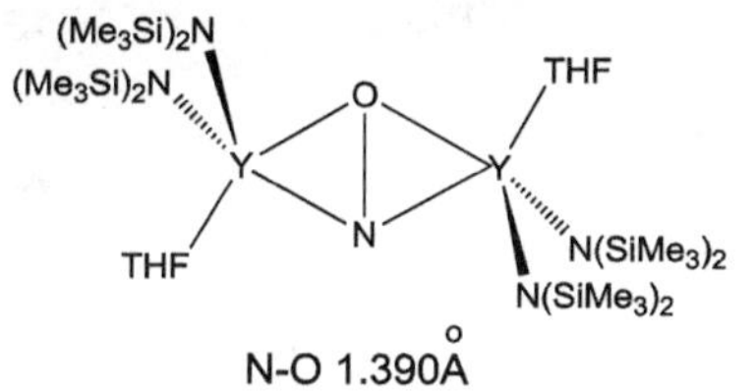

22

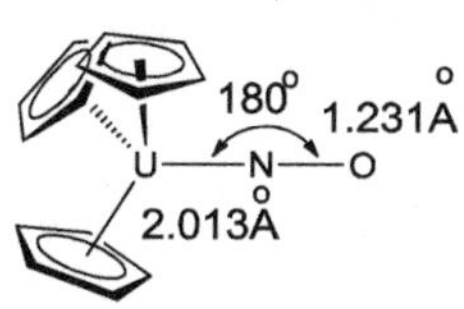

23

2.2 *Vibrational Spectroscopy*

In the gas phase, nitric oxide shows a single infrared stretching mode at 1,860 cm^{-1}, and the NO^+ ion has a stretching mode at 2,200 cm^{-1} and NO^- at 1,470 cm^{-1}. On coordination to a transition metal, the nitrosyl stretching frequency is generally observed to lie between 1,900 and 1,300 cm^{-1} [16, 63, 72, 73]. Figure 2 illustrates typical frequency ranges for transition metal-nitrosyl complexes [16]. The infrared stretching frequency *ν*(NO) is the most widely quoted spectroscopic fingerprint for metal-nitrosyl complexes, but it does not independently provide a definitive parameter for distinguishing either linear and bent nitrosyl or terminal and bridging nitrosyl ligands because the observed frequency ranges for these complexes overlap (see Fig. 2). The lowest reported frequency for a linear nitrosyl is 1,450 cm^{-1}, and this is significantly lower than those observed for many bent nitrosyl complexes, viz., 1,710–1,600 cm^{-1}. Haymore and Ibers [74] have proposed corrections which try to factor in the position of the metal in the periodic table, the metal coordination number and the electron-donating properties of the spectator ligands. This does provide a more definitive way of distinguishing linear and bent complexes, but since there are many variables, it has not been widely adopted.

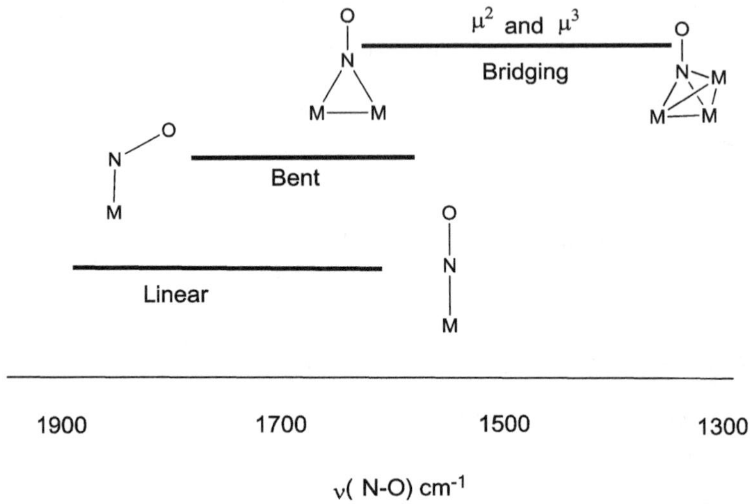

Fig. 2 Range of frequencies observed in nitrosyl complexes

Recently De La Cruz and Sheppard [26] have collected together the vibrational data for a large number of nitrosyl complexes which conform to the effective atomic number rule. They note that the paucity of binary nitrosyl compounds has made it difficult to establish suitable reference compounds, but the neutral effect of carbonyl ligands on *ν*(NO) has enabled them to use carbonyl-nitrosyl compounds as references. The wide range of *ν*(NO) for linear nitrosyl complexes have been corrected for the charge on the complex and the electron donating/withdrawing effects of the ligands. The following corrections have been proposed:

1. A positive charge on the complex is corrected by +100 cm^{-1} and a negative charge by −140 cm^{-1}.
2. The following ligand corrections in cm^{-1} are proposed: halides (+30), bridging halides (+15), PF_3 (+10), $P(OPh)_3$ (−30), PPh_3 (−15), $P(alkyl)_3$ (−70), η-C_5H_5 (−60) and η-C_5Me_5 (−80).

These lead to the following corrected ranges for linear complexes:

First-row transition metals – 1,750 (Cr)–1,840 (Ni) cm^{-1}
Second row – 1,730 (Mo)–1,845 (Pd) cm^{-1}
Third row – 1,720 (W)–1,760 (Ir) cm^{-1}

The narrower ranges when combined with isotope shifts for ^{14}N and ^{15}N (NO) have been used to distinguish linear and bent nitrosyl complexes, and it was noted that isotope shift differences are more discriminating than isotope frequency ratios. The review also analyses the data for bridging nitrosyl and analyses environmental and solvent effects. Infrared spectroscopy has proved particularly useful for identifying complexes which have structural isomers in the solid state. For example, infrared data have indicated isomeric forms of $[CoCl_2(NO)(PR_3)_2]$, and the significant frequency difference is consistent with linear and bent nitrosyl in the isomeric forms. The interconversion of linear and bent nitrosyls on the same metal has also been studied using vibrational data [75].

The biological importance of nitrosyl complexes and particularly heme nitrosyls has stimulated the development of more sophisticated spectroscopic tools for characterising nitrosyl complexes [76]. The vibrational spectra of 5C ferrous heme nitrosyls show very characteristic vibrational features associated with the Fe–N–O unit. The N–O stretching mode, ν(N–O), is observed at 1,670–1,700 cm^{-1} (by IR spectroscopy). The Fe–NO stretch, ν(Fe–NO), appears at 520–540 cm^{-1} and is most conveniently observed by nuclear resonance vibrational spectroscopy (NRVS) but is also detected quite straightforwardly with resonance Raman spectroscopy using Soret excitation. NRVS probes vibrational energy levels using synchrotron radiation and is applicable to nuclei which respond to Mössbauer spectroscopy, e.g. iron. The method exploits the high resolution offered by synchrotron light sources, which enables the resolution of vibrational fine structure and especially those vibrations which are coupled to the Fe centre(s). The in-plane Fe–N–O bend is most elusive but has been definitively assigned by NRVS to a feature around 370–390 cm^{-1}. These assignments were initially based on isotope labelling in conjunction with normal coordinate simulations of the obtained data. More recently, orientated single-crystal NRVS data for the model complex $[^{57}Fe(OEP)(NO)]$ have been reported that further confirm these assignments [77]. Excitingly, these NRVS measurements allow for the determination of the direction of iron motion within the molecular framework. These results are discussed in more detail in the chapter contributed by Lehnert, Scheidt and Wolf [22, 23, 23, 30–33].

2.3 Nitrogen NMR Spectroscopy

^{15}N NMR spectroscopy has proved to be a particularly useful technique for clarifying the coordination mode of the nitrosyl ligand in diamagnetic transition metal complexes. Both ^{14}N and ^{15}N nuclei have low sensitivities, but the latter has the advantage that $I = ½$, and this leads to sharper spectra which overcome the disadvantage of its low natural abundance. Generally it is necessary to isotopically enrich the samples in order to obtain satisfactory spectra. The low sensitivity of ^{15}N NMR spectroscopy may be improved by low temperatures, high fields and wide bore spectrometers and proton decoupling if NOE is unfavourable. The ^{14}N has a high natural abundance but is quadrupolar [16]. Mason and her co-workers [78–86] have extensively reviewed the chemical shifts of inorganic and nitrogen-containing organic molecules and noted the dominance of the second-order paramagnetic term arising from the mixing in of excited states with the ground state under the perturbing influence of the magnetic field. These second-order effects are mediated by the magnetic dipole operator, resulting from changes in orbital angular momentum associated with the rotation of electronic charge around the nucleus. Such electron density currents interact with the magnetic dipole of the nucleus, giving rise to changes in the effective field experienced by that nucleus, which are measured as changes in the shielding. For nitrogen compounds with lone pairs, the $n_N \rightarrow \pi^*$ electronic transition dominates the paramagnetic term and creates a very large chemical shift difference. For example, $[NO]BF_4$ in liquid sulphur dioxide has a chemical shift (relative to nitromethane) of −7.5 ppm, whereas the corresponding adduct $[Me_2SONO]^+$ has a upfield chemical shift of 612.6 ppm. The first nitrogen ^{14}N NMR studies on linear nitrosyls were undertaken in the 1960s, and in 1979 ^{15}N NMR studies on cyclopentadienyl nitrosyl complexes of Cr, Mo and W [87] established the chemical shifts are 49.0 to 16.5 ppm relative to nitromethane. The first ^{15}N NMR spectra of bent nitrosyls were reported by Larkworthy et al. in 1983 using 99% ^{15}N-enriched samples, and these showed the bent nitrosyls to be deshielded by several hundred ppm and up to 700 ppm (in square-pyramidal complexes of cobalt). The technique has subsequently been extended to a wide range of complexes including some with bridging nitrosyls. Mason, Larkworthy and Moore summarised [86] the relevant literature in 2002. A recent example of the usefulness of this technique for distinguishing linear and bent nitrosyls can be found in Milstein's research on four and five coordinated rhodium nitrosyls with pincer ligands [88].

Variable temperature effects have been very useful for studying the dynamics of the interconversion of linear and bent nitrosyl ligands on the same metal centre. It is also possible to measure the ^{15}N NMR spectra of metal-nitrosyl complexes in the solid-state cross-polarisation magic-angle spinning techniques (CPMAS) and in addition to the large chemical shift differences between linear and bent nitrosyls, the difference in rotational symmetry for the linear and bent complexes is reflected in the chemical shift shielding tensors. The technique may be used to define dynamic processes in the solid. In particular the swinging and spinning of the bent nitrosyl ligands in cobalt porphyrin complexes has also been studied using this technique [86].

Solution and solid-state ^{15}N NMR in combination with infrared studies of ^{15}N-enriched isotopomers have proved particularly useful for studying dynamic processes involving linear and bent nitrosyls. Solid-state NMR studies have established that the linear and bent nitrosyls in **9** do not exchange in the solid state. In solution it has been suggested that the square-pyramidal isomer with linear and bent nitrosyls is in equilibrium with a trigonal-bipyramidal isomer with essentially linear nitrosyls as shown in Scheme 4b. Indeed for the corresponding osmium compound the stable isomer has bond of 169 and 171°. Both complexes undergo a rapid fluxional process which makes the linear and bent forms equivalent on the NMR time scale (see Scheme 4b). The difference in zero-point energies between the isotopomers with ^{15}NO-linear: ^{14}NO-bent and ^{14}NO-linear:^{15}NO-bent leads to a significant equilibrium isotope effect. The $^{14,15}N$ equilibrium isotope effect for bent and linear nitrosyl is observable because of the significant difference in stretching frequencies and large difference in chemical shifts. Both complexes occur in solution as a mixture of square-pyramidal and trigonal-bipyramidal forms (73 and 27% for ruthenium and <50 and >50% for osmium). The interconversion may occur either through a Berry

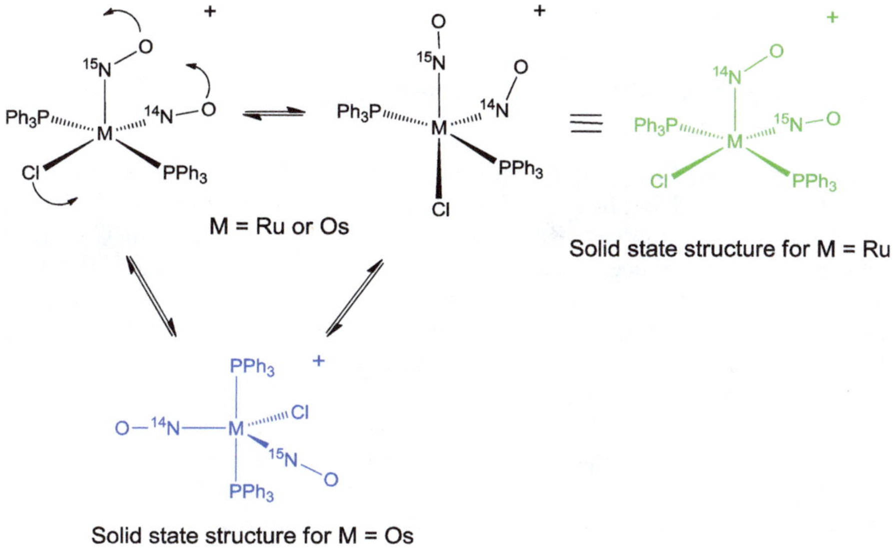

Scheme 4b The dynamic processes which have been proposed to account for the ^{15}N NMR spectra of $[MCl(NO)_2(PPh_3)_2]^+$ M = Ru or Os

pseudo-rotation involving the intermediate trigonal-bipyramidal complex or via a turnstile mechanism involving the two nitrosyls and the chloride as shown in Scheme 4b [83–85].

2.4 Other Spectroscopic Techniques

Photoelectron [89], Mössbauer [90, 91] and electron paramagnetic resonance spectroscopies [92–95] were widely used from the early days to obtain structural and bonding data on metal-nitrosyl complexes. The biological importance of these complexes has led to these techniques being used more extensively on model complexes as well as for direct studies of biologically active materials. The application of these techniques to nitrosyl complexes of the transition metals has been reviewed in some detail by Lewandowska in a subsequent chapter. A recent paper by Wieghardt et al. has demonstrated elegantly how the application of the majority of the spectroscopic and structural techniques may be used effectively to solve a particular problem. This paper also describes the application of XANES, EXAFS and UV and visible spectroscopy [96]. The biological importance of nitric oxide has also required the development of techniques for studying it in vivo and stimulated research into developing metal-based fluorescent sensors for the detection of NO in living cells. A fluorescein complex of copper described by Lippard et al. has proved to be valuable lead compound for these studies, and their results are described in a recent review [59–61].

3 Bonding in Nitrosyl Complexes

3.1 Nitric Oxide: A Non-Innocent Ligand

The assignment of formal oxidation states to the metal in complexes containing ligands such as Cl, NH_3, SCN and CN is generally uncontroversial and leads to a single designation (although Kaim has recently suggested that even CN^- may behave in a non-innocent manner in some low oxidation state solid-state compounds) [97–100]. Other ligands, e.g. bipyridyl, dithiolene, tropolone and π-bonded organic ligands, may be represented by resonance forms which result in the assignment of alternative formal metal oxidation states. For example, the dithiolene ligand may be represented either as a neutral **24** or dianionic **25** ligand, which results in formal metal oxidations states X^{n+} and $X^{(n+2)+}$. It is also possible to form a wide range of complexes with dithiolene as a radical monoanion [97, 98, 100, 101] which would result in a formal metal oxidation state of X^{n+}. In the 1960s, Jørgensen introduced the concept of *innocent and non-innocent ligands* which sought to resolve these ambiguities [101, 102]. NO and SO_2 are also non-innocent ligands, and as discussed in previous sections, this may lead to alternative X^{n+} and $X^{(n+2)+}$ oxidation states depending on the assignment of the charge on the ligand [97–100].

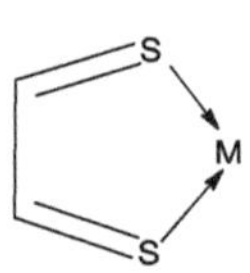

24

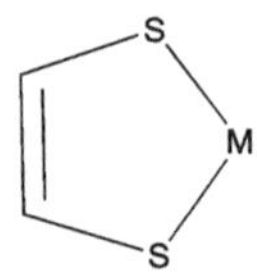

25

The oxidation state ambiguities for NO as a non-innocent ligand led Feltham and Enemark [103] to propose a widely adopted and cited notation (nearly 600 citations to date), which made no assumptions concerning the formal charge on the ligand or the metal. In this $\{MNO\}^x$ notation, the electrons occupying the π^*(NO) orbitals are added to the *x electron count* for the remainder of the complex containing the metal and the "innocent" spectator ligands. For example, $\{MNO\}^x$ for $[Mn(CN)_5NO]^{3-}$ assumes that the cyanide ligands are "innocent" and are associated with a charge of −1, each giving −5 total. To balance the fragment's overall charge, the charge on {MnNO} is +2 (i.e. −3 = −5 + 2). Since Mn^{2+} has 5d electrons and neutral NO· has one electron in π^*(NO), the total electron count x is 6, i.e. it is a $\{MNO\}^6$ complex. Since CN^- is a strong field ligand, the resultant complex would be expected to have a low-spin configuration. The advantage of this notation is its simplicity and that the *d*-electron count is the same no matter whether the nitrosyl ligand is formulated as NO^+, NO or NO^-. Wieghardt has recently commented [96] that the Enemark–Feltham description often seems to be purposefully ambiguous. When NO acts as a non-innocent ligand and antiferromagnetic coupling between spins on the metal and NO is indicated by the most accurate calculations, then it is important to know the oxidation state of the metal and NO and their spin states. He concluded that over a 15-year period, a range of bonding pictures had been proposed for many key systems of biological importance, but a consensus had not been achieved. In its initial inception, Enemark and Feltham viewed the M–N–O moiety as a strongly bonded primary fragment in the molecule and considered the spectator ligands as a relatively minor electronic perturbation. This is not necessarily the case for complexes where antiferromagnetic coupling rather than covalent bonding is indicated and many organometallic compounds where the nitrosyl and spectator ligands work together to ensure that the compounds conform to the 18- or 16-electron rules [16].

The following limitations of the Enemark–Feltham notation have become apparent over the last 40 years:

1. Enemark and Feltham noted that "it is quite misleading to describe all linear complexes as derivatives of NO^+ and all bent complexes as derivatives of NO^-," but did not provide a notation for indicating the M–N–O geometry and did not connect the notation to the 8 and 18 EAN rules.
2. From an effective atomic number perspective, the complex $[RuCl(NO)_2(PPh_3)_2]^+$ **9** is compatible with two linear three-electron nitrosyl ligands, but in the solid state, one is linear and the other bent, and in solution the two ligands exchange on the ^{15}N NMR time scale [83–85]. Enemark and Feltham's classification describes the complex as $\{M(NO)_2\}^8$ and does not provide an indicator to distinguish the bonding modes of the nitrosyls. It has also been noted that the $\{MNO\}^x$ is the same as d^x derived from the NO^+ formalism regardless of whether a nitrosyl ligand is linear, intermediate or bent.
3. Enemark and Feltham's notation deliberately does not specify the oxidation state of the complex or its coordination number. However, recent studies based on independent spectroscopic and structural data may provide compelling evidence for a specific metal oxidation state, which has implications for the metal-nitrosyl bonding. Some authors therefore see a need to communicate this information.
4. The suggestion of antiferromagnetic coupling between nitrosyl unpaired electrons and unpaired electrons on the metal undermines the fundamental assumptions of the Enemark–Feltham model since it implies weaker interactions within the MNO moiety. Furthermore, the antiferromagnetic coupling models require a definition of the oxidation states and spin states of the metal and the nitrosyl ligand.
5. Finally the Enemark–Feltham notation has not found wide applicability for polynitrosyl complexes.

Feltham and Enemark believed that the transformation of linear and bent nitrosyls represented a specific example of a more general phenomenon which they described as the *stereochemical control of valency* [103]. This term was meant to indicate that the linear-bent preferences were controlled primarily by the total number of electrons, x. As the years have gone by, the completion of nearly 3,000 structures has shown that the problem is more complex than originally envisaged. The fact that there are many complexes with intermediate M–N–O bond angles has undermined the aspiration of control over the valency, and it has become more apparent that the M–N–O angle also depends on the number of the spectator ligands, their steric and electronic characteristics and the metal's coordination geometry. The situation is certainly an order of magnitude more complex than the Walsh generalisations for 16-, 17- and 18-electron triatomic main group molecules. Past experience [66] has shown that electron-counting rules work best when the covalent bonds are strong and may be accurately represented by single Slater determinental wave functions. If the interactions are weaker, then it becomes a moot point whether an electron pair bond exists or whether the observed diamagnetism is the result of antiferromagnetic coupling by unpaired spins on the two atoms. The electron-counting rules also depend on all the valence orbitals contributing significantly to the bonding [66]. For transition metals, the radial distribution functions of the d orbitals are very different from those of $(n + 1)$s and $(n + 1)$p [104], and this not only leads to different overlap integrals but also creates higher electron-electron repulsion energies for those electrons more localised in the metal *d* orbitals. The latter makes it necessary to consider not only orbital occupations but also spin states and undermines the applicability of the *aufbau principle*.

3.2 An Alternative Notation

Since the vast majority of nitrosyl complexes conform to the 18- and 16-electron rules, it seems reasonably to focus the notation on these parameters rather than the modified *d*-electron count proposed by Enemark and Feltham. Furthermore, the more routine nature of single-crystal X-ray measurements these days and the possibility of accurately estimating the M–N–O bond angle from spectroscopic data means that this parameter can be incorporated in the notation using the short hand introduced in Table 1, i.e. 180–160° *l (linear)*, 140–160° *i (intermediate)*, 110–140° *b (bent)*.

The general form of the proposed alternative notation is:

$$[L_mM(NO)_n l/i/b]^y$$

where $n + m$ = coordination number of complex including the nitrosyls and y = EAN count with NO donating 3 electrons for *l* and 1 electron for *b* and either 3, 2 or 1 for *i (intermediate)*. The electron-donating capabilities for linear, intermediate and bent nitrosyl complexes follow directly from the valence bond and molecular orbital depictions of bonding in Schemes 1 and 2 given above. For nitrosyl complexes with M–N–O angles between 140 and 160° (intermediate *i*)), it is necessary to bring other factors into consideration – these depend on additional spectroscopic, magnetic and structural data before making a final assignment. This point is best illustrated by reference to some specific examples.

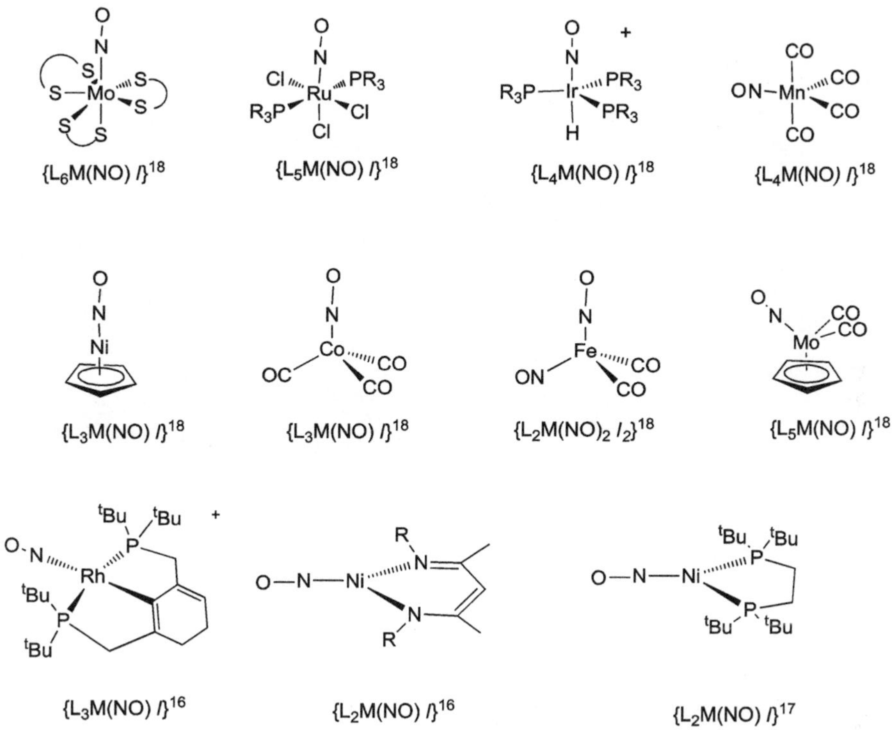

Fig. 3 EAN notation for complexes with linear nitrosyls

Examples. Figure 3 illustrates the application of the new notation to a range of coordination compounds, which conform to the 18- or 16-electron EAN rule and have linear nitrosyls. This geometry defines NO as a 3-electron donor according to Schemes 1 and 2. Figure 3 provides specific examples of 7, 6, 5 and 4 coordinate complexes which conform to the 18-electron rule and organometallic complexes with π-bonded carbocyclic ligands. For cyclopentadienyl complexes, the ligand is assumed to occupy three coordination sites in defining the number of spectator ligands. The notation is readily adapted to polynitrosyls, and the iron dinitrosyl complex shown in Fig. 3 provides a specific example of its use. The figure provides an example of a 16-electron square-planar complex (a common coordination geometry for d^8 complexes) with a linear nitrosyl [88]. The figure also illustrates examples of 16- and 17-electron trigonal nickel nitrosyl complexes. The latter departs from the usual 16-electron count for a trigonal planar d^{10} complex, and the localisation of the additional electron has to be defined by spectroscopic data or density functional calculations. The notation is applicable to $[M(NO)(CN)_5]^{x-}$ which have octahedral geometries and filled or partially filled t_{2g} orbitals and according to the new notation are described as $\{L_5M(NO)\ l\}^{18\text{-}14}$.

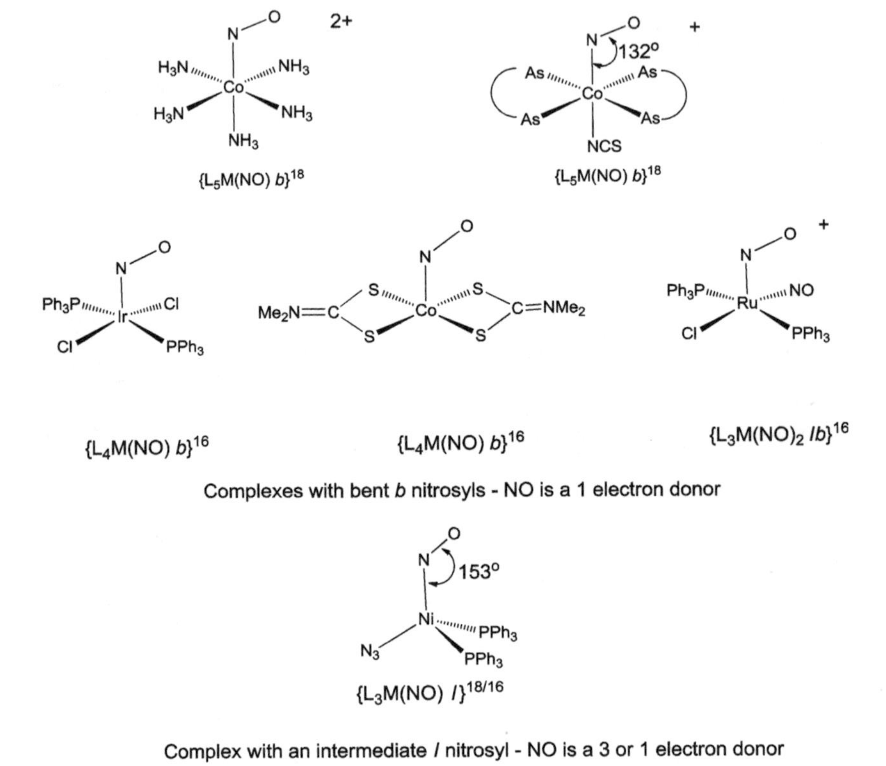

Fig. 4 EAN notation for complexes with bent nitrosyls

Figure 4 provides examples of complexes with bent nitrosyls which fall into either 18- or 16-electron EAN categories and have either octahedral (18-electron) or square-pyramidal (16-electron) geometries. Following the depictions in Schemes 1 and 2, the bent nitrosyl is defined as a 1-electron donor in these examples. The last example in the figure shows a tetrahedral nickel complex which has a 153° Ni–N–O bond angle and therefore falls into the intermediate *i* category. As noted above, this ambiguous category requires additional considerations in order to decide whether the NO is acting as a 1-, 2- or 3-electron donor which is reflected in the two alternative electron counts shown. The 17-electron count is excluded by the diamagnetism of the complex. The structures of several related tetrahedral nickel complexes with alternative phosphines have been studied and shown to have linear geometries, and therefore, it may be classified as an $\{LM(NO)\ i\}^{18}$ complex. [37] When the complex contains both linear and bent nitrosyls, then this may be indicated using both *l* and *b* as shown for the ruthenium complex in Fig. 4. **26** and **27** illustrate the structures of two closely related osmium complexes which differ by one having Cl and the other OH as innocent spectator ligands. In the Enemark–Feltham notation, they are both

$\{M(NO)_2\}^8$ complexes, but in the new notation, the different nitrosyl geometries are indicated and the resultant electron count differentiates them [83–85].

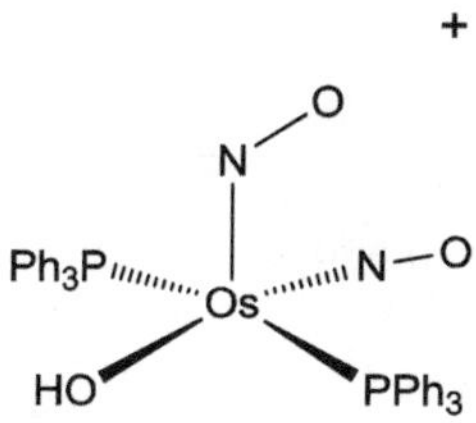

$\{L_3Os(NO)_2\ l,b\}^{16}$

26

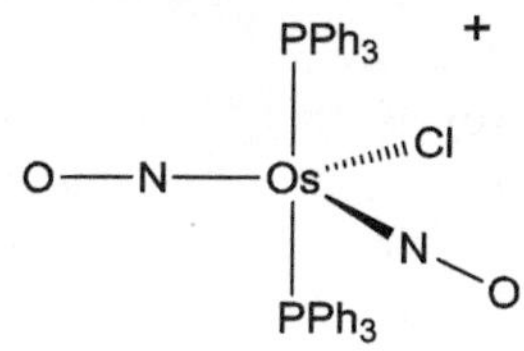

$\{L_3Os(NO)_2\ l,l\}^{18}$

27

In common with Enemark–Feltham, the new notation makes no attempt to define the formal charges on the nitrosyl ligand and the formal metal oxidation state but focuses attention on the geometry of the nitrosyl, the metal's coordination number and the total electron count. As De La Cruz and Sheppard have recently pointed out [26] in their extensive analysis of the vibrational data for nitrosyl complexes, the great majority of them conform to 18- and 16-electron rules, and therefore, this parameter establishes whether the molecule has a closed shell. The total electron count has important chemical implications since it indicates whether the compound is likely to undergo electrochemical conversion or nucleophilic addition in order to achieve an 18-electron configuration.

In view of the many disputes which have arisen in the literature regarding the assignment of a formal charge to NO and the formal oxidation state or valency of the metal, it is preferable to have a notation which is not dogmatic and which specifies only the total number of valence electrons. There is ample structural, spectroscopic and theoretical evidence that the M–N–O bonding is a highly delocalised multiply bonded unit which conforms to the electroneutrality principle and the charges on M, N and O do not diverge greatly from zero. Section 3.3 gives some specific data from DFT calculations which underline this view.

If, and only if, there is "compelling" spectroscopic or theoretical evidence that the metal has a clearly defined oxidation state, then this information may be incorporated in the notation, e.g. $\{L_5Ru^{II}(NO^+)\ l\}^{18}$ $\{L_5Co^{III}(NO^-)\ b\}^{18}$ (two examples taken from Figs. 3 and 4), and this addition confirms that these two complexes have the low-spin d^6 configurations common for octahedral complexes. Similarly the extended notation for the rhodium complex with a pincer ligand at the bottom of Fig. 3 is $\{L_3Rh^I(NO)\ l\}^{16}$, and this immediately suggests a d^8 square-planar geometry. However, this addition should be made in the full knowledge that it represents a formalism and not a rigorous statement regarding the actual charge distribution in the complex.

For complexes where metal oxidation states and spin states have been designated on the basis of reliable theoretical or spectroscopic data and it is felt necessary to indicate that the bonding is best represented by antiferromagnetic coupling, then this may be indicted in the following manner [73, 105, 106]:

trans-[Fe(NO)(cyclam)Cl]$^+$ is designated as {$L_5Fe^{III}(NO^-)$ *l* ($S = 1/2$}^{19}, suggesting an intermediate spin Fe(III) centre ($S = 3/2$) antiferromagnetically coupled to NO^- ($S = 1$).
trans-[Fe(NO)(cyclam)Cl]$^{2+}$ is designated as {$L_5Fe^{IV}(NO^-)$ *l* ($S = 1$}^{18}, suggesting Fe(IV) ($S = 4/2$) coupled antiferromagnetically to NO^- ($S = 1$).
trans-[Fe(NO)(cyclam)Cl] is designated as {$L_5Fe^{III}(NO^{2-})$ *b* ($S = 0$}^{18}, suggesting a low-spin Fe(III) ($S = 1/2$) antiferromagnetically coupled to what is formally described as NO^{2-} ($S = 1/2$).

The number of unpaired electrons in the complex, determined by magnetic measurements, are indicated by $S = 1/2, 1, 3/2,\ldots$ and a multiplicity of $2S + 1$.

Figure 5 shows some five coordinate complexes of iron and cobalt which would have been described as {M(NO)}^{6,7,8} according to the Enemark–Feltham notation and have M–N–O bond angles which are classified as linear, intermediate and bent according to the data in Table 1. The [Fe(NO)(oep)]$^+$ complex has a linear MNO geometry and is classified as a 16-electron square-pyramidal complex. The related complex [Co(NO)(oep)] has a bent MNO (123°) and the NO donates a single electron since an electron pair is localised on N, and this results in an EAN count of 16. The intermediate [Fe(NO)(oep)] complex has a intermediate (143°) geometry, and the NO may function as a 3-, 2- or 1-electron donor. The preferred designation will depend on other chemical considerations and supplementary spectroscopic data [107, 108]. For example, if ESR data suggests that the unpaired electron, which is consistent with $S = 1/2$ ground state, is localised on the nitrogen, then NO effectively functions as a 2-electron donor and the appropriate notation is {$L_4Fe^{II}(NO^{\cdot})$ $S = 1/2$ *i*}^{16}. In Fig. 4, the odd electron is indicated as a partial lone pair, and this is reflected in the large Fe–N–O angle. This assignment is consistent with the simplified bonding models presented in Schemes 1 and 2 for the nitric oxide radical. This series of OEP complexes are therefore all classified as 16-electron complexes, and the additional electrons are located primarily on nitrogen. The figure also illustrates a pair of dithiocarbamate complexes of iron and cobalt which have very different MNO bond angles which are reflected in their EAN notations. In this instance, the localisation of a partial lone pair on nitrogen is less compelling because the angle is very close to linearity and the odd electron is probably more localised on the metal. The iron *salen* complexes [109] shown on the right-hand side have been isolated in two isomeric forms with $S = 3/2$ and $S = 1/2$. The $S = 3/2$ isomer has MNO = 147° and would be described as intermediate, and the $S = 1/2$ isomer has MNO = 127° and is clearly bent. Their EAN notations are {$L_4Fe^{III}(NO^-)$ $S = 3/2$ *i*}^{17} and {$L_4Fe^{II}(NO^{\cdot})$ $S = 1/2$ *b*}^{15}, respectively [110].

The Feltham and Enemark and the alternative notation described above is most reliable when the ground state of the resultant nitrosyl complex is adequately described by a single determinant wave function but will show limitations when the wave function is no longer amenable to a simple Hartree–Fock analysis and electron correlation effects have to be incorporated. For transition metal complexes

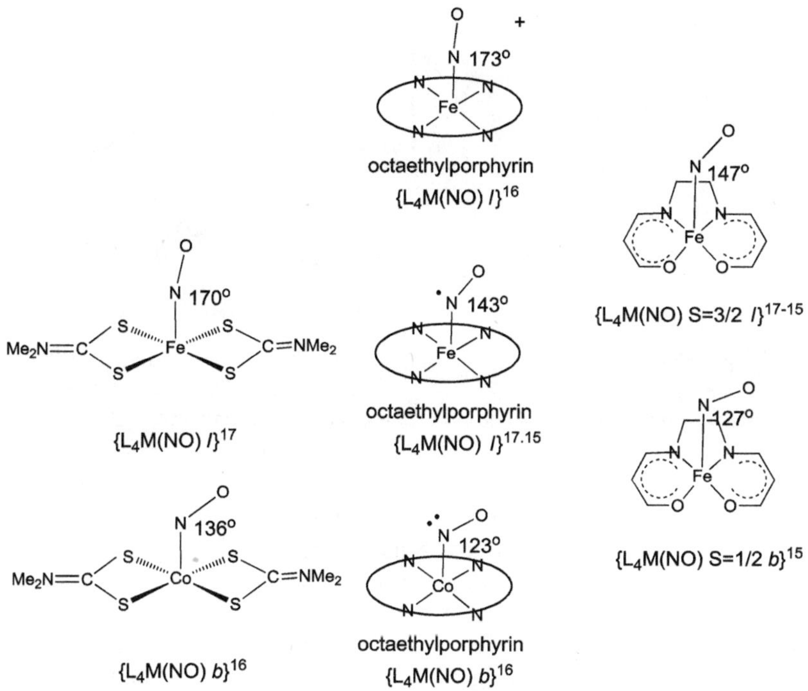

Fig. 5 EAN notation for complexes with intermediate and bent nitrosyls [73]

with open-shell configurations, the high electron–electron repulsion energies associated with d electrons means that electron correlation effects are significant and have to be built into the multi-determinant wave function. In addition, the large exchange effects associated with the d electrons favour configurations with electron pairs with parallel spins. Antiferromagnetic interactions between unpaired electrons on the NO ligand and the transition metal *d* electrons replace simple covalent bond pair descriptions for many complexes of biological relevance. One usually associates antiferromagnetic interactions as rather weak; however, this prejudice need not be transferrable to wave functions used to correct for correlation and exchange energy effects. Currently density functional molecular orbital calculations are extensively used to throw light on these bonding problems, and many examples of this methodology are described in the subsequent chapters of this pair of volumes. Historically less accurate molecular orbital calculations were used initially to define the geometric and spectroscopic properties of nitrosyl complexes, and these results are discussed in the next section.

Wieghardt et al. [96] have recently reported a very detailed and thorough theoretical analysis (DFT and TD-DFT) of the Tp^* complexes of cobalt, copper and nickel shown in Fig. 6. The predominant wave function for the diamagnetic 18-electron nickel complex is based on a Ni(II)/NO^- valence bond structure rather than the traditional Ni(0)/NO^+ or Ni(IV)/NO^{3-} proposed by Harding et al. [111]. The

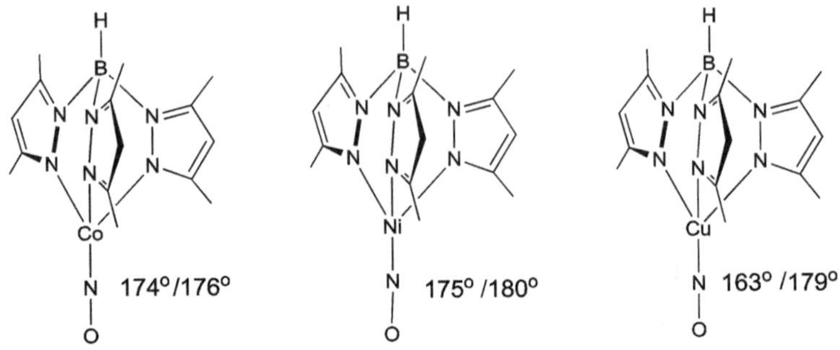

$\{L_3Co(NO)l\}^{17}$ $\{L_3Ni(NO)l\}^{18}$ $\{L_3Cu(NO)l\}^{19}$

Fig. 6 EAN notation for some cobalt, nickel and copper nitrosyl complexes

ground state is best described as an antiferromagnetically coupled $S = 1$ Ni(II) centre coupled to a $S = 1$ NO^- anion leading to $S = 0$ for the complex. The 17-electron cobalt complex is similarly described as antiferromagnetically coupled $S = 3/2$ Co (II) ion coupled to $S = 1$ NO^-. The interaction between $d_{xz,yz}$ and π^*(NO) has low ionicity, and the Co and NO orbitals contribute almost equally to bonding and antibonding combinations, but the energy separation is surprisingly small at ca. 8 kcal mol^{-1}. This results in significant contributions to the final wave functions from singly and doubly excited states resulting from population of the antibonding $d_{xz,yz} - \pi^*$(NO) orbitals. The electron hole in the complex is associated with the d_{xy} orbital of the Co(II) manifold. The 19-electron copper complex is described as $S = 0$ Cu (I) antiferromagnetically coupled to $NO^{\cdot}$ ($S = 1/2$), and the additional electron is thought to reside in the antibonding $d_{xz,yz} - \pi^*$(NO) orbital, which undergoes a Jahn-Teller distortion involving a slight bending of the nitrosyl ligand. The multiconfigurational nature of the wave functions for this series of complexes is an important general characteristic, and the ground state contributes only 80% in the copper complex, 52% in the nickel complex and 75% in the cobalt complex. In the proposed notation, these complexes may be described as

$$\{L_3Co^{II}(NO^-)l(S = 1/2)\}^{17}\{L_3Ni^{II}(NO^-)l(S = 0)\}^{18}\{L_3Cu^{I}(NO^{\cdot})l(S = 1/2)\}^{19}$$

This description incorporates the essential bonding information derived from the extensive calculations and spectroscopic analyses at least as the important contributors to the ground state.

The antiferromagnetic description of the bonding in metal-nitrosyl has become more common in recent years and is a notable feature of DFT calculations on biomimetic complexes of the first-row transition metals and model complexes which closely resemble biological active sites. It does, however, carry some contradictions which in my opinion have not been completely resolved. On the one hand, there is ample structural evidence that the metal-nitrosyl bonding is highly covalent and the multiple bonding is strong. Indeed bond length comparisons discussed above suggest that the multiple bonding between M and NO, approaches that in metal nitrido complexes. The satisfactory modelling of the electron correlation and exchange effects in DFT calculations reproduce these short bond lengths but present a picture which requires the bonding to be described in terms of antiferromagnetic interactions, which one usually associates with weak bonding and energy separations between π and π^* which are calculated to be less than 10 kcal mol^{-1}. The adequate representation of the correlation and exchange effects leads to the polarisation of one of the spins on the metal and the other on the nitrosyl. This may distort the usual view of bonding, and in my opinion, the circle has not been squared and a contradiction remains.

3.3 *Initial Detailed Molecular Orbital Calculations*

The first semiempirical molecular orbital calculations on nitrosyl complexes were reported by Harry Gray and Dick Fenske and their co-workers in the 1960s [112, 113]. The calculations were used to interpret the spectroscopic properties of first-row transition metal-nitrosyl pentacyano complexes. The calculations confirmed that NO was superior to CO and CN^- as a π-acceptor and noted that the antibonding π*(NO) orbitals had similar energies to the manifold of *d* orbitals [112, 113]. When the first examples of bent nitrosyl complexes were reported, some qualitative symmetry analyses were proposed [114–118] which attempted to rationalise the linear to bent transformation in nitric oxide complexes and the bent geometry in dioxygen complexes [116]. The absence of bent carbon monoxide and dinitrogen complexes provided an interesting contrast. These analyses emphasised the proximities of the π*(NO) and metal d orbitals and drew attention to the symmetry implications of the Walsh type analysis [115]. The first semiempirical molecular orbital analysis of the problem was reported by Mingos [43, 44] in 1973, and in 1974 Hoffmann, Mingos and their co-workers [119] analysed the bonding in five coordinate metal-nitrosyl complexes, which at that time were unique in their ability to form both linear and bent nitrosyl complexes. In the Walsh diagram analysis illustrated in Fig. 7 for octahedral $[Co(NO)(NH_3)_5]^{2+}$, the d_{z^2} and $d_{x^2-y^2}$ (e_g) components of the parent octahedron are readily identifiable and they are strongly metal-ligand antibonding, and the t_{2g} orbitals lose their degeneracy because of the strong π interactions between π*(NO) and d_{xz}, d_{yz}. The antibonding component of this interaction lies between the e_g and t_{2g} orbitals of the parent octahedron. For low-spin $\{L_5M(NO)\}^{12\text{-}18}$ complexes (based on the formalism described above), the linear M–N–O geometry is favoured because the $e(d_{xz,yz} + \pi^*(NO))$ bonding orbitals become less bonding as M–N–O decreases. Specifically the π overlap between the d_{xz} and π_x*(NO) orbitals is reduced on bending (in the xz plane). The calculations confirmed for the first time that the bent geometry was associated with the steep descent of one component of the $e(\pi^*(NO)- d_{xz,yz})$ antibonding

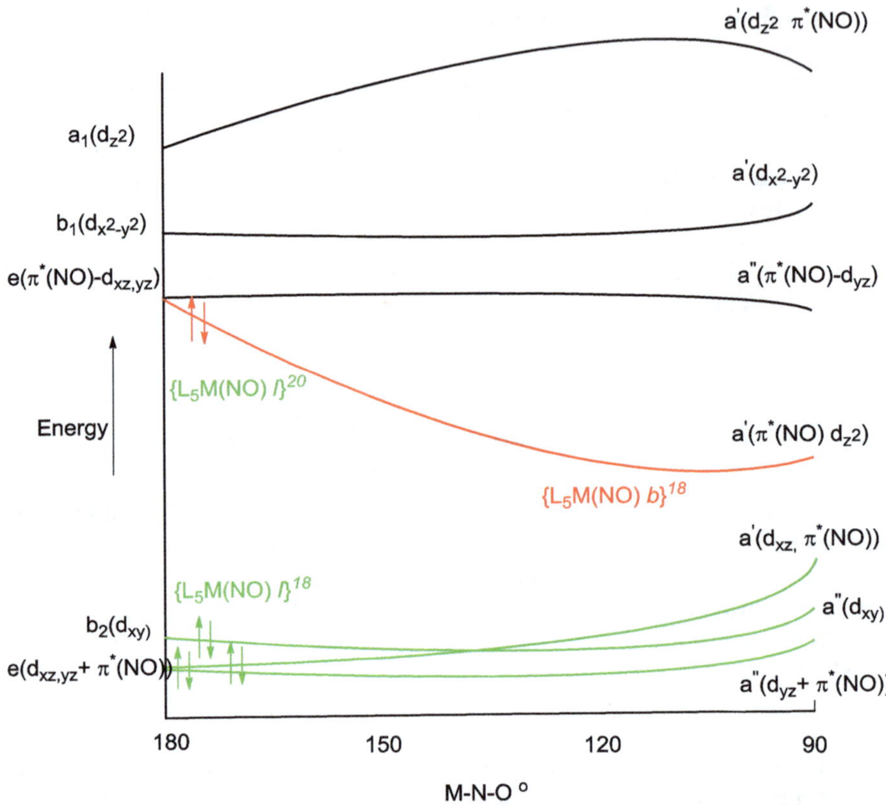

Fig. 7 Calculated Walsh diagram for $[Co(NO)(NH_3)_5]^{2-}$. The orbital primarily responsible for the bending distortion is shown in *red*. It corresponds to a 20-electron count for the linear geometry (shown in *green*) and 18-electron count for the bent geometry (shown in *red*)

orbitals (see Fig. 7) [119]. This orbital is occupied for low-spin $\{L_5M(NO)\}^{18-22}$ complexes. For low-spin $\{L_5M(NO)\}^{18,19}$ complexes, the additional electrons occupy the (π*(NO):d_{xz}) which greatly favours the bent geometry, because the energy of the orbital falls steeply as the M–N–O angle decreases. The d_{z^2} component of the original octahedral e_g set points directly at the lone pair of the NO ligand and has a similar energy to the antibonding degenerate pair of orbitals derived from the out-of-phase combination of π*(NO) and $d_{xz,yz}$. As the linear nitrosyl bends, the d_{z^2} orbital mixes extensively with d_{xz}^{-} π^*_x, which also has the same symmetry for the bent geometry, and the intimate mixings of these orbitals shown in **28** and **29** make these MOs diverge dramatically. In the more stable component **28,** the nitrogen atom orbital mixings result in an outpointing hybrid which resembles a lone pair orbital on nitrogen, and at the metal a stabilising interaction is turned on between d_{z^2} and π_{xz}*(NO). The bending distortion in **29** rehybridises the orbitals towards the metal and increases the antibonding interaction with d_{z^2}. The bending distortion is especially favoured for NO (and O_2) because the metal and π*(NO) orbitals have similar energies and less favoured for CO and N_2 because the π^* orbitals of these ligands are higher lying.

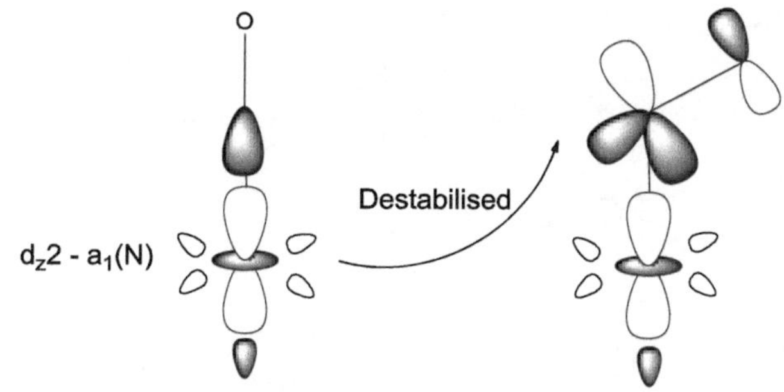

29

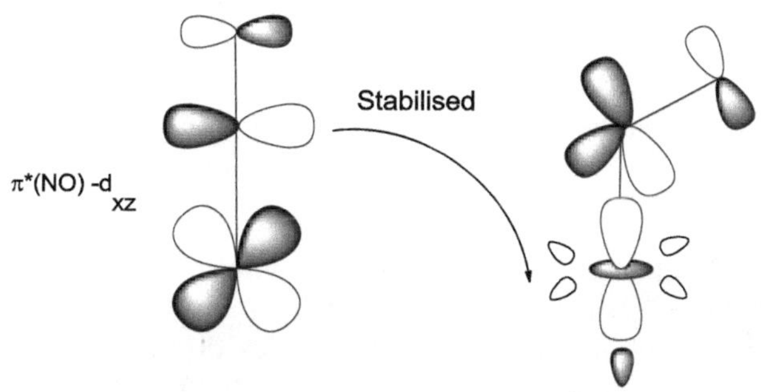

28

The calculations accounted for the following:

$\{L_5M(NO)\}^{12-18}$linear M-N-O ($\sim$180°)

$\{L_5M(NO)\}^{19}$intermediate M-N-O (140 – 160°) longer M-N bond

$\{L_5M(NO)\}^{20}$bent M-N-O ($\sim$120°)longer M-N bond

$\{L_5M(NO)\}^{20-22}$bent M-N-O ($\sim$120°) longer M-N bond

Specific examples of these complexes are illustrated in Scheme 5. According to the new notation described above, the bending process is represented by the following change $\{L_5M(NO)\ l\}^{20}$ to $\{L_5M(NO)\ b\}^{18}$ (see Fig. 7) and corresponds formally to a relocation of an electron pair from the metal to a molecular orbital which resembles a lone pair on nitrogen (see **28**). The two examples of $\{L_5M(NO)\}^{19}$ complexes in Fig. 7 have quite different M–N–O bond angles which places them in the bent and intermediate categories and underlines the effect of subtle changes in the metal and ligands. The lengthening of the M–N bond on bending arises because of the diminishing π overlap between d_{xz} and π_x*(NO) and this is

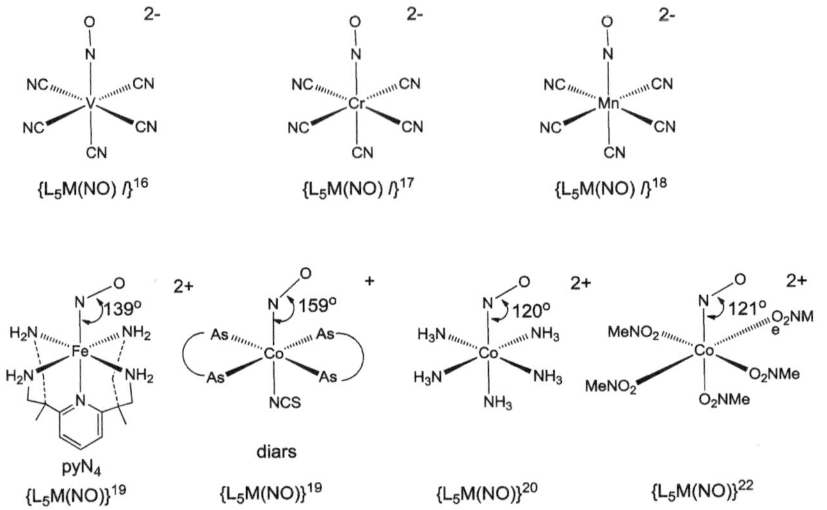

Scheme 5 Geometries of octahedral nitrosyl complexes. The notation is based on NO acting as a 3-electron donor in all examples in order to emphasise the effect of populating the orbitals in Fig. 7 in an *aufbau* fashion

accentuated for 20-electron species. The only example of a 22-electron complex is $[Cu(NO_2Me)_5(NO)]^{2+}$ which has a Cu–N–O bond angle of 121°, but it does not have a spin-paired ground state. The Jahn–Teller distorted structure, and long Cu–NO bond suggests a formulation of Cu(II) and NO [13, 14].

The energies of the metal d_{z^2} and d_{xz} relative to π*(NO) play a major role in determining the slopes of the molecular orbitals as the M–N–O angle becomes smaller. The energies are influenced by the position of the metal in the periodic table and the donor/acceptor properties of the spectator ligands and particularly the *trans* ligand. Specifically strong π-donor ligands raise the energy of the metal orbitals d_{xz}^{-} π_x* and favour bending.

Octahedral complexes containing a bent nitrosyl ligand show a significant *trans* influence, and for nitrogen ligands, the *trans* ligand bond is at least 0.2 Å longer than the *cis* ligand bonds. The bent nitrosyl ligand shows a much larger *trans* influence than linear nitrosyl ligands and in that respect resembles the nitrido ligand more closely. The *trans* influence has its origins in the orbital mixings illustrated in **28** [43]. The d_{z^2} orbital not only is involved in bonding with the nitrosyl ligand but is antibonding with respect to the amine ligands and particularly the *trans* amine ligand. Therefore, the orbital mixings responsible for the bending also increase the antibonding interaction to the *trans* ligand, and this causes the observed bond lengthening. In the limit lengthening, a *trans* bond leads to the dissociation of the ligand and related pairs of octahedral (18-electron) and square-pyramidal (16-electron) complexes result. The biological significance of the larger *trans* influence of NO compared to CO was recognised in 1976 by Perutz and co-workers [45] who used to it interpret differences in changes of the quartenary structures of haemoglobin adducts of these ligands. The biological consequences of this *trans* influence in porphyrin complexes of NO have been amplified particularly by Lehnert and Scheidt [33, 120, 121].

If the ligand *trans* to the nitrosyl has a strong *trans* influence, then this may result in structural consequences which do not follow the simplified generalisations discussed above. Specifically if strongly electron-donating ligands such as C_6H_4F or thiol ligands are located *trans* to nitrosyl in octaethylporphyrin $\{L_5Fe(NO)\}^{18}$ complexes, an intermediate nitrosyl (Fe–N–O = 157°) geometry is observed rather than the usually observed linear nitrosyl. The structural studies on these complexes have been determined by Lehnert and Scheidt et al. [33, 120–122], and Ghosh [123] has provided a DFT-based theoretical analysis.

The Walsh methodology was subsequently applied to five coordinate metal-nitrosyl complexes. Although the relative energies of the metal d_{z^2} and e{π*(NO)–d_{xz}, d_{yz}} are reversed, they still mix extensively on bending, and the components diverge in a similar manner to that noted above for $[Co(NO)(NH_3)_5]^{2+}$. The detailed analysis for five coordinate complexes suggested that the bent geometry is favoured for d^7 and d^8 square-pyramidal complexes. The methodology provided a molecular orbital interpretation of the following structural features of nitrosyl complexes.

1. The better the σ- or π-donating capability of the basal ligands, the more likely is the nitrosyl to bend.
2. In compounds of the type $ML_2L'_2(NO)$, L *trans* to L the nitrosyl group should bend in the plane containing the poorer donors.
3. In a compound of the type $ML_2DA(NO)$, D = π-donor trans to A = π-acceptor, if the NO group bends in the D–M–A plane, then it should bend towards the acceptor.
4. The nitrosyl is less likely to bend in the equatorial position of a trigonal–bipyramid than the apical site of a square pyramid.
5. If a nitrosyl in the equatorial position of a trigonal bipyramid bends, then it would prefer to do so in the axial plane rather than equatorial one.
6. Nitrosyl ligands in axial positions in a trigonal bipyramid and equatorial positions of a square pyramid prefer to be linearly coordinated.
7. In $ML_4(NO)$ species, if L are strong π-acceptors, a trigonal bipyramid with an equatorial nitrosyl will be preferred. If the ligands are strong π-donors, a range of geometries is possible, from a strongly bent nitrosyl to a less bent trigonal bipyramid.
8. A bent nitrosyl will move its nitrogen off the coordination axis in the direction of π-coordination. The movement of axis represents the initial stages of a distortion towards a π-bonded geometry.

These generalisations and detailed analysis of subtle geometric details demonstrated that the extended Hückel methodology could be used effectively to interpret the bonding in a wide range of coordination and organometallic compounds. The analysis also demonstrated the utility of the fragment approach for analysing a wide range of bonding problems in transition metal chemistry and led to the articulation of the *isolobal* analogy which united wide areas of main group and transition metal structural chemistries and provided effective bridges between organic and organometallic chemistry.

3.4 More Recent DFT Molecular Orbital Calculations

The broad generalisations associated with the Walsh diagram approach provide an excellent pedagogical tool, but there are cases where a more sophisticated approach is required. In recent years, DFT calculations have proved to be sufficiently accurate to reproduce the ground and excited states of metal complexes [123–126]. They may also be used to provide an accurate tool for the analysis of spectroscopic data. The increased study of complexes relevant to the biological role of nitric oxide has made the accurate description of the bonding in NO complexes more essential. Recent studies on a number of complexes relevant to the biological role using a combination of spectroscopic and computational techniques have provided an impetus for a bonding model which recognises the non-innocent nature of NO as a ligand and which proposes antiferromagnetic coupling between metal- and ligand-based unpaired electron spins rather than strong covalent bonds.

Ligand non-innocence involving antiferromagnetic coupling is generally described by a classical singlet diradical bonding description which depends on the relative orbital energies of metal and ligand and the energetic of the spin interactions. However, the situation is complicated for nitrosyls (and particularly first-row transition metal nitrosyls), because of static correlation and exchange energy effects, which favour singlet diradical states. Currently density functional molecular orbital calculations are extensively used to throw light on these bonding problems, and many examples of this methodology are described in the subsequent chapters of this pair of volumes. The DFT calculations, if the appropriate functionals are used, are able to reproduce the geometries of closed shell molecules very well (see, e.g. **29** and **30** and **31** which compare the results from alternative methods [38–40]) and with some additional effort may also be applied to complexes of the first-row transition metals where electron-electron repulsion and correlation effects are important. They may also be used to explore reaction coordinates.

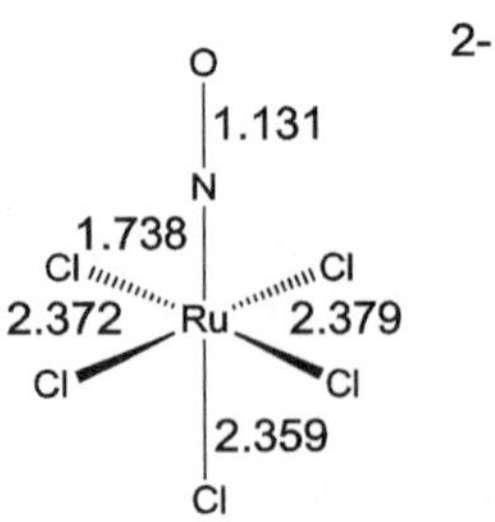

Experimental (Å)

29

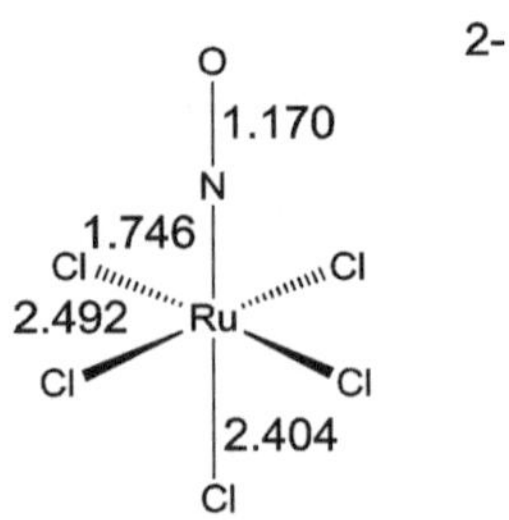

B3LYP
Calculated (Å)

30

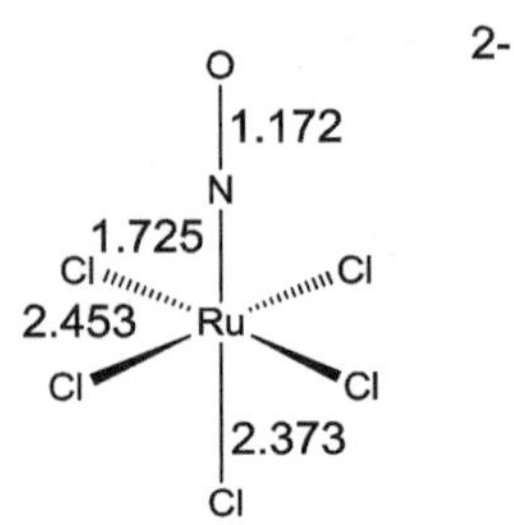

SVWN5
Calculated (Å)

31

Frenking et al. [127, 128] have calculated the geometries and charges in a series of pentacyanonitrosyl complexes and have concluded from the computed charges on nitrogen and oxygen and the energy decomposition analysis that the electronic structures are most accurately represented by $[M(CN)_5]^{q-}$ and NO even when the nitrosyl is bent. The charges (in electrons) are summarised below:

	$[Cr(NO)(CN)_5]^{4-}$	$[Mn(NO)(CN)_5]^{3-}$	$[Fe(NO)(CN)_5]^{2-}$	$[Co(NO)(CN)_5]^{3-}$
q(N)	0.22	0.34	0.36	0.03
q(O)	−0.52	−0.35	−0.20	−0.33
q(N) + q(O)	−0.30	−0.01	+0.16	−0.30

For the first three examples which are $\{L_5M(NO)\ l\}^{18}$, the charge varies from slightly less than zero to slightly more than zero across the periodic row and becomes more negative for $\{L_5Co(NO)\ b\}^{18}$ which has a bent M–N–O geometry. The effect represents a very attenuated version of the NO^+ to NO^- formalism reflecting once again the reliability of the electroneutrality principle. When related compounds which have the central metal atom belonging to different rows of the periodic table are compared, the changes are less marked:

	$[Fe(NO)(CN)_5]^{2-}$	$[Ru(NO)(CN)_5]^{2-}$	$[Os(NO)(CN)_5]^{2-}$
q(N)	0.36	0.27	0.21
q(O)	−0.20	−0.20	−0.23
q(N) + q(O)	0.16	0.07	−0.02

The charge on the nitrosyl becomes more negative as the group is descended probably reflecting more covalency and more effective back donation for the heavier metals. Frenking et al. [127–130] have noted the high degree of covalency in the metal-nitrosyl bonding and estimated that the π-bonding contributes 70% of the total. The DFT calculations have corroborated the essential features of the Walsh diagram approach to the linear to bent distortion, and the calculated bond lengths and angles have put on a more quantitative basis the qualitative changes in bond lengths and angles proposed in the initial calculations. For example, in $[Co(NO)(CN)_5]^{3-}$, the computed Co–N–O bond angle is calculated to be 121.7°, and the following bond lengths N–O 1.223, Co–NO 1.916, Co–CN_{trans} 2.077 and Co–CN_{cis} 1.920–1.945 Å reproduce the experimental data pretty well and in particular the changes in N–O and Co–N bond lengths on bending and the *trans* influence of the bent NO ligand [127, 128].

DFT calculations have also been used to contrast and compare the bonding and *trans* influences in related nitrido and nitrosyl complexes [38–40] and to explore reaction coordinates. The increasing accuracy of DFT calculations has also resulted in its more routine use for interpreting spectroscopic properties of nitrosyl complexes. Scheidt et al. [33, 120, 121] have noted that DFT calculations are very useful for interpreting the vibrations of Fe-ligand systems in porphyrin complexes which are particularly sensitive to electronic structures. The calculations do reproduce many structural features and provide a quantitative basis for understanding torsional dynamics of the Fe–NO unit. However, neither B3LYP nor BP86 functionals reproduce all the observed features satisfactorily, because they fail to completely describe the axial bonding interactions. Initially attention was focussed on calculating infrared frequencies of nitrosyl complexes, but increasingly more ambitious applications of DFT to ESR, NMR and visible and UV spectroscopies have been published, and TD-DFT calculations have been used to predict Co-K edge XANES spectra [131–142]. For [Tp*Co(NO)], the linear response DFT calculations were poor at reproducing the significant g-anisotropy of the $S = 1/2$ ground state, and it was necessary to perform CASSCF/MRCI computations in conjunction with quasi-degenerate perturbation theory for estimating the spin-orbit coupling in order to more accurately model the molecular g-tensor [96].

Recently the DFT methodology has been applied to the more challenging task of understanding the biological activation of small molecules [127–144]. These studies provide deeper insights into the mechanisms of these processes and interpreting the structural and spectroscopic properties of these important systems. The reliability of the calculations remains an issue, and the use of mainstream functionals may sometimes lead to opposing conclusions on the key aspects of the structures of a biological intermediates or the mechanism of its biological function. At one time, it was thought that the errors may not be larger than 1–5 kcal mol^{-1}, but recent experience suggests that this may be a serious underestimate [139]. Therefore, clearly more work is required in this area.

4 Summary

This review has provided a historical backdrop to the chemistry of nitric oxide and its complexes and demonstrated what an extremely versatile and interesting molecule it is. It is hardly surprising to learn that *Science* described it as the molecule of the year in 1992. It is certainly a non-innocent ligand capable of adopting alternative coordination geometries and stabilising metal complexes in a variety of oxidation states, and its secrets have gradually been unravelled by the application of a wide range of structural and spectroscopic techniques. The study of the reactions of its complexes has revealed that it participates in important catalytic and photochemical reactions. However, it was the discovery 30 years ago that nitric oxide has an important biological role as the endothelium-derived relaxation factor which has provided the impetus for the majority of the current research. Nitric oxide acts as a secondary messenger by activating soluble guanylyl cyclase (sGC) introducing a downstream pathway which leads to vascular smooth muscle relaxation. The mode of action of nitric oxide requires coordination of a heme iron at the active site of sGC. The remaining chapters in these volumes of *Structure and Bonding* provide a detailed account of these recent developments by experts in the field. Not surprisingly many of the chapters deal with the coordination chemistry of heme and related model compounds; however, there is also considerable interest in the reactions of nitric oxide with iron-sulphur clusters which result in the disassembly of the clusters and the formation of dinitrosyl iron compounds (DNIC) and derivatives of Roussin's black salt and red ester which were described in the first part of the review. So in historical terms, we have almost gone full circle, but in the process, I hope we have learned some important lessons.

Acknowledgement I would like to dedicate these volumes to Professor Jim Ibers who encouraged my interest in the structures of nitrosyl complexes when I held a Fulbright Fellowship at Northwestern University from 1968 to 1970 and Professor Roald Hoffmann who invited me to join him for a month at Cornell University in the summer of 1973, where we spent many hours talking about the bonding in transition metal compounds and developing the bonding picture for linear and bent nitrosyl complexes and the *isolobal* analogy. I would also like to thank Professor Karl Wieghardt, Professor Nicolai Lehnert, Professor John McGrady and Professor Gerard Parkin for helpful comments on parts of the manuscript and Dr Rene Frank for completing the analysis of nitrosyl structures from the Cambridge Crystallographic Data Base.

References

1. Butler A, Nicholson R (2003) Life death and nitric oxide. Royal Society of Chemistry, Cambridge
2. Bauer G (1963) Handbook of preparative inorganic chemistry. Academic, New York
3. Sharpe AG (1976) The Chemistry of cyano-complexes of the transition metals. Academic, London
4. Rucki R (1977) Anal Profiles Drug Subst 6:487
5. Rucki R (1986) Anal Profiles Drug Subst 15:781
6. Toledo JC, Ohara A (2012) Chem Res Toxicol 25:975
7. Wilkinson G, Gillard RD, McCleverty JA (1987) Comprehensive coordination chemistry. Pergamon, Oxford
8. Lipscomb WN, Wang FE, May WR, Lippert EL (1961) Acta Cryst 14:1100
9. Dulmage WJ, Meyers EA, Lipscomb WN (1953) Acta Crys 6:760
10. Halland A (2008) Molecules and models – the molecular structures of main group compounds. Oxford University Press, Oxford, 298
11. Jorgensen WL, Salem L (1973) The organic chemists book of orbitals. Academic, New York, 80
12. Mingos DMP, Yau J (1994) J Organometal Chem 479:C16
13. Wright AM, Wu G, Hayton TW (2010) J Amer Chem Soc 132:14336
14. Wright AM, Wu G, Hayton TW (2011) Inorg Chem 50:11746
15. Cotton FA, Wilkinson G (1962) Advanced inorganic chemistry. Wiley, New York, 333
16. Sherman DJ, Mingos DMP (1989) Adv Inorg Radiochem 34:293
17. Walsh AD (1953) J Chem Soc 2260, 2266, 2296, 2301, 2306
18. Hodgson DJ, Payne NC, McGinnety JA, Pearson RG, Ibers JA (1968) J Amer Chem Soc 90:4487
19. Pierpont CG, Eisenberg R (1972) Inorg Chem 12:1088
20. Snyder DA, Weaver DL (1970) J Chem Soc
21. Coppens P, Novozhilova I, Kovalovsky A (2002) Chem Rev 102:61
22. Lehnert N, Berto TC, Sage J, Silvernail N, Scheidt WR Alp EE, Sherhan W, Zhao J (2013) Coord Chem Rev 257:244
23. Praneeth VKK, Nether C, Peters G, Lehnert N (2006) Inorg Chem 45:2795
24. Lehnert N, Scheidt WR, Wolf A (2013) Struct Bond 155
25. Wells AF (1975) Structural inorganic chemistry, 4th edn. Oxford University Press, Oxford, 651
26. De La Cruz C, Sheppard N (2011) Spectrochim Acta A Mol Biomol Spectrosc 78:7
27. Cambridge Crystallographic Data Base, Cambridge. www.ccdc.cam.ac.uk/products/csd/
28. Lichtenberger DL, Gruhn NE, Renshaw SK (1997) J Mol Struct 405:709
29. Green JC (2006) In: Crabtree RH, Mingos DMP (eds) Comprehensive organometallic chemistry III. Elsevier, Oxford, 1:381
30. Dodle B, Kanjenpillar R, Blaque O, Berke H (2011) J Amer Chem Soc 133:8168
31. Dodle B, Kanjenpillar R, Blaque O, Berke H (2013) Struct Bond 152
32. Berto TC, Praneeth VKK, Goodrich IC, Lenhert N (2009) J Amer Chem Soc 131:17116

33. Lenhert N, Sage JT, Silvernail NJ, Scheidt WR, Alp EE, Sturhahn W, Zhao J (2010) Inorg Chem 49:7197
34. Hursthouse MB, Motavelli M (1979) J C S Dalton Trans 1362
35. Pandey KK (1990) J Coord Chem 22:307
36. Manoharan PT, Gray HB (1966) Inorg Chem 5:823
37. Landry VK, Pan K, Quan SM, Parkin G (2007) J Chem Soc. Dalton Trans 820
38. Lyne PD, Mingos DMP (1995) J Chem Soc Dalton Trans 1635
39. Sivova OV, Lyubimova OO, Sizov VV (2002) Russ J Gen Chem 74:317
40. Bright D, Ibers JA (1969) Inorg Chem 8:709
41. Akashi H, Yamauchi T, Shibahara T (2004) Inorg Chim Acta 357:325
42. Pratt CS, Coyle BA, Ibers JA (1971) J Chem Soc A 2146
43. Mingos DMP (1974) Inorg Chem 10:1479
44. Mingos DMP (1980) Trans Amer Crystallog Assoc 16:17
45. Perutz MF, Kilmartin JV, Nagai K, Szabo A (1976) Biochemistry 15:378
46. Li H, Bonnet D, Bul E, Neese F, Weyhermuller F, Blum N, Sellmann D, Wieghardt K (2002) Inorg Chem 41:3444
47. Mingos DMP, Ibers JA (1971) Inorg Chem 10:1035
48. Mingos DMP, Robinson WT, Ibers JA (1971) Inorg Chem 10:1043
49. Ibers JA (1971) Acta Cryst B27:250
50. Formitchev DV, Furlani TR, Coppens P (1998) Inorg Chem 37:1519
51. Lynch MS, Cheng M, van Kuikan BE, Khalil (2011) J Amer Chem Soc 133:5255
52. Tocheva EI, Rosssell FI, Mauk AG, Murphy MP (2007) Biochemistry 46:12366
53. Tocheva EI, Rosssell FI, Mauk AG, Murphy MP (2004) Science 304:867
54. Merkle AC, Lehnert N (2009) Inorg Chem 58:11504
55. Li L, Enright GD, Preston KF (1994) Organomet 13:4686
56. Li L, Enright GD, Preston KF (2013) Struct Bond 155
57. Mingos DMP, Ibers JA (1970) Inorg Chem 9:1105
58. Barybin HV, Young VG, Ellis JE (1999) Organomet 18:2744
59. Tonzetich ZJ, McQuade LE, Lippard SJ (2010) Inorg Chem 49:6338
60. Tonzetich ZJ, Heroguel F, Do LH, Lippard SJ (2011) Inorg Chem 50:1570
61. Pluth MD, Lippard SJ (2012) Chem Commun 48:11981
62. Mingos DMP (1979) J Organometal Chem 179:C29
63. Hayton TW, Legzdins P, Sharp WB (2002) Chem Rev 102:935
64. Hsieh C-H, Darensbourg MY (2010) J Amer Chem Soc 122:14119
65. Zong-Sian L, Tzung-Wen C, Kuan-Yu L, Chang-Chih H, Jen-Shiang YK, Wen-Feng L (2012) Inorg Chem 51:10092
66. Mingos DMP (1984) Acc Chem. Res 17:311
67. Berringhelli T, Ciani G, d'Alfonso G, Molinan H, Sironi A, Freni M (1984) JCS Chem Commun 1327
68. Norton JR, Collman JP, Dolcetti R, Robinson WR (1972) Inorg Chem 11:382
69. Calderon JL, Fontana S, Fraundhiofer E, Day VW (1974) J Organometal Chem 64:C10
70. Siladke NA, Meihas KR, Ziller SW, Fang M, Furche F, Long JR, Evans JW (2012) J Amer Chem Soc 134:1243
71. Evans WJ, Fang M, Bates JE, Furche F, Ziller JW, Kiesz MD, Zink JI (2010) Nature Chem 2:644
72. Adams DM (1968) Metal ligand and related vibrations. St Martin's, New York
73. McCleverty JA (2004) Chem Rev 104:403
74. Haymore BL, Ibers JA (1975) Inorg Chem 14:3060
75. Collman JP, Farham P, Dolcelli G (1971) J Amer Chem Soc 93:1788
76. Lehnert N, Sage JT, Silvernail NJ, Scheidt WR, Alp EE, Sturhahn W, Zhao J (2010) Inorg Chem 49:7197
77. Pavlik JW, Barabanschikov A, Oliver AG, Alp EE, Sturhahn W, Zhao J, Sage JT, Scheidt WR (2010) Angew Chem Int Ed 49:4400

78. Bell LK, Mason J, Mingos DMP, Tew DG (1983) Inorg Chem 22:3497
79. Evans DH, Mingos DMP, Mason J, Richards A (1983) J Organometal Chem 249:293
80. Bultide J, Larkworthy LF, Mason J, Povey DC, Mason J, Sandell B (1984) Inorg Chem 23:3629
81. Bell LK, Mingos DMP, Tew DG, Larkworthy LF, Sandell B Povey PC, Mason J (1983) J C S Chem Commun 125
82. Mason J, Mingos DMP, Sherman DJ, Schaffer J, Stejskal EO (1985) J C S Chem Commun 444
83. Mingos DMP, Sherman DJ (1984) J C S Chem Commun 1223
84. Mingos DMP, Sherman DJ (1987) Trans Met Chem 12:400
85. Mingos DMP, Sherman DJ (1987) Trans Met Chem 12:897
86. Mason J, Larkworthy LF, Moore EA (2002) Chem Rev 102:913
87. Botto RE, Kolthammer I, Legzdins P, Roberts JD (1979) Inorg Chem 18:2049
88. Gaviglio C, Ben-David Y, Shimon LJW, Dotorovisch F, Milstein D (2009) Organomet 28:1917
89. Finn P, Pearson RK, Hollander JM, Jolly WL (1971) Inorg Chem 10:378
90. Gary AN, Goel PS (1971) Inorg Chem 10:1314
91. Greatrex R, Greenwood NN, Kaspi P (1971) J C S Dalton 1873
92. Wayland BB, Olsen LW (1974) Inorg Chim Acta 11:L23
93. Gibson J (1962) Nature (Lond) 1962:64
94. Manoharan PT, Kuska HA, Roger MT (1967) J Amer Chem Soc 89:4564
95. Krzystek J, Sevenson, DC, Zuyagu SA, SmirnovD, Ozarowski A (2010) J Amer Chem Soc 132:5241
96. Tomson NC, Crimm MR, Petrenko T, Roseburgh LE, Sproules S, Boyd WC, Bergman RG, DeBeer S, Toste FD, Wieghardt K (2011) J Amer Chem Soc 133:18785
97. Kaim W (2011) Angew Chem Int Ed 50:10498
98. Kaim W (2011) Inorg Chem 50:9752
99. Sproules S, Wieghardt K (2010) Coord Chem Rev (2010) 254:1358
100. Sproules S, Weyhermueller T, De Beer S, Wieghardt K (2010) Inorg Chem 49:5241
101. Jorgenson CK (1962) Coord Chem Rev 1:164
102. Schonherr T (2004) Struct Bond 106:1
103. Enemark JH, Feltham RD (1974) Coord Chem Rev 13:339
104. Mingos DMP (1998) Essential trends in inorganic chemistry. Oxford University Press, Oxford
105. Westre CE, Di Cicco, Filiponi A, Natoli CR, Hedman B, Solomon EI, Hodgson KO (1994) J Amer Chem Soc 116:6757
106. Wylie RA, Scheidt WR (2002) Chem Rev 102:1067
107. Elison MK, Schulz CE, Scheidt WR (2000) Inorg Chem 39:5102
108. McCleverty JA, Atherton NM, Locke J, Wharton EJ, Winscom CJ (1967) J Amer Chem Soc 89:6082
109. Wells FV, McCann SW, Wickman HH, Kessel SL, Hendrickson DN, Feltham RD (1982) Inorg Chem 21:2306
110. Fitzsimmons BW, Larkworthy LF, Rogers KA (1980) Inorg Chim Acta 44:L53
111. Harding DJ, Harding P, Adams H, Tuntulani T (2007) Inorg Chem 360:3335
112. Manoharan PT, Gray HB (1966) Inorg Chem 5:873
113. Fenske RF, DeKock RL (1972) Inorg Chem 11:437
114. Wayland BB, Olson LW (1974) J Amer Chem Soc 96:6037
115. Mingos DMP (1971) Nature Phys Sci (Lond) 229:193
116. Mingos DMP (1971) Nature Phys Sci (Lond) 230:154
117. Pierpont CG, Eisenberg R (1971) J Amer Chem Soc 93:4905
118. Feltham RD, Enemark JH (1974) J Amer Chem Soc 97:5002
119. Hoffmann R, Chen MML, Elian M, Rossi AR, Mingos DMP (1974) Inorg Chem 13:2666
120. Berto TC, Praneeth VKK, Goodrich IC, Lenhert N (2009) J Amer Chem Soc 131:17116

121. Richter-Addo GB, Wheeler RA, Hixson CA, Chen L, Khan MA, Ellison MK, Schulz CE, Scheidt WR (2001) J Am Chem Soc 123:6314
122. Goodrich LE, Paulat F, Praneeth VKK, Lehnert N (2010) Inorg Chem 49:6293
123. Ghosh A, Hopmann KH, Conradie J (2009) Electronic structure calculations: transition Metal–NO complexes. In: Solomon EI, King RB, Scott RA (eds) Computational inorganic and bioinorganic chemistry. Wiley, Chichester
124. Kaltsoyanis N, McGrady JE (2004) Principles and applications of density functional theory I & II. Struct Bond 112:1;113:1
125. Mingos DMP, Putz MV (2013) Applications of density functional theory to chemical reactivity. Struct Bond 149:1
126. Mingos DMP, Putz MV (2013) Applications of density functional theory to biological and bioinorganic chemistry. Struct Bond 150:1
127. Lyubinova C, Sizova OV, Losden C, Frenking G (2008) J Molecular Struct Theo Chem 865:28
128. Caramori GF, Kunitz AG, Andriaru KF, Doro FG, Frenking G, Thiouni E (2012) Dalton Trans 41:7327
129. Scheidt WR, Barabanschikov A, Paolek JW, Silvernail NJ, Sage JT (2010) Inorg Chem 49:6240
130. Scheidt WR (2008) J Porphyrins Phthalocyan 12:979
131. Lehnert N, Praneeth VKK, Paulat F (2006) J Comput Chem 27:1338
132. Silaghi-Dumitrescu R (2013) Struct Bond 150:97
133. Ghosh A (2005) Acc Chem Res 38:943
134. Ghosh A (2006) Biol Inorg Chem 11:712
135. Radon M, Pieloot KJ (2008) J Phys Chem 112:11824
136. Olah J, Harvey JN (2009) J Phys Chem 113:7338
137. Ghosh A (2009) J Inorg Biochem 253:523
138. Conradie J, Ghosh A (2007) J Phys Chem 111:12621
139. Conradie J, Ghosh A (2006) J Inorg Biochem 100:2069
140. Ghosh A (2006) J Biol Inorg Chem 11:671
141. Blomberg MRA, Siegbahn PEM (2012) Biochemistry 51:5173
142. Riplinger C, Neese F (2011) Chem Phys Chem 12:3192
143. Bykov D, Neese F (2012) J Biol Inorg Chem 17:741
144. Paulat F, Lehnert N (2007) Inorg Chem 46:1547

Struct Bond (2014) 153: 45–114
DOI: 10.1007/430_2013_102

Published online: 26 June 2013

Coordination Chemistry of Nitrosyls and Its Biochemical Implications

Hanna Lewandowska

Abstract A comprehensive overview is presented of the biologically relevant coordination chemistry of nitrosyls and its biochemical consequences. Representative classes of metal nitrosyls are introduced along with the structural and bonding aspects that may have consequences for the biological functioning of these complexes. Next, the biological targets and functions of nitrogen (II) oxide are discussed. Up-to-date biochemical applications of metal nitrosyls are reviewed.

Keywords Biological action · Electronic structure · Iron proteins · Metal nitrosyls · Nitric oxide · Non-innocent ligands

Contents

H. Lewandowska (✉)
Institute of Nuclear Chemistry and Technology, Centre for Radiobiology and Biological Dosimetry, 16 Dorodna Str., 03-195 Warsaw, Poland
e-mail: h.lewandowska@ichtj.waw.pl

Abbreviations

5C	Five-coordinate
AAA+	ATPases associated with diverse cellular activities
ATP	Adenosine triphosphate
Bax	Bcl-2-associated X protein
BIPM	2,2′-(phenylmethylene)*bis*(3-methylindole)
Bu	Butyl
Cbl	Cobalamin
cGMP	Cyclic guanosine monophosphate
Cp	Cyclopentadienyl
CPMAS	Cross polarization magic angle spinning
CuFL	Fluorescein-copper (II) complex
Cys	Cysteine
DETC	Diethyl thiocarbamate
DFT	Density functional theory
DMSO	Dimethyl sulfoxide
DNDGIC	Dinitrosyl–diglutathionyl–iron complex
DNIC	Dinitrosyl iron complex
DTC	Dithiocarbamate
EDRF	Endothelium-derived relaxing factor
ENDOR	Electron nuclear double beam resonance
eNOS	Endothelial nitrogen (II) oxide synthase
EPR	Electron paramagnetic resonance
Et	Ethyl
FL	Fluorescein
GAF domain	cGMP-specific phosphodiesterases, adenylyl cyclases, and FhlA characteristic domain
GC	Guanylate cyclase
GSH	Glutathione
GSNO	Nitrosoglutathione
GST	Glutathione transferase
GTP	Guanosine triphosphate
H_2bpb	*N,N′-bis*(bipyridine-2-carboxamido)-1,2-diaminobenzene
Hb	Hemoglobin
HiPIP	High-potential iron–sulfur protein
His	Histidine
HOMO	Highest occupied molecular orbital
HTH	Helix–turn–helix
iNOS	Cytokine-inducible nitrogen (II) oxide synthase

IR	Infrared
IRE	Iron responsive element
IRP	Iron regulatory protein
IUPAC	International Union of Pure and Applied Chemistry
LIP	Labile iron pool
LMW	Low molecular weight
LUMO	Lowest unoccupied molecular orbital
Me	Methyl
metHb	Methemoglobin
MNIC	Mononitrosyl iron complex
MNIP	4-Methoxy-2-(1*H*-naptho[2,3-*d*]imidazol-2-yl)phenol
MorDTC	Morpholyldithiocarbamate
MRP1	Multidrug resistance-associated protein1; iron regulatory protein1
MT	Metallothionein
MTP1	Ferroportin1
NAD	Nicotinamide adenine dinucleotide
naphth-enH_2	2-Hydroxy-1-naphthaldehyde and ethylenediamine
naphth-mphH_2	4-Methyl-*o*-phenylenediamine
naphth-phH_2	*o*-Phenylenediamine
NHase	Nitrile hydratase
NHE	Normal hydrogen electrode
NIR	Nitric oxide reductase
NMR	Nuclear magnetic resonance
nNOS	Neuronal nitrogen (II) oxide synthase
NO	Nitrogen (II) oxide
NorR	Anaerobic nitric oxide reductase transcription regulator
NOS	Nitrogen (II) oxide synthase
NRVS	Nuclear resonance vibrational spectroscopy
OEP	Octaethylporphyrin
$PaPy_3H$	*N,N-bis*(2-pyridylmethyl)amine-*N*-ethyl-2-pyridine-2-carboxamide
PeT	Photoinduced electron transfer
Ph	Phenyl
porph	Porphyrin
$PPDEH_2$	Protoporphyrin IX diester
PPh_3	Triphenylphosphine
PPN	3-Phenyl-2-propynenitrile
RBS	Roussin's black salt
RRE	Roussin's red salt ester
RRS	Roussin's red salt
RSNO	*S*-nitrosothiol
S_2-o-xyl	Dianion of 1,2-phenylenedimethanethiol
SCE	Saturated calomel electrode
Ser	Serine

sGC	Soluble guanylate cyclase
SNP	Sodium nitroprusside
SoxR	Superoxide response DNA-binding transcriptional dual regulator
SoxS	Activator of superoxide stress genes
TCPP	meso-*tetrakis* (4-carboxyphenyl) porphyrin
TMPyP	meso-*tetrakis* (4-N-methylpyridinium) porphyrin
TPE	Two photon excitation
TPP	Tetraphenylporphyrin
TPRR′	*tris*(pyrazolyl) borato ligand
TTMAPP	meso-*tetrakis* [4-(N,N,N-trimethyl) aminophenyl] porphyrin
Tyr	Tyrosine
UTR	Untranslated region of RNA
UV–VIS	Ultraviolet–visible spectroscopy

1 Introduction

The investigation of transition metal nitrosyl complexes has gained in importance, since the role of nitric oxide (NO) in various physiological processes was discovered. Non-innocent nature of nitric oxide as a ligand and the associated ambiguity in assignment of metal and NO–ligand oxidation states in metal nitrosyls complicate their classification and the interpretation of their electronic spectra and redox properties, chemical behavior, and biological activity. The chemistry of nitrosyl compounds is important, because of the structural, synthetic, and mechanistic implications. This review focuses attention particularly on those nitrosyl complexes that are of direct relevance to biology and medicine. The structural aspects of representative classes of metal nitrosyls of biological significance are reviewed, with particular emphasis on their biological functions and sources, as well as some of their biochemical applications.

2 Structural Aspects for Metal Nitrosyls

NO is a diatomic, stable free radical with an N–O bond length of 1.154 Å and a $^2\Pi$ ground state [1]. Simple molecular orbital theory predicts a bond order of 2.5, consistent with its bond length between those of N_2 (1.06 Å) and O_2 (1.18 Å). The singly occupied MO is a π^* orbital, but polarized toward nitrogen in a manner opposing the polarization of the lower energy π^b orbitals. The result is a relatively nonpolar diatomic molecule; consequently, the ν(NO) stretching vibration, at 1,875 cm^{-1} (15.9 mdynes/cm), has a very low intensity in the infrared absorption spectrum. Because of spin–orbit coupling of the unpaired electron with its π-orbital (~121 cm^{-1}), NO exhibits P, Q, and R rotational branches in its gas-phase vibrational spectrum [2, 3]. The increase in the population of the π^* orbital in the row:

NO^+, NO^0, NO^- leads to the increase in the bond length (1.06 Å, 1.154 Å, 1.26 Å, respectively) and to the decrease in the IR stretching frequencies of ν(NO) from 2,377 for NO^+ to 1,470 cm^{-1} for NO^- [4, 5]. More information on the NO molecule bonding and electronic structure can be found in the chapter by Mingos.

2.1 Bonding in Metal Nitrosylates

Bonding of NO to transition metals usually occurs via the nitrogen atom. Other types of bonding occur in some rare cases and include: twofold and threefold nitrosyl bridges σ/π-dihaptonitrosyls (so-called side-on nitrosyls) and isonitrosyls (M–O–N bound), as reviewed in [6].

Table 1 illustrates a variety of possible NO bonding patterns for which structural data are available. It also gives the IUPAC notations and other ones encountered in the literature [23]. The typical values of the most important structural parameters for the species listed in Table 1 are quoted according to [6]. Structural aspects for each of the group of compounds have recently been reviewed and can be found in the cited reference.

Apart from the above, in biological and biomimetic systems, metals usually bind one or two terminal NO groups. The M–N–O bond angles may be essentially linear or bent, up to ca. 120°. The bridging mode of NO coordination is rarely found in biomimetic systems but is encountered in organometallic and cluster chemistry. In the *molecular orbital* approach, the bonding of NO to a metal can be generally presented as the superposition of two components: The first one involves donation of electron density from the N atom of NO to the metal, involving a σ molecular orbital; the second component is back-donation from metal dπ orbitals to the π* orbitals of NO (π backbonding[1]). The bond order within the ligand is decreased by this process, while the metal–ligand bond order is increased [24]. It is often formally assumed that the NO ligands of metal–nitrosyl moiety are derivatives of the nitrosyl cation, NO^+. The nitrosyl cation is isoelectronic with carbon monoxide; thus, the bonding between a nitrosyl ligand and a metal follows the same principles as the bonding in carbonyl complexes. Yet, the formal description of nitric oxide as NO^+ does not match certain measureable and calculated properties. In an alternative description, nitric oxide serves as a 3-electron donor, and the metal–nitrogen interaction is a triple bond. Both these approaches are presented in the literature. Despite the general analogy between metal carbonyls and nitrosyls, it is noteworthy that NO is even stronger π-acceptor than CO. This manifests itself in the relatively

[1] According to the definition of IUPAC: A description of the bonding of π-conjugated ligands to a transition metal which involves a synergic process with donation of electrons from the filled π-orbital or lone electron pair orbital of the ligand into an empty orbital of the metal (donor–acceptor bond), together with release (back-donation) of electrons from an *n*d orbital of the metal (which is of π-symmetry with respect to the metal–ligand axis) into the empty π*-antibonding orbital of the ligand [7].

Table 1 Classification of metal nitrosyls in terms of the bonding mode of nitrosyl group; according to [6]

Structure diagram	Nomenclatures [7]	Typical structural parameters (according to [6])	Exemplary complex (reference)
M–N–O	*Linear nitrosyl* IUPAC: σ-NO M(NO)	r(NO) = 1.14–1.20 Å r(MN) = 1.60–1.90 Å α(MNO) = 180–160 Å	Manganium porphyrinates [8, 9] $\{Fe(NO)_2\}^9$, e.g., $[Fe(NO)_2(CO)_2]$ $[Co(NO)(CO)_3]$ [10] $[PPN][S_5Fe(NO)_2]$ [11]
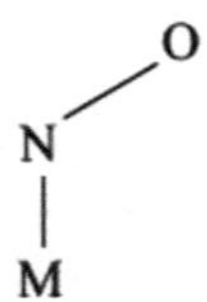	*Bent nitrosyl* IUPAC: σ-NO M(NO)	r(NO) = 1.16–1.22 Å r(MN) = 1.80–2.00 Å α(MNO) = 140–110 Å	Nitrosyl halides, alkanes, and arenes, four-coordinate complexes of Co [12], Os [13], $Fe(NO)_4$ (two of four NO groups) [14] $[Pt(NO)(NO_2)_2(Cl)_3][K]_2$ [15]
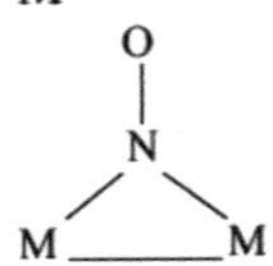	*Normal twofold nitrosyl bridge* IUPAC: μ_2-NO $M_2(\mu_2$-NO) M_2(NO)	r(MM) = 2.30–3.00 Å r(NO) = 1.18–1.22 Å r(MN) = 1.80–2.00 Å α(MNM) = 90–70 Å	$[Mn_2(NO)(X)(\mu_2\text{-}NO)_2(Cp)_2]$ (X=Cp [16] or NO_2 [17]) $Ru_3(CO)_{10}(NO)_2$ [18]
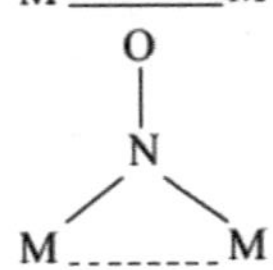	*Long twofold nitrosyl bridge* IUPAC: μ_2-NO $M_2(\mu_2$-NO)	r(M···M) = 3.10–3.40 Å r(NO) = 1.20–1.224 Å r(NM) = 1.90–2.10 Å α(MNM)= 130–110 Å	$Cp_2Mn_2(\mu\text{-}NO)_2(NO)_2$ [19]
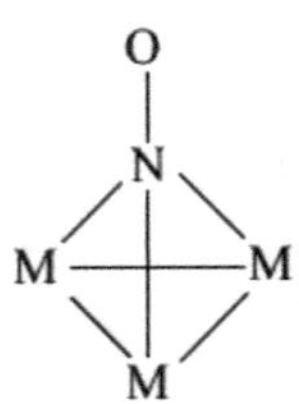	*Threefold nitrosyl bridge* IUPAC: μ^3-NO $M^3(\mu^3$-NO) M^3(NO)	r(NO) = 1.24–1.28 Å r(MN) = 1.80–1.90 Å	$Cp^3Co^3(\mu^3\text{-}NO)^2$ [20]

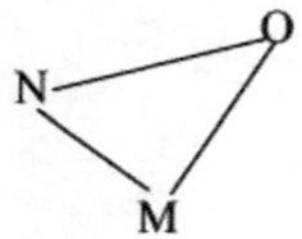	*σ/π-Dihaptonitrosyl* IUPAC: η^2-NO σ/π-NO M(η^2-NO)	r(MN) = 1.70–1.90 Å r(MO) = ca. 1.90–2.00 Å	
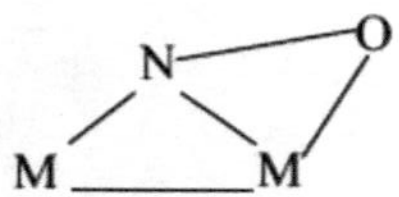	*σ/π-dihaptonitrosyl bridge* IUPAC: η^2, μ_2-NO M_2(η^2, μ_2-NO) M(NO)/M		Metastable states of $2H_2O\cdot[Fe(\eta^2\text{-}NO)(CN)_5][Na]_2$ (sodium nitroprusside) [21]
M – O – N	*Linear isonitrosyl* IUPAC: σ-ON M(ON)		BrON (metastable state, UV irradiation of BrNO) [22]
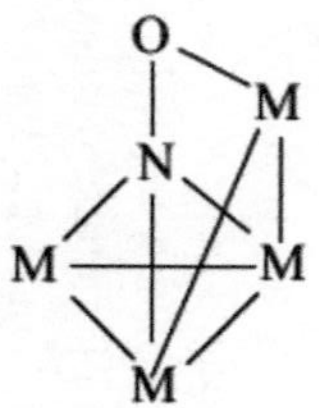 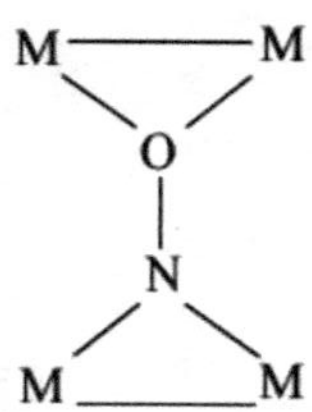	*Isonitrosyl bridges* IUPAC: η^2, μ^4-NO M_4(η^2, μ^4-NO) M_3(NO)M IUPAC: η^2, μ^4-NO M_4(η^2, μ^4-NO) M_2(NO)M_2		Metastable states of $2H_2O\cdot[Fe(\eta^2\text{-}NO)(CN)_5][Na]_2$ (sodium nitroprusside) [21]

PPN 3-Phenyl-2-propynenitrile, *Cp* Cyclopentadienyl

strong metal–nitrogen bonding, lowering of the bond order within the ligand, and hence the high degree of electron density delocalization in this triatomic fragment. Therefore, in accordance with the Enemark and Feltham notation, MNO is treated as a covalently bound functional group [25] represented as $\{M(NO)_x\}^n$, in which n denotes the total number of electrons associated with the metal d and π^* (NO) orbitals. The structure becomes disturbed by the coordination of additional ligands. Energies of π^* orbitals of NO are close to d orbitals of transition metals. Therefore, the relative charge distribution in M–N bonding orbitals may vary to a great extent for the two isoelectronic complexes, and very small changes in metal or ligand properties may cause substantial changes in the character of the nitrosyl moiety (NO^+, NO, or NO^-), and consequently, in the NO-donating properties of the complex. Despite the difficulty in assigning formal oxidation states to the metal and the NO in nitrosyl complexes, the charge distribution within this residue, which is comprised between $\{M^{z-1}(NO^+)\}$ $\{M^z(NO)\}$ and $\{M^{z+1}(NO^-)\}$ remains an important issue, particularly with respect to the electronic and magnetic behavior of complexes, as reflected in the chemical properties of the nitrosyls.

A good indicator of the backbonding and bond polarization degree is IR spectroscopy; the NO vibrations are usually observed in the broad region: ~1,300–1,900 cm^{-1}, the higher for the linear, the medium for bent and the lower being characteristic for the bridging mode of NO coordination (see Table 1), as summarized in the chapter on spectroscopy by Lewandowska and in the chapter of Mingos. As noted by McCleverty [5], these three ranges overlap significantly and are strongly dependent on the multiple electronic factors, thus no simple correlation can be seen between the position of ν(NO) and the M–N–O bond angle, unless these factors are allowed for a certain adjustments [6, 13] (see spectroscopic chapter by Lewandowska). The typical N–O and M–N bond distances and M–N–O bond angles for the linear and bent nitrosyls are: 1.14–1.20 Å/1.60–1.90 Å/180–160 Å and 1.16–1.22 Å/1.80–2.00 Å/140–110 Å, respectively [6]. Apart from the vibrational spectroscopy, among other tools applied in the determination of the structure of nitrosyls especially useful are: Mössbauer spectroscopy, nuclear resonance vibrational spectroscopy, and in case of the paramagnetic forms, EPR and ENDOR. Some easily available information on the electronic structure and the geometry of electronic transitions can be found by UV–VIS and magnetic circular dichroism. For even-electron MNO fragments also NMR chemical shifts can give some additional information on the electron charge distribution [26–32]. However, in order to obtain definitive data on the structure it is necessary to complement the results with theoretical methods. Conclusions on nitrosyl structures, which can be drawn from the spectroscopic data will be discussed in the next chapter by Lewandowska.

Comprehensive analyses of the electron charge of metal nitrosyls based on theoretical methods including spectroscopic data have been carried out by Ghosh and coworkers [33–38]. The results of their work are summarized in the chapter by Mingos. In the search of new tools that would allow simple and unambiguous way to determine the electronic nature of transition metal nitrosyl complexes, it has recently been proposed by Sizova et al. [39] to use the quantum chemical bond order indices deconvoluted into σ-, π-, and δ-contributions, for some organometallic nitrosyls of ruthenium and rhodium with bent and linear configurations of the MNO groups. The new (as regards nitrosyls) approach, considering bond order a good characteristic of electron distribution in molecules, implemented in parallel with the classical structural theory proved useful for structural analysis of nitrosyl complexes. The charges of the NO moieties and the π- and σ-contributions to the bond order were calculated. Unusually large values of through-atom indices confirmed the delocalization of the π-electron density over the linear Ru–N–O fragment. Changes in the electron charge distribution in a series of $\{Ru\text{–}N\text{–}O\}^{6\text{–}7\text{–}8}$ complexes were analyzed in terms of the bond order, the metal charge, and the nature of the nitrosyl moiety. This approach allowed the authors [39] to explain the mechanistic aspects of the metal–metal bond destruction in the paddlewheel-type dimetal complexes $M_2(L)_4(NO)_2$ caused by the coordination of nitric oxide.

2.2 *Linear Versus Bent Nitrosyl Ligands*

Typical ranges for the values of internuclear N–O and M–N bond distances and M–N–O bond-angles are: 1.14–1.20 Å/1.60–1.90 Å/180–160 Å for linear nitrosyls and 1.16–1.22 Å/1.80–2.00Å/140–110 Å, for bent nitrosyls. It should be noted that in some early papers mononitrosyl compounds containing linear M–N–O groups were assumed to contain bound NO^+, whereas those having bent M–N–O arrangements were regarded as containing NO^-, but this has subsequently been found misleading [5]. When applying a *valence bond* approach to a linear M–N–O arrangement, it is convenient to regard the N and O atoms in the NO^+ group as being sp hybridized.

According to the Enemark-Feltham approach, the factor that determines the bent vs. linear NO ligands in octahedral complexes is the sum of electrons of π-symmetry. Complexes with π-electrons in excess of 6 tend to have bent NO ligands. Thus, $[Co(ethylenediamine)_2(NO)Cl]^+$, with seven electrons of π-symmetry (six in t_{2g} orbitals and one on NO), adopts a bent NO ligand, whereas $[Fe(CN)_5(NO)]^{3-}$, with six electrons of π-symmetry, adopts a linear nitrosyl [40].

According to Scheme 1, the electronic configuration for $\{M(NO)\}^6$ will be $(e_1)^4(b_2)^2(e_2)^0$. In such an arrangement, or in any other with fewer metal d electrons (for six-coordinated complex), the population of the strongly bonding (e_1) and nearly nonbonding metal centered (b_2) orbitals, and the vacancy at the antibonding e_2 orbital explain the multiple bond order along the linear MNO moiety. An addition of one more electron, giving the configuration $(e_1)^4(b_2)^2(e_2)^1$ would result in occupation of a totally antibonding π-type orbital. This enforces bending according to Walsh's rules, distortions of the M–N–O bond angle, and a change in symmetry. Mixing of the a_1 and the x component of the previously designated e_2 orbital (see Scheme 1b) affords a more bonding a' level, mainly π*(NO) admixed with d_{z2}, and an equivalent antibonding level which is mainly d_{z2} in

Scheme 1 Arrangement of molecular orbitals in six-coordinate $\{M(NO)\}^n$ when M–N–O is (**a**) 180° and (**b**) 120°. The MNO bond defines the *z*-axis [5]

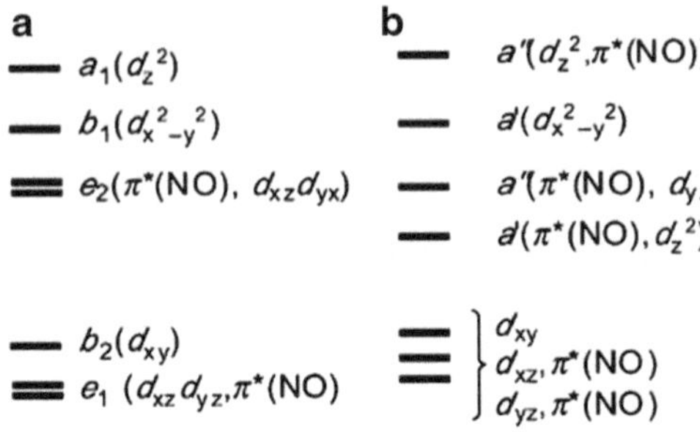

character. Consequent on the bending of the M–N–O bond, the electronic configuration of the frontmost orbitals in $\{M(NO)\}^7$ will be $(a')^1(a'')^0$. This is equivalent to describing the coordinated nitric oxide as NO•. In $\{M(NO)\}^8$, the electronic configuration in this MO system will be $(a')^2(a'')^0$, representing the coordination of singlet NO^-. The molecular orbital scheme shown in Scheme 1 assumes a relatively strong ligand field, in which the separation of the "t_{2g}" and "e_g" levels is significant [5].

Linear and bent NO ligands can be distinguished using infrared spectroscopy. Linear M–N–O groups absorb in the range 1,650–1,900 cm^{-1}, whereas bent nitrosyls absorb in the range 1,525–1,690 cm^{-1} (please note that these regions overlap). The differing vibrational frequencies reflect the differing N–O bond orders for linear (triple bond) and bent NO (double bond). Application of amendment factors to ν(NO) values proposed by De La Cruz et al. [6], depending on the environment of the central ion and electron configuration, allows a more unambiguous differentiation of the spectral characteristics of nitrosyls (see also the spectroscopic chapter by Lewandowska).

3 Representative Classes for Metal Nitrosyls of Biological Significance

3.1 Homoleptic Nitrosyl Complexes

In contrast to metal carbonyls, homoleptic nitrosyls are rare and include $[M(NO)_4]$ (where M= Cr, Mo, W, the premier member being $Cr(NO)_4$ [25]), $[Co(NO)_3]$, $[Rh(NO)_3]$, $[Ir(NO)_3]$, as well as transient copper and iron nitrosyls [41]. The structure of the latter has been solved just recently, in the work of Lin et al. [14]. The iron teranitrosyl displays three nitrosyl bands, at 1,776, 1,708, and 1,345 cm^{-1} in KBr, and the band at 1,345 cm^{-1} is assigned to the bent N–O vibration. The slightly lower-energy NO bands of complex 1 shifted by ~4 cm^{-1} from those of $[(NO_2)_2Fe(NO)_2]$ (1,782, 1,712 cm^{-1} in KBr) reflect the similar electron-donating ability and

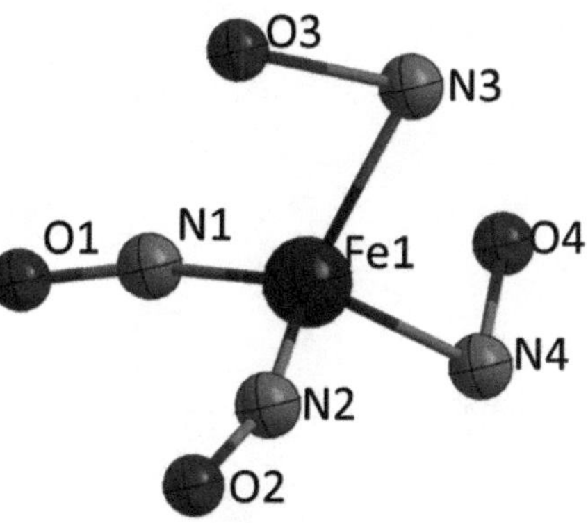

Fig. 1 Spatial structure of dinitroxyldinitrosyl iron. Reprinted from [14], with the permission of the American Chemical Society, copyright 2012

the binding affinity of NO^- and $NO^{2\bullet-}$ ligands toward the $\{Fe(NO)_2\}^9$ motif. The various character of the two coordination modes is explicitly seen in the X-ray structure as the difference between the M–N–O angles (see Fig. 1). Its electronic structure thus is best described as a $\{Fe(NO)_2\}^9$ motif coordinated by two nitroxyl (NO^-) ligands. The infrared and Raman spectra of homoleptic nitrosyls together with DFT calculations are helpful in solving nitrosyl structures of biological importance, as will be shown in the chapter on spectroscopy by Lewandowska.

3.2 *Roussin's Red and Black Salts*

Roussin's red and black salts were among the earliest synthesized nitrosyl complexes. These iron–sulfur nitrosyls were discovered by Roussin while studying the action of sulfur on solutions of sodium nitroprusside and described in 1858 [42].

Roussin's red salt, $K_2[Fe_2S_2(NO)_4]$ is a historically important complex of a great interest for biochemists, being the first synthetic model of iron–sulfur cluster, analogical to those present widely in proteins (*vide infra*). Moreover, it is already a nitrosylated form of such a cluster, and the cluster nitrosylation just recently has been widely studied, since this process is associated with the regulatory action of many non-heme iron–sulfur enzymes.

Roussin's red salt anion is an edge-shared bitetrahedron, wherein a pair of $Fe(NO)_2$ units are bridged by a pair of sulfide ligands (Scheme 2). The Fe–NO bonds are close to linear indicating that NO is acting as a three-electron donor. The diamagnetic compound obeys the 18-electron rule, and each iron center is assigned the oxidation state of Fe(−I). It is formed during the nitrosylation of the dinucleated Rieskie-type clusters in proteins [43].

Roussin's black salt has a more complex cluster structure. The $[Fe_4(NO)_7S_3]^-$ displays an incomplete cubane geometry of the anion cluster and consists of a tetrahedron of iron atoms with sulfide ions on three faces of the tetrahedron (see Scheme 3). The point group symmetry of the anion is C_{3v}. Three iron atoms

Scheme 2 Roussin's red salt

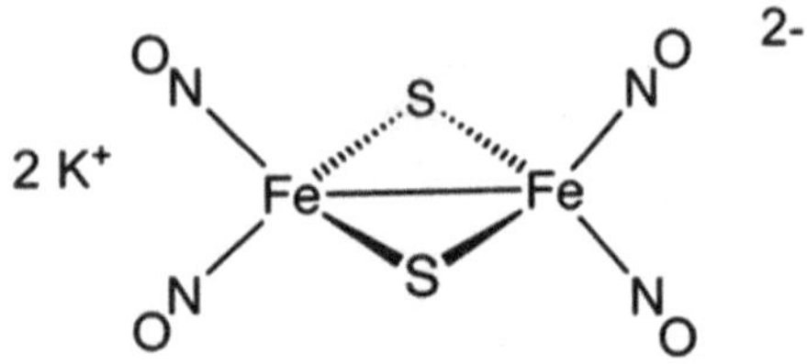

Scheme 3 Roussin's black salt

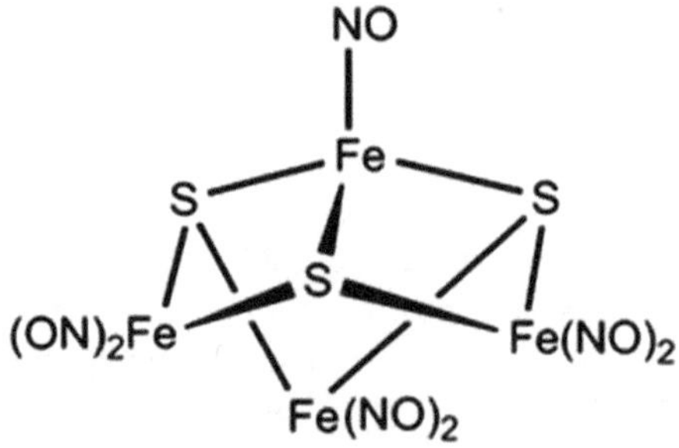

are bonded to two nitrosyl groups. The iron atom on the threefold symmetry axis has a single nitrosyl group which also lies on that axis. Roussin's red and black salts can interconvert among each other upon the change of the pH, according to the reaction given below:

Roussin's Black Salt with the formula $\mathbf{NaFe_4S_3(NO)_7}$ displays an incomplete cubane geometry of the anion cluster. The point group symmetry of the anion is C_{3v} and possesses one $\{FeNO\}^7$ and three $\{Fe(NO)_2\}^9$ units. RBS and RRS interconvert among each other, depending on the pH, from RBS in acidic, to RRS in alkaline solution. Roussin's red salt and its esters (RREs, $Fe_2(SR)_2(NO)_4$, where "R" is an alkyl group) are dinuclear forms of dinitrosyl iron complexes (DNICs, *vide infra*). They are diamagnetic and EPR silent due to the antiferromagnetic coupling between the dinuclear irons. The esters are being investigated as nitric oxide donors in biology and medicine, due to the relatively low toxicity and good stability. In addition, Roussin's salts are discussed in the fields of microbiology and food science due to their mutagenic properties. Their bactericidal effect on the food-spoilage bacteria was demonstrated (see, e.g., [44]). It is known that RREs act as promoters for the carcinogenic properties of other substances for a long period of time (see, e.g., [45]). At the same time attempts to use them against melanoma cancer cells have proven promising [46]. Photolysis of the compounds induced the release of NO, thereby sensitizing target cells to exposure to radiation. Both black and red roussinate anions undergo photodecomposition in aerobic solution to give, eventually, ferric precipitates plus NO, the red roussinate anion being more photoactive. This property of roussinates enables developing photochemical strategies for delivering NO to biological targets on demand [47, 48].

3.3 Mononitrosyl Iron Complexes

Mononuclear mononitrosyl iron complexes (MNIC), designated as $\{Fe(NO)\}^7$ in the Enemark-Feltham notation, have an $S = 3/2$ ground state in contrast to heme-type $\{Fe(NO)\}^7$ species for which $S = 1/2$. EPR spectra pointing to formation of MNIC are obtained upon NO treatment of various non-heme iron proteins, such as rubredoxins and mammalian ferritins; yet, these nitrosyls are of transitory character and readily disproportionate into DNICs and Fe(III) thiolates [49]. Reaction of aquated Fe^{2+} salts with NO affords a six-coordinate iron nitrosyl, $[Fe(NO)(H_2O)_5]^{2+}$, an $\{Fe(NO)\}^7$ species. This complex also is formed in the so-called brown ring test for NO_2^- used in simple qualitative analysis [50]. Four-coordinate mononitrosyl iron complexes such as $[Fe(NO)(SR)_3]^-$ are rare but have been known for several decades [51, 52]. Nitrosylation of the biomimetic reduced- and oxidized-form rubredoxin was shown to give the extremely air- and light-sensitive mononitrosyl *tris*(thiolate) iron complexes. Transformation of $[Fe(NO)(SR)_3]^-$ into dinitrosyl iron complexes (DNICs) $[(RS)_2Fe(NO)_2]^-$ and Roussin's red ester $[Fe_2(\mu\text{-}SR)_2(NO)_4]$ occurred rapidly under the addition of 1 equivalent of NO(g) and $[NO]^+$, respectively [53]. Chemistry of the Fe(III) thiolate, $[Fe(SR)_4]^-$, with NO also proceeds to the DNIC through the MNIC, like in case of the Fe(II) complex [53] (Scheme 4). The reactivity of these iron thiolates demonstrates that RS− ligands play a role of reductants during the process of transformation of Fe(II) or Fe(III) to $\{Fe(NO)_2\}^9$ species.

3.4 Dinitrosyl Iron Complexes

An important and representative group of biologically significant nitrosyls are dinitrosyl iron complexes (DNICs). Their occurrence has been observed in many kinds of organisms and in a wide spectrum of physiological conditions associated with inflammation, Parkinson's disease, and cancer. Accumulation of DNICs coincides with intensified production of nitric oxide in macrophages, spinal cord, endothelial cells, pancreatic islet cells, and hepatocytes [55–57]. DNICs are also important in NO-dependent regulation of cellular metabolism and signal transduction [58–61]. Dinitrosyl iron complexes (DNICs) and *S*-nitrosothiols (RSNO) have been known to be two possible forms for storage and transport of NO in biological systems [61]. Depending on the micro-environment, the low molecular weight (LMW) DNICs can provide at least two types of nitrosylating modification of proteins, forming either protein-*S*-nitrosothiols or protein-bound DNICs. Also, DNICs were shown to mediate the iron-catalyzed degradation and formation of *S*-nitrosothiols [53, 62]. Abstracting from the subtitle of this section it is worth to note that the reversible degradation of S-nitrosothiols also was shown to be catalyzed by copper ions [63]. The interaction of NO with $Cu(DTC)_2{\cdot}3H_2O$ (DTC: dithiocarbamate) was studied and the formation of two stable nitrosyl complexes, $Cu(DTC)_2NO$

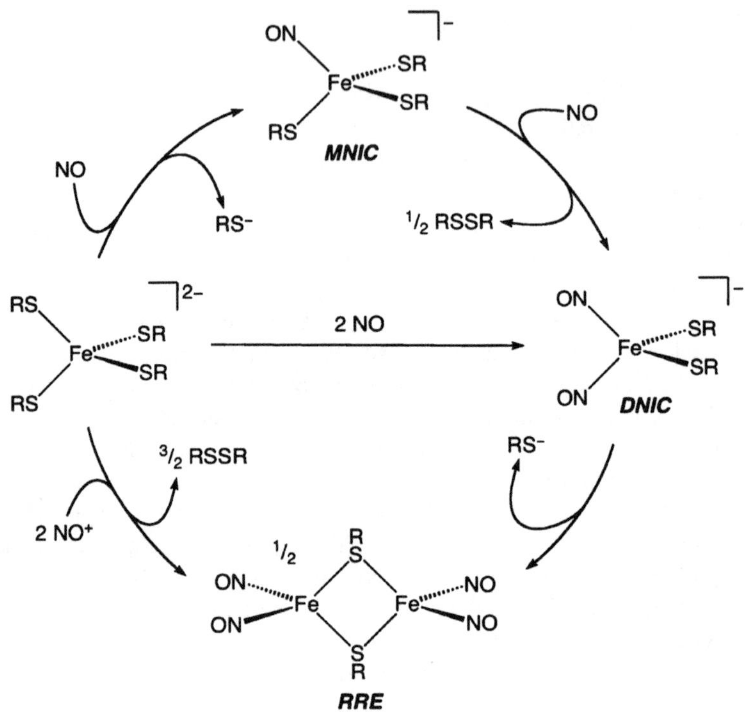

Scheme 4 Redox interconversions between MNIC, DNIC, and RREs. Reprinted from [54], with permission of the American Chemical Society, copyright 2010

and $Cu(DTC)_2(NO)_2$, was proven. The complexes were air-stable and were not disrupted by purging the solution with an inert gas. $Cu(MorDTC)_2NO{\cdot}3H_2O$ was isolated in the solid state and its NO (IR) band was observed at 1,682 cm^{-1}. The latest results by Lim et al., concerning the mechanism of reduction of the 2,9-dimethyl-1,10-phenanthroline copper(II) complex by NO, support an inner sphere mechanism, where the first step involves the formation of a copper–nitrosyl (Cu(II)–NO or $Cu(I)–NO^+$) adduct, and argue against an outer sphere electron transfer pathway, which was assumed earlier [64]. LMW-DNICs may release NO to various targets. As observed in cells or tissues, LMW-DNICs exerting cyclic GMP-independent effects were attributed to the nitrosylating modification of proteins via transfer of NO or $Fe(NO)_2$ unit, thus yielding protein *S*-nitrosothiols or protein-bound DNICs [65, 66]. The transfer of the $Fe(NO)_2$ motif of LMW-DNIC resulting in the formation of protein-bound DNIC serving as an NO-storage site also was demonstrated in isolated arteries, although the persistent *S*-nitrosylation of protein is another mechanism of formation of releasable NO-storage site in arteries [67–70]. LMW-DNICs elicited an NO release-associated relaxant effect in isolated arteries [71]. Also, it has been proposed that free thiols/thiolates can displace the proteins of the protein-bound DNICs via thiolate exchange to LMW-DNICs [72, 73]. Protein-bound DNICs are stable for hours and can accumulate in high concentrations in tissues, their stability being lower in the presence of LMW-thiols, due to transnitrosylation [72–74]. The complex relationship between iron and NO and the putative biological role of DNIC have been reviewed by Richardson and Lok [75] and Lewandowska et al. [76]. Although DNICs chemistry has been studied since early 1960s, determination of the electronic structure of the core complex is still a matter of research and discussion [77]. Among the very few dinitrosyl complexes of well-defined structural and spectral parameters are metal dinitrosyl halides [78, 79], $[Fe(NO)_2(CO)_2]$ [80], osmium and ruthenium phenylophosphinates [13, 81–83].

The ON–M–NO angles for DNICs lie between 180° in the *trans* case and 90° in the *cis* case. This has further spectroscopic implications, as further discussed in this volume, in the chapter on spectroscopy. When the two equivalent bent NO groups are attached to the same metal atom in four-coordinate complexes, they can conform either with the NO moieties slightly bent towards each other, or shooting out in opposite directions, in other words N–O–M–O–N fragment can adopt either the *attracto* or the *repulso* conformation. Martin and Taylor [84] concluded that the *attracto* form is characteristic for those dinitrosyls, where N–M–N angle is less than ~130°, and *repulso* form occurs in complexes with α(NMN) greater than ~130°.

Regarding the electronic structure of the $\{Fe(NO)_2\}$ core present in DNICs, the known stable DNICs can be classified into three groups: the paramagnetic mononucleated $\{Fe(NO)_2\}^9$ DNICs, the dimerized forms of $\{Fe(NO)_2\}^9$, which are diamagnetic due to electron pairing, and the diamagnetic $\{Fe(NO)_2\}^{10}$ DNICs. Ligands contained in $\{Fe(NO)_2\}^9$ type complexes are (1) thiol groups of amino acids or inorganic sulfur (II) atoms in sulfur clusters, (2) imidazole rings present, e.g., in histidine and purines, (3) pyrole rings of heme-type prosthetic groups [58, 61, 68, 70, 85]. The $\{Fe(NO)_2\}^{10}$ DNICs are coordinated by CO, PPh_3, and N containing ligands [86]. Prototypical DNICs, which contain a single iron atom, are paramagnetic low spin $S = ½$ $\{Fe(NO)_2\}^9$ species. The doublet ground state gives rise to a characteristic axial EPR signal at $g_{av} = 2.03$ [85]. This common spectroscopic signature was a hallmark of DNICs occurrence in both synthetic and biological contexts, dating back several decades to early studies on cancerous liver samples [87]. DNICs typically take the form of an $[Fe(NO)_2(X)_2]^-$ anion, where, in the case of biological milieu, X is a ligand such as a protein-based cysteinate residue or an LMW thiol, like glutathione. The nature of the non-nitrosyl ligands in tissue-derived DNICs is uncertain, and it is assumed that beside thiolate ligands, also imidazole rings present in residues, such as histidine or purines, can form LMW-DNICs. In high molecular weight complexes, thiols are responsible for DNIC formation in non-heme iron proteins, whereas histidine participates in the formation of iron complexes containing heme ligands [76]. A single example of a crystallographically characterized, protein-bound DNIC contains a tyrosinate ligand [88]. Thus, the similarity of the EPR spectra of thiol- and histidine-derived dinitrosyl complexes in aqueous media precludes unequivocal assignment of their biological role [57, 77, 89]. While dinitrosyl-dithiol iron (II) complexes are well-characterized species in which iron is coordinated by two sulfur atoms and two NO molecules [57], the biological importance of complexes of histidine, NO, and iron has scarcely been investigated. In the very early publication of Woolum [90], it was proposed that the N7 atom of histidine imidazole ring is responsible for coordination of iron and DNIC formation in non-thiol proteins.

According to several authors, the toxicity of DNIC's components seems to be mutually dependent on each other, but the reported results give an ambiguous picture. Bostanci et al. [91, 92] reported attenuation of iron-induced neurotoxicity by nitric oxide synthase inhibitors. On the other hand, the presence of NO donors was reported to protect against iron-induced nephrotoxicity [93, 94]. The reported ability of iron ions to rescue tumor cells from the pro-apoptotic effects of NO is in line with these results, showing a mutual interrelationship of nitric oxide and iron toxicity [95]. Another effect is attributed to coinciding active transport of iron and glutathione outside the cells [96] that was proven to be dependent on MRP1 [97]. The dependence of iron release on glutathione provides evidence that intracellular iron might be depleted via MRP1 in the form of DNIC. It was also shown that depletion of glutathione rendered the cells vulnerable to NO donors [98]. Recently more and more data have been collected on the regulatory functions that DNICs may have in proteins. In addition to the positive regulatory roles of DNICs that are described in the following sections, DNICs are also known for their toxic effects [99]. In particular, diglutathionyl DNIC is a potent and irreversible inhibitor of glutathione reductase [100, 101]. At the same time the formation of DNICs may play a protective role due to their higher stability than that of NO [102]. Denninger et al. found that the formation of DNICs decreases the labile iron pool (LIP[2]) and therefore makes the cell less susceptible to oxidative stress [105]. Several works suggested that LIP is a target for •NO complexation [97, 106, 107]. These findings imply the role of DNICs in the cellular mechanisms of protection against a labile iron surge during inflammation.

DNICs can exist in equilibrium with their dimeric analogs, Roussin's red esters [108]. Several factors including solvent polarity, concentration, and the nature of the thiolate ligand can influence which species predominates in solution [49, 68, 109–112]. In the case of LMW thiolate DNICs, an excess of ligand (usually ca. 20-fold) is required to maintain the mononucleated form. If the concentration of thiol compound is low, the ions are promptly condensed into a binuclear RRE structure via formation of two RS-bridges and release of other two RS-ligands, according to Eq. (1).

[2] The labile iron pool has recently been defined by Cabantchik et al. [103] as the pool of iron labilly bound to low-molecular complexes available for redox reactions. Typical LIP concentration in the cell does not exceed 1 μM [104].

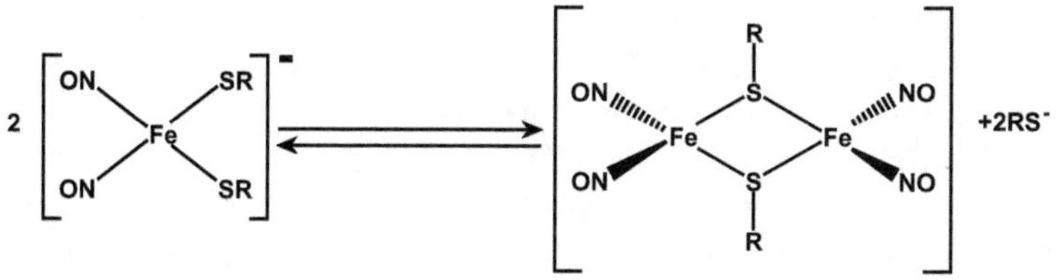

(1)

The above reaction is, in fact, an electrochemically reversible one-electron reduction corresponding to the $\{Fe(NO)_2\}^{9/10}$ couple [113]. In the presence of a different thiol, its S-nitrosation, yielding nitrosothiol and other decomposition products, can also occur [68, 114]. This behavior is responsible for the role of $[(RS)_2Fe(NO)_2]^-$ in the NO storage and NO-transport occurring in vivo [114]. Theoretical explanation of the fact that DNICs tend to form bi-nucleated complexes was presented by Jaworska and Stasicka [115]. According to these authors, the effective overlap of the spatially extensive HOMO orbital (derived from sulfur orbitals) and the LUMO orbital of the d type between two molecules can result in the formation of the RS-bridge. This reaction can be further facilitated by solvent-enhanced polarization of the S–C bond, a contribution of S orbitals to HOMO, and distorted tetrahedral geometry of the complex. A considerable contribution of π^*_{NO} to LUMO and polarization of the NO bond enable the RS nucleophile attack followed by the Fe–NO bond cleavage. Not surprisingly, one-electron reduction of RRE derivatives also gives rise to the corresponding rRREs. Reduced Roussin's red esters have been detected in nitrosylated protein samples that have been subjected to reduction [116]. The $\{Fe(NO)_2\}^9$ units can also occur as a component of several different structures, including a product of [4Fe–3S] cluster nitrosylation known as Roussin's black salt (see above) [51].

As to the geometry of the coordination sphere in DNICs, the anisotropy of the *g* values, determined from the electron spin resonance spectra of frozen solutions, varies considerably from complex to complex. The results are consistent with the view that all these complexes have a distorted tetrahedral geometries, but the extent of the distortion depends on the ligands. As a result of this variation there are changes in the nature of the spin-containing d orbital. Ligands containing hard, nonpolarizable donor atoms such as oxygen or fluorine produce a distortion towards a planar geometry, placing the odd electron in a predominantly d_{x2-y2} orbital, while those containing softer donor atoms such as phosphorus or sulfur give complexes with a different type of distortion, leading to placement of the odd electron in a predominantly d_{z2} orbital. Nitrogen and halide donor ligands produce smaller distortions, leading to spin-containing molecular orbitals with contributions from a mixture of d orbitals. Bryar et al. [117] suggested that complexes of this type have a structure based on a trigonal bipyramid with a missing ligand. Costanzo et al. [110], based on spectroscopic data, determined the geometry of those complexes as distorted tetrahedral (the electronic configuration $\{Fe\ (NO)_2\}^9$). This conclusion was based on a comparison of the molar extinction coefficients of absorption bands for d–d transitions in typical tetrahedral and octahedral complexes of iron. For

DNICs the extinction coefficients are much higher than those for octahedral complexes and suggest a tetrahedral environment. Since 2001 these predictions have been confirmed in numerous other papers [6, 118–121].

Three electronic states of $\{Fe(NO)_2\}$ are recognized, i.e. paramagnetic $\{Fe(NO)_2\}^9$, diamagnetically coupled $\{Fe(NO)_2\}^9$–$\{Fe(NO)_2\}^9$ ligand-bridged RREs, and $\{Fe(NO)_2\}^{10}$ species [122]. The electronic and geometric structures of paramagnetic iron dinitrosyl complexes were investigated using electron spin resonance, infrared spectroscopy, and X-ray crystallography. It was concluded that these compounds are best described as 17 electron complexes with a d^9 configuration, the unpaired electron being localized mainly around the iron atom. NO groups have partial positive charges (bent nitrosyl groups) and the metal atom charge is close to zero [115]. The earliest conclusion that the electron configuration of iron dinitrosyl complexes is best described as d^9 was formulated in the study of Bryar and Eaton [117], yet proposals for other electronic configurations emerged as well [89], due to the fact that DNICs in solution give characteristic EPR signals implying axial symmetry, while X-ray analyses of DNIC crystals reveal a distorted tetrahedral structure. This inspired Vanin et al. [89, 123, 124] to conclude that in solution d^9 complexes transform into square-planar structures with d^7 configuration, with the unpaired electron localized on the d_{z2} iron orbital. Consequently, taking into consideration the low-spin state of DNICs, d^7 configuration would result in square-planar DNIC spatial structure, whereas the d^9 would imply tetrahedral structure. Crystallographic studies of DNICs with some LMW thiols support the assumption of the deformed tetrahedral configuration with C_{2v} symmetry [110, 111] supporting this suggestion, but it should be noted that (as stated above) the symmetry and coordination of crystalline compounds changes in the solution where their occurrence and metabolism can be observed [89, 123, 124]. Initially it was assumed that DNICs with thiol ligands have octahedral geometries, in which additional ligands, or solvent molecules, are located axially to the plane of the four remaining ligands [123]. The infrared spectroscopy revealed the existence of the inter-isomeric redox equilibrium of the linear and bent NO groups of two forms of $[Ru(NO)_2(PPh_3)_2(Cl)]^+$ [82]. A trigonal bipyramidal form with two linear NO groups has been proposed as the transition state for this process ($E_a < 35\ \mathrm{J\ mol^{-1}}$). This is not a unique phenomenon in solution, as shown by the fluxional interconvertibility between the linear NO group of the trigonal bipyramidal structure and the bent NO group of the tetragonal pyramidal structure of a series of compounds of general formulae $[Co(NO)\{P(XY_2)_2\}(Cl)_2]$ [where (X,Y) = (Et/Et), (Bu/Bu), (Me/Ph), (p-Me-C_6H_4) or (Ph/Ph)] [82, 125] (see also spectroscopic chapter by Lewandowska).

The final and decisive conclusion on the electronic configuration and its interchanges among the different isomeric dinitrosyl iron complexes has recently been reviewed by Lu et al. [126]. Their comprehensive research and analysis has unambiguously identified the reduced DNIC as a bimetallic species containing antiferromagnetically coupled $\{Fe(NO)_2\}^9-\{Fe(NO)_2\}^{10}$ centers. This electronic configuration gives rise to an S= ½ ground state and an axial EPR signal centered at $g_{av} = 1.99$. The series of papers by Liaw and coworkers [127–132] presents a thorough and comprehensive analysis of the changes in the electronic structure that occur during the syntheses of DNICs from mononitrosyl complexes, the transformations between four-coordinated and six-coordinated DNICs, as well as the $\{Fe(NO)_2\}^9$–$\{Fe(NO)_2\}^{10}$ interconversions. Changes in the structural properties are illustrated by the spectral shifts and appropriate electronic states have been assigned. An exceptionally interesting example of how the EPR signal geometry changes from axial to rhombic along with the shift between the six-coordinated and four-coordinated $\{Fe\ (NO)_2\}^9$ complex is presented in the paper of Tsai et al. [133]. In contrast to tetrahedral $\{Fe(NO)_2\}^9$ DNICs with an EPR g value of 2.03, the six-coordinate $\{Fe(NO)_2\}^9$ DNIC displays an EPR signal with $g_{av} = 2.013$ (in THF solution). The temperature-dependent reversible transformation occurs between both DNICs. The addition of 2 equivalent of PPh_3 into the hexacoordinate complex promotes O-atom transfer of the chelating nitrito ligand to generate $OPPh_3$, the neutral EPR-silent $\{Fe(NO)_2\}^{10}$ tetracoordinate complex, and NO. Recently, the group of Liaw presented a series of papers on the reversible interconversions between d^9 and d^{10} forms of DNICs. Cyclic voltammetry experiments indicated pseudo-reversibility of the $\{Fe(NO)\}^9 \longleftrightarrow \{Fe(NO)\}^{10}$ redox process, suggesting that the $\{Fe(NO)\}^{10}$ DNICs with thiolate coordination are not as stable as their $\{Fe(NO)\}^9$ counterpart [86]. This finding implies that higher electron density around the Fe center in the $\{Fe(NO)\}^{10}$ DNICs no longer favors an electron-rich coordination sphere by anionic thiolate ligation. According to the EPR Fe K-edge spectroscopic results, the electronic structure of the dimeric d^{10} $[\{Fe(NO)_2\}]_2L_2$ complex is best described as $\{Fe^{II}(NO^-)_2\}^{10}$. The binding affinity of $[SR]^-$ type ligands is greater than that of $[OPh]^-$ in the examined $\{Fe(NO)_2\}^{9/10}$ type complexes, which explains the fact that the most of the DNICs and RREs in living organisms are bound to proteins through cysteinate side chains. As indicated in Scheme 5, the dinuclear DNICs can be classified according to oxidation levels and configurations: the EPR-silent neutral and diamagnetically coupled $\{Fe(NO)_2\}^9\{Fe(NO)_2\}^9$ RRE [134, 135], the EPR-active neutral $\{Fe(NO)_2\}^9\{Fe(NO)_2\}^9$ RRE containing two separate $\{Fe(NO)_2\}^9$ motifs [136], the EPR-silent anionic $\{Fe(NO)_2\}^9\{Fe(NO)_2\}^9$ RRE containing mixed thiolate–sulfide-bridged ligands [137], the EPR-silent $\{Fe(NO)_2\}^9\{Fe(NO)_2\}^9$ Roussin's red salt [138], the EPR active $\{Fe(NO)_2\}^{10}\{Fe(NO)_2\}^9$ reduced RREs [122, 126], and the EPR-silent $\{Fe(NO)_2\}^{10}\{Fe(NO)_2\}^{10}$ dianionic reduced RREs [128]. A direct interconversion between $\{Fe(NO)_2\}^{10}$ DNIC and the $\{Fe(NO)_2\}^9$ DNIC was demonstrated. On the basis of IR ν(NO) stretching frequencies, Fe–N (N–O) bond distances and Fe K-edge pre-edge energy values, the electronic structure of $\{Fe(NO)_2\}^{10}$ core is best described as $\{Fe^{II}(NO^-)_2\}^{10}$. Interconversions were shown to be driven to a specific pathway by the site selective interactions. The distinct S K-edge pre-edge absorption energies and patterns can prove efficient tools for the characterization of the various oxidation-state dinuclear DNICs. The reviewed results indicate that protein-bound $\{Fe(NO)_2\}^{10}$ DNICs dianionic reduced RREs may exist in living organisms. Also, the cysteine-containing $\{Fe(NO)_2\}^{10}$ DNICs, $\{Fe(NO)_2\}^{10}\{Fe(NO)_2\}^{10}$ RREs, and $\{Fe(NO)_2\}^{10}\{Fe(NO)_2\}^9$ reduced RREs may be regarded as the potential species derived from nitrosylation of [Fe–S] proteins. The diversity of DNICs/RREs in

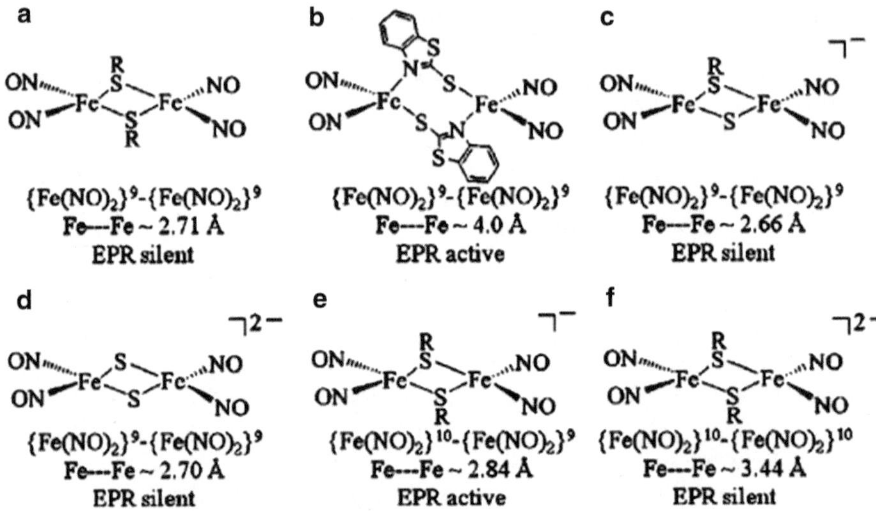

Scheme 5 The various forms of DNICs. Reprinted from [128], with the permission of the American Chemical Society, copyright 2012

biological evolution and bioavailability of a given DNICs/RREs during repair of the modified [Fe–S] proteins remain open questions.

3.5 Iron–Sulfur Cluster Nitrosyls

Iron–sulfur [Fe–S] clusters are ensembles of iron and sulfide centers ubiquitously present in proteins, where they function as prosthetic groups. Fe–S clusters are evolutionary ancient and are involved in sustaining fundamental life processes. The group of known regulatory proteins that contain an iron–sulfur cluster cofactors is growing both in number and in importance, with a range of functions that include electron transfer, sensing of molecular oxygen, stress response, substrate binding/activation, iron regulation and storage, regulation of gene expression. They are involved in a number of enzymatic activities and as well possess structural functions. In some cases, the cluster is required for the protein to attain its regulatory form, while in others the active form requires loss or modification of the cluster [139]. Three main types of [Fe–S] clusters present in biosynthetic systems are presented in Scheme 6. Various synthetic iron sulfur clusters have been synthesized [54, 141].

Most of the research studying the binding of NO to synthetic FeS clusters has been aimed at molecular and/or electronic structural issues and to facilitate the understanding of the reactivity of the coordinated NO group. The reaction of synthetic iron–sulfur clusters (both [2Fe–2S] and [4Fe–4S]) with nitric oxide was

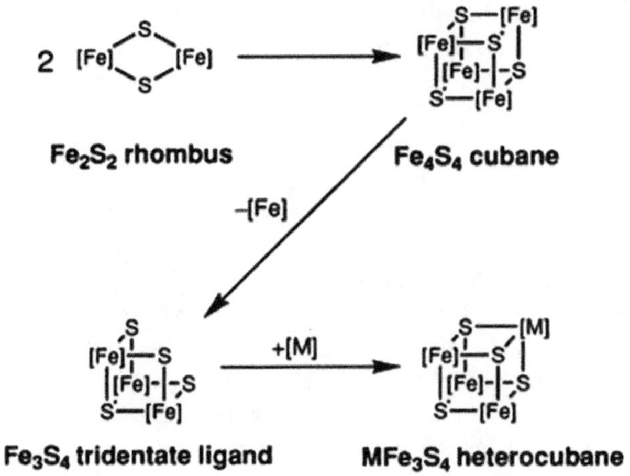

Scheme 6 Some of the important iron–sulfur cluster units found in metalloenzymes. [2Fe–2S] rhombus cluster is characteristic of [2Fe–2S] ferredoxins and Rieskie proteins, [4Fe–4S] cubane – e.g., in [4Fe–4S] ferredoxins, aconitase; [3Fe–4S] clusters are present in the inactive form of aconitase, [3Fe–4S] ferredoxins. The iron vertices, designated as [Fe], have high-spin tetrahedral FeS_4 coordination. Reprinted from [140], with the permission of Elsevier, copyright 2000

first communicated in 1985 [49]. These studies demonstrated the propensity for cluster disassembly by NO and NO^{2-}. Subsequently, it was shown that this process involves the formation of DNICs. The mechanism of cluster nitrosylation was shown to proceed with modification of the sulfur ligands, not thiolates, as in case of homoleptic iron thiolates. An exemplary reaction of that type, for the oxidized synthetic [2Fe–2S] type complex, is given by Eq. (2).

$$\left[Fe_2S_2(SR)_4\right]^{2-} + 4NO \rightarrow 2\left[Fe(NO)_2(SR)_2\right]^{-} + 2S^0 \qquad (2)$$

The elemental sulfur that is formed by destruction of the [2Fe–2S] cluster can further react with the DNIC, resulting in oxidation of the thiolate ligands to disulfide and formation of the Roussin's black salt in the reaction of unknown stoichiometry, which can be formally formulated as follows [Eq. (3)].

$$4\left[Fe(NO)_2(SR)_2\right]^{-} + 4S^0 \rightarrow 4RSSR + NO^{-} + S^{2-} \qquad (3)$$

The described reaction proceeds slowly for aryl-thiolate DNICs and is fast for alkyl-thiolate complexes. Trapping of the elemental sulfur atoms by phosphine inhibits oxidation [137]. The fate of the sulfide ligands in reactions of biological iron–sulfur clusters with NO remains unknown, although analogically it is expected that sulfur by-products must be sequestered from the vicinity of the iron atoms in order for the DNICs to remain stable. It was shown that the cluster reassembly proceeds in vivo in aerobically growing *Escherichia coli* cells in the presence of L-cysteine and cysteine desulfurase (thus, via sulfur atom transfer), which is consistent with results using synthetic clusters [142]. The synthetic cluster can be regenerated from the DNIC, in the presence of S_8 and Fe(1,2-benzenedithiolate)$_2$, which serves as a trap for NO [86] or via formation and subsequent reduction of Roussin's red salt in the transnitrosylation reaction with $[Fe(SR)_4]^-$ (this being, however, not relevant to a possible model of [2Fe–2S] repair in vivo). The described processes are affected by O_2, due to nitric oxide oxidation. Consequently, it was shown that a synthetic Rieske-type [2Fe–2S] cluster, [Fe2(μ-S)$_2$(BIPM)(S_2-*o*-xyl)]$^{2-}$(BIPM = 2,2′-(phenylmethylene)*bis*(3-methylindole), S_2-o-xyl = dianion of 1,2-phenylenedimethanethiol), could be nitrosylated to the corresponding *N*-bound and *S*-bound DNICs [143–146]. Unlike DNICs containing thiolate ligands, DNICs containing nitrogen ligands might be expected to undergo redox reactions without decomposition because of a lesser propensity for ligand oxidation. Unlike DNICs, RRE derivatives are EPR silent due to antiferromagnetic coupling between the two $\{Fe(NO)_2\}^9$ units, which gives rise to a diamagnetic ground state. Consequently, tracking the emergence of these species in vivo is more difficult. Presently, a complementary use of UV–vis, Raman, Mössbauer, and nuclear resonance vibrational spectroscopy (NRVS) is employed to aid the identification of protein-bound RREs [147, 148]. A detailed structural characterization of model DNIC-type complexes and mechanistic data on their interconversions are presented by the group of Liaw [14, 53, 86, 122, 126, 127, 136, 149–152]. Of special interest, as regards nitrosylation of proteins, is their recent work on the peptide-bound DNICs and RREs/rRREs [130]. Using aqueous IR, UV–vis, EPR, CD, XAS, and ESI-MS the authors provide data on the dynamic equilibria between high- and LMW DNICs, thus giving insight into the possible mechanisms of interaction between DNICs in nitrosylated proteins. Reaction of [4Fe–4S] clusters from aconitase [116], HiPIP [147], and endonuclease [153] with nitric oxide leads to formation of protein-bound DNICs. As with [2Fe–2S] clusters, the repair of [4Fe–4S] clusters by DNICs has been demonstrated in both protein and synthetic systems [54]. According to the recent results of Crack et al. [145, 154, 155] and Smith et al. [156], the [4Fe–4S] cluster nitrosylation (in *M. tuberculosis* WhiD/WhiB1 proteins, transcription factors in *Actinobacteria*, required for differentiation and sporulation) proceeds via a complex, multiphasic reaction, with a rate at least ~4 orders of magnitude greater than for the reaction with O_2. The first step of the [4Fe–4S] nitrosylation was shown to be first order with respect to NO, and it was concluded that during that process one NO molecule binds to the cluster. This increased the accessibility to further NO binding in the two following steps: these reactions could be observed spectrophotometrically, at 360 nm and 420 nm, respectively. Both were first order for NO, indicating either that a single NO is involved in each one or that NO binding to different irons of the cluster occurs independently, giving an overall first-order dependence. The fluorescence titration data revealed a stable intermediate at a stoichiometry of ~4 NO molecules per [4Fe–4S] cluster. The last step of the reaction was again first order in NO leading to an EPR-silent octa-nitrosylated product(s) with the overall stoichiometry Fe:NO = 1:1.

3.6 *Nitrosylated Porphyrins and Porphyrin Analogues*

The reactions of NO with heme are of great biological significance. The first known physiological target of NO was the soluble guanylate cyclase (sGC). This intracellular NO-sensing enzyme involved in vasodilation is a protein receptor for NO; see Sec. 4.2.1. The generation of NO in vivo is catalyzed by the nitric oxide synthase (NOS) class of enzymes, which belong to the family of heme-thiolate enzymes [157]. NO binds to the ferrous heme in sGC and releases the heme-ligating histidine, resulting in a heme Fe^{2+}–NO complex formation. This reaction triggers a change in heme geometry and a subsequent conformational change of the protein to an enzymatically active form [105]. Hemoglobin nitrosyl complexes are formed with both reduced ($Hb(Fe^{2+})$) and oxidized ($Hb(Fe^{3+})$ – metHb) hemoglobin. The iron atom is most often in the reduced state and paramagnetic heme-Fe^{2+}(NO) is formed. The ferrous porphyrin has a greater affinity for NO. In that case the nitrosylation reaction is reversible (see Table 2 for kinetic data). Another heme-type target for NO is cytochrome c oxidase (CcOx, *aliter* Complex IV), the last enzyme in the respiratory electron transport chain of mitochondria (or bacteria) located in the mitochondrial (or bacterial) membrane. Nitrosyl complex formation inhibits cytochrome *c* peroxidase activity due to the impediment of H_2O_2 access to heme Fe. Besides nitrosyl complex formation, cytochrome *c* can participate in redox reactions with NO. Reduction of cytochrome *c* with NO is well known and is analogous to that of Hb [Eqs. (4)–(6)] [167].

$$\text{cyt}\, c^{3+} + \text{NO} \leftrightarrow \text{cyt}\, c^{3+} - \text{NO} \tag{4}$$

$$\text{cyt}\, c^{3+} - \text{NO} \leftrightarrow \text{cyt}\, c^{2+} - \text{NO}^{+} \tag{5}$$

$$\text{cyt}\, c^{2+} - \text{NO}^{+} + 2\text{OH}^{-} \leftrightarrow \text{cyt}\, c^{2+} + \text{NO}_2{}^{-} + \text{H}_2\text{O} \tag{6}$$

The cyt *c* $^{3+}$–NO complex can be spontaneously destroyed, and in this case practically the entire NO is converted into nitrite [166]. Besides that, NO can be reduced to nitroxyl anion [NO^-, Eq. (7)], and this reaction is characterized by the rate constant of 200 M^{-1} s^{-1} [166].

$$\text{cyt}\, c^{2+} + \text{NO} \rightarrow \text{cyt}\, c^{3+} + \text{NO}^{-} \tag{7}$$

Both Hb and cytochrome *c* produce stable nitrosyl complexes that in some cases serve as an NO depot in the organism [168–170]. The described nitrosyl complexes are photosensitive and can be destroyed upon irradiation with visible light.

Due to the many biological functions of ferrous heme nitrosyls, many corresponding model complexes have been synthesized and structurally and spectroscopically characterized, viz. tetraphenylporphyrin ($TPPH_2$), octaethylporphyrin ($OEPH_2$) and protoporphyrin IX diester ($PPDEH_2$), meso-*tetrakis* (4-carboxyphenyl) porphyrin (TCPP), meso-*tetrakis* [4-(N,N,N-trimethyl) aminophenyl] porphyrin (TTMAPP),

Table 2 Kinetic and thermodynamic constants for nitric oxide binding to different forms of ferrous and ferric iron

Compound	On rate ($M^{-1}\ s^{-1}$)	Off rate (s^{-1})	Dissociation constant (M)	Reference
Ferrous compounds				
Hemoglobin (R-state)	2×10^7	1.8×10^{-5}	0.9×10^{-12}	[158]
Hemoglobin (T-state)	2×10^7	3×10^{-3}	1.5×10^{-10}	[158, 159]
Myoglobin	1.7×10^7	1.2×10^{-4}	0.7×10^{-11}	[160, 161]
Cytochrome *c*	8.3×10^0	2.9×10^{-5}	3.5×10^{-6}	[162, 163]
Ferric compounds				
Hemoglobin	4×10^3	1×10^0	2.5×10^{-4}	[144]
Hemoglobin α-chain	3.3×10^3	2.1×10^0	1.4×10^{-4}	[144]
Hemoglobin β-chain	1.3×10^4	3×10^0	2.3×10^{-4}	[144]
Myoglobin (sperm whale)	5.3×10^4	14×10^0	2.6×10^{-4}	[162, 164, 165]
Myoglobin (elephant)	2.2×10^7	40×10^0	1.8×10^{-6}	[144]
Microperoxidase	1.1×10^6	3.4×10^0	3.1×10^{-6}	[165]
Cytochrome *c*	1×10^3	3×10^{-2}	3×10^{-5}	[162, 164, 166]

Adopted from [74], with the permission of Elsevier, copyright 1999

and meso-*tetrakis* (4-N-methylpyridinium) porphyrin (TMPyP) [3, 171–173] (for a more detailed review on the porphyrinate nitrosyls, see also chapter by Lehnert et al.). In addition to these porphyrin-type ligands, there are model chelators mimicking porphyrin binding and geometry designed to obtain certain required spectral or dynamic features. Examples are five-coordinate (5C) Schiff base-type tetradentate macrocyclic ligands. These complexes have structural and vibrational features, which make them good spectroscopic models for 5C ferrous heme-nitrosyls by giving less obscure electronic spectra due to absence of very intense $\pi \rightarrow \pi^*$ transitions of the heme macrocycle [174]. Importantly, vibrational spectroscopy has always been a key technique in examination of heme nitrosyl model complexes, because their vibrational properties are very sensitive to the electronic and spin state of the metal, its coordination number, etc. In the case of ferrous heme nitrosyls, it has been demonstrated that binding of an axial ligand (usually an N- or S-donor) in position *trans* to the NO moiety weakens the Fe–NO and N–O bonds in comparison with the corresponding 5C species. In this way, it increases the amount of radical character (spin density) on the coordinated NO by a *trans*-effect [28, 175]. Other examples of the five-coordinated mononitrosyl Schiff base complexes are those prepared by condensation of 2-hydroxy-1-naphthaldehyde and ethylenediamine (naphth-enH_2), *o*-phenylenediamine (naphth-phH_2), and 4-methyl-*o*-phenylenediamine (naphth-mphH_2) with Co(II) and Fe(II) (in methanol) forming the tetragonal pyramidal mononitrosyls [176]. The iron complexes have temperature-independent magnetic moments corresponding to the high-spin electronic structure ($S = 3/2$). The cobalt complexes were shown to be diamagnetic $\{Co(NO)\}^8$ species and were shown to contain bent Co–N–O bonds (ca. 125°), as shown by crystallographic studies, and by $^{14/15}N$ NMR spectroscopy in solution and in the solid state [176, 177]. These diamagnetic complexes, like their porphyrinato analogues, are subject to photolysis of the Co–NO bond.

Vibrational methods applied to proteins and model complexes include IR, resonance Raman, and NRVS [173, 178–180]. The interpretation of these vibrational data combined with DFT studies on the reaction mechanisms of heme proteins is frequently based on the porphine approximation, i.e. all porphyrin ring substituents are neglected. Whereas this is intuitively a good approximation for D_{4h}-symmetric macrocycles (e.g., TPP complexes), it is noteworthy that the biologically observed hemes all contain asymmetric substitution patterns on the porphyrin ring [181].

Iron nitrosyl porphyrins may be five- or six-coordinate, the sixth ligand being usually either an N-heterocycle, water, alkyl/aryl, or NO_2^-. The metal–NO bond angles in the diamagnetic $\{Fe(NO)\}^6$ group of complexes lie between 169° and 180°, i.e. are essentially linear, and the Fe–N(O) distances (in the range 1.63–1.67 Å) are generally independent of coordination number. Occasionally, a slight off-axis tilt of the Fe–N–O bond system was observed. The NO stretching frequencies of this group lie between 1,830 and 1,937 cm^{-1}. In model systems, the NO dissociation rates for the complexes of the same heme-type ligand are usually higher for six-coordinated than for five-coordinated species, in line with the negative trans-effect rule. Nevertheless, the electron donating/withdrawing properties of the substituents to the porphyrin ring are critical for the k_{off} rates [160, 173, 182]. Weakening of the FeNO bond resulting from the axial trans base coordination is associated with a lowering of the ν(NO) frequency. Distortions of the porphyrinato ligand do not have any significant effect on the Fe–N–O bond angle [173, 183]. There is a group of complexes consisting major exception to the structural generalities for $\{Fe(NO)\}^6$ porphyrinates referred to above. The most cited are [Fe(NO)(OEP)(C_6H_4F-*p*)], (T(p-OMe)TPP)Ru(NO)Et. In addition to the unexpected bending of the MNO, the nitrosyl group is also tilted off the normal to the porphyrin plane. Analysis of the crystal structure excludes the possibility of steric factors. The tilting of the nitrosyl group was also observed in other nitrosyl metalloporphyrinates. For example, the high-resolution crystal structure for [Co(OEP)(NO)] shows that the axial Co–N (NO) vector is tilted from the normal to the porphyrin plane. The off-axis tilt is correlated with an asymmetry in the equatorial Co–Np bond distances [184]. [Fe(NO)(OEP)(C_6H_4F-*p*)] has an Fe–N–O bond angle of 157° and the Fe–N(O) distance is 1.73 Å, both dimensions more similar to $\{Fe(NO)\}^7$ species, and the Fe–N–O group is tilted significantly off-axis [185]. It was shown early on that this species and its analogues [Fe(NO)(OEP)R] (R = Me, C_6H_5, etc.) are diamagnetic, thus the d^7 configuration is excluded [186]. The NO stretching frequency (1,791 cm^{-1}) and the Mössbauer isomer shift ($\acute{a}$) 0.14 mm s^{-1} of this complex are significantly different from those of other six-coordinate $\{Fe(NO)\}^6$ species. However, DFT calculations show that the structure of this unusual species represents a minimum energy form. The bending and tilting of the nitrosyl group was proposed to result from the interaction of the NO π^* orbital with mixed metal orbitals $d_{x2\text{-}y2}$ and d_{xz}. In the $\{Fe(NO)\}^7$ group, the Fe–N–O bond angles are 140–150° and the Fe–N(O) distances 1.72–1.74 Å. The NO stretching frequencies range from 1,625 to 1,690 cm^{-1} and are dependent on the coordination number and the nature of the *trans* axial ligand in the six-coordinate species. These species are paramagnetic ($S = 1/2$), and the Mössbauer spectral isomer shifts vary from 0.22 to

0.35 mm s^{-1} [187]. A comparison of the structures of five-coordinate metal nitrosyl porphyrinato complexes of Mn, Co, Fe, Ru, and Os, based on the $\{MNO\}^6$, $\{MNO\}^7$, and $\{MNO\}^8$, cores reveals that the M–N–O bond angle changes from essentially linear in $\{MNO\}^6$, as exemplified by [Mn(NO)$^-$(porph)] and [Fe(NO)(porph)]$^+$, through ca. 143° for $\{FeNO\}^7$, to ca. 122° in $\{MNO\}^8$, as in [Co(NO)(porph)] [6, 172, 188]. The M–N–O bond lengths also progressively lengthen, but the displacement of the metal atom out of the N_4 porphyrin plane decreases.

Cobalt nitrosyl porphyrinato-complexes containing the $\{CoNO\}^8$ core are five-coordinate and essentially square pyramidal, the Co–N–O bond angle falling close to 120° [189], although that in [Co(NO)(TPP)] is unexpectedly large (ca. 135°; this may be due to the quality of the X-ray data). This anomaly may be related to the ability of the NO group to swing or rotate about the Co–N–O bond in the solid state, an effect detected by CPMAS NMR spectral studies (see above) [189]. The Co–N(O) bond distances average 1.84 Å, and ν(NO) falls in the range 1,675–1,696 cm^{-1} (KBr or Nujol). Porphyrinato and related (chlorins, isobaterio-chlorins) cobalt nitrosyls can be electrochemically oxidized and reduced [190]. In general, oxidation is primarily associated with electron loss from the macrocyclic ligands, affording metal (nitrosyl)-stabilized porphyrin ð-radical cations. Reduction affords mono- and dianionic species which are unstable and readily lose NO. Cobalamin, a precursor of vitamin B_{12}, possessing a corrinoid prosthetic group was shown to inhibit deleterious and regulatory effects of nitric oxide [191–197]. The reduced form of aquacobalamin binds NO under physiological conditions yielding a diamagnetic six-coordinate product with a weakly bound α-dimethylbenzimidazole and a bent nitrosyl coordinated to cobalt at the β-site of the corrin ring. It has also been described as $Co^{III}NO^-$, on the basis of UV–vis, ^1H, ^{31}P and ^{15}N NMR data [198].

Porphyrinato ruthenium complexes contain the $\{Ru(NO)\}^6$ group and are six-coordinate. In ruthenium chemistry, the $\{Ru–NO\}^6$ configuration is generally accepted as NO^+ bound to an Ru(II) center. This is largely based on the high NO stretching frequencies (ν_{NO} = 1,820–1,960 cm^{-1}) noted with $\{Ru–NO\}^6$ nitrosyls, versus that of either free NO (~1,750 cm^{-1}) [199, 200] or bound NO• (1,650–1,750 cm^{-1}) in $\{Ru–NO\}^7$ species [201–205]. Mössbauer [206, 207] and K-edge X-ray absorption spectroscopic data on $\{Fe–NO\}^6$ nitrosyls with similar NO stretches have unequivocally established their formal {Fe(II)–NO^+} description. Although similar data on $\{Ru–NO\}^6$ species have not been reported, the {Ru(II)–NO^+} formulation best describes most of the $\{Ru–NO\}^6$ nitrosyls. Ruthenium nitrosyls with $\{Ru–NO\}^7$ exhibit lower νNO values (1,650–1,750 cm^{-1}) and a characteristic S = ½ EPR signal of the bound NO• radical near g ≈ 2; their properties, however, have not been fully characterized [200]. With two exceptions, all six-coordinate ruthenium nitrosyl porphyrinato complexes, [Ru(NO)(porph)L]$^+$ (L = neutral ligand), contain an essentially linear Ru–N–O bond angle, the Ru–N(O) bond distance falling in the range 1.74–1.77 Å. The exception to this general rule is [Ru(NO)(porph)(C_6H_4F-*p*)] which, like its iron analogue, has a bent Ru–N–O bond angle, 152–155°. The NO stretching frequencies of those species containing linear Ru–N–O range from 1,790 to 1,856 cm^{-1} (KBr), whereas those in the alkyl or aryl species are significantly lower (1,759–1,773 cm^{-1}).

Once again, the *trans* σ-bonding alkyl or aryl ligand exerts a powerful influence on the M–N–O bond angle. Studies on transient intermediates during photolysis of $\{Ru\text{–}NO\}^6$ nitrosyls have revealed different modes of binding (and dissociation) of coordinated NO at the ruthenium centers. Metastable NO linkage isomers have been observed for $\{MNO\}^6$ (M = Fe, Ru, Os) and for $\{MNO\}^{10}$ complexes of Ni, as well as for $\{FeNO\}^7$ iron nitrosyl porphyrins [208–212].

3.7 Copper Complexes

Copper-containing enzymes play a central role in denitrification. To date, there have been several types of copper nitrite reductases discovered [213]. These CuNIR are found in many different plants and bacteria. NO_3^- and NO_2^- serve as terminal electron acceptors, ultimately producing NO, N_2O, and/or N_2. All CuNIR contain at least one type 1 copper center in the protein. A copper–NO species has been proposed as a key intermediate in biological nitrogen oxide reduction. The site binding NO_2^- or NO has a distorted trigonal planar molecular geometry that is stabilized by two weaker interactions with a methionine sulfur and a peptide oxygen [214]. The nitrosyl adduct, which has the configuration $\{Cu(NO)\}^{10}$, is described as Cu(I) bound by NO^+. The crystal structure of an intermediate (type 2 copper)-nitrosyl complex formed during the catalytic cycle of CuNIR reveals an unprecedented side-on binding mode in which the nitrogen and oxygen atoms are nearly equidistant from the copper cofactor [215, 216]. The first well-characterized copper nitrosyl coordination compound was derived from a dinucleating ligand (2,6-*bis*[*bis* (2-pyridylethyl)aminomethyl] phenolate) and contains a bridging NO group [217]. The Cu–N–O bond angles are 130°, the N–O bond distance is 1.18 Å, typical of NdO, and ν(NO) = 1,536 cm^{-1}. The coordination around each copper atom is distorted square pyramidal, and the Cu–Cu distance is 3.14 Å, thus nonbonding. Complexes that can be treated as models for the active site of nitrite reductases have been obtained using sterically hindered *tris*(pyrazolyl) borato ligands (TPRR′). The obtained model mononuclear copper–nitrosyl complexes of hydro*tris*(pyrazolyl) borate ligands, with various substituents to the pyrazole ring [Cu(NO)TPRR′], display a pseudotetrahedral coordination geometry, with a terminal N end-on nitrosyl, possessing virtually linear (163.5°–176.4°) M–N–O groups with a short Cu–N(O) bond (1.76 Å). NO addition was reversible, and the complexes reacted with oxygen to form $[Cu^{II}(O_2NO)TPRR']$. It was suggested that these $\{Cu(NO)\}^{11}$ species should be described as Cu(I) coupled to NO• ($S = 1/2$), and so they have one more electron than the proposed active site in nitrite reductase. However, it showed up that very similar structures, varying in the substituted R groups had different magnetic properties. NO(g) dismutation in model complexes is strongly influenced by the nature of the substituents R and R′ on the TPRR′ ligand, as given in the recent review by Gennari [218]. There are various hypotheses on the structure of active complex formed with the copper center during nitrite transmutation to NO: A hypothesis proposed by Averill [219, 220] assumes an initial reduction of the metal center and a subsequent binding of nitrite to Cu(I) by N atom coordination (analogical binding in model complexes supports this hypothesis). Subsequent protonation of the substrate would give water and an unstable $\{CuNO\}^{10}$ terminal N-bonding nitrosyl. This high-energy complex would promptly release NO and Cu(II), allowing the catalytic cycle to continue. An alternative hypothesis to Averill's proposed mechanism for the catalytic cycle of CuNIR [221–224] considers the initial binding of nitrite to the oxidized copper(II) form, which has the effect of increasing the reduction potential of Cu(II) [225, 226]. This favors the subsequent electron transfer (and dehydration) to give a $\{CuNO\}^{10}$ species that, to minimize structural rearrangements, would display a side-on η^2-NO coordination. For this hypothesis there is that the Cu(II)-nitrito form of CuNIR exhibits η^2-O,O′ coordination of NO_2^- to copper [222, 226] stabilized by hydrogen bond interactions with the protein-matrix side chains. This type of coordination was shown also for some (TPRR′)Cu(II)-nitrito complexes, and it is noteworthy that bulky substituents to the TPRR′ drove an asymmetric coordination mode of NO_2 moiety, like in the active site of CuNIR. Recently it has been discovered that the nitrite ion is bound in a tridentate fashion in the reduced form of CuNIR, with its oxygen atoms coordinated to Cu(I) and an additional weak Cu–N interaction [227], a binding mode promoting minimal structural rearrangements from haptocoordinated NO_2^- towards side-on coordinated NO ($\{CuNO\}^{10}$) species. Side-on coordination of NO to copper was found in the crystal structure of reduced $\{CuNO\}^{11}$ CuNIR [215]. However, it was suggested that this peculiar coordination may be an artifact of the solid state and could not reflect the real situation in solution [228]. The spectroscopic properties of the model complexes, in which NO is end-on bound to the metal, were compared to those of CuNIR [209]. This has allowed for the determination of the binding mode of the $\{CuNO\}^{11}$ species in protein solutions as strongly bent end-on (with a Cu–N–O angle of ~135°), but not side-on [229]. For the detailed review on the copper complexes of NO see the chapter by Tolman et al. in this volume of Structure and Bonding.

4 Nitrosyls in Biology

Probably the first evidence for the formation of nitrosyl complexes of iron in living cells was provided by Commoner and coworkers in 1965 [87]. Using electron paramagnetic resonance spectroscopy, they recorded the signal of the *g* factor 2.04 in the livers of rats with chemically induced cancerous changes [90, 230]. Similar discoveries made simultaneously by Vanin et al. [231, 232] and Maruyama et al. [233]. The EPR signal was correctly interpreted as originating from the complex of iron with two molecules of nitric oxide and other ligands containing sulfhydryl groups. In 1978 Craven proved that nitrosyl–heme complex was involved in activation of guanylate cyclase. Little attention was paid to these findings until 1986, when nitrogen (II) oxide has been identified as an endothelium-derived relaxing factor (EDRF), that is, a chemical responsible for the expansion of the smooth muscle of blood vessels and thus regulating blood pressure [234–237]. In naturally occurring circumstances nitric oxide is synthesized from L-arginine and oxygen by various NO synthases and by reduction of inorganic nitrate. Formation of DNICs in biological material under the laboratory conditions is observed after the addition of various NO donors, the review on which is presented by Yamamoto et al. [238].

4.1 The Biological Functions of Nitrogen (II) Oxide

Nitrogen (II) oxide, historically referred to as nitric oxide, is now recognized as an important signaling molecule that impacts a wide range of physiological responses, including blood pressure regulation, insulin secretion, airway tone, peristalsis, neurotransmission, immune response, apoptosis; it is also involved in angiogenesis and in the development of nervous system [199]. It is believed to function as a retrograde neurotransmitter and hence is likely to be important in the process of learning [239]. In the living cells NO is enzymatically synthesized from L-arginine by nitric oxide synthases (NOSs) by the stepwise oxidation of L-arginine to citrulline, which is accompanied by the generation of one molecule of NO. Arginine-derived NO synthesis has been identified in mammals, fish, birds, invertebrates, and bacteria [240]. The NO-forming reaction occurs in the oxidase domain of eNOSs, which contains a heme responsible for L-arginine hydroxylation. There is abundant evidence for NO signaling in plants, but plant genomes are devoid of enzyme homologs found in other kingdoms [241]. In mammals, three distinct genes encode NOS isozymes: neuronal (nNOS), cytokine-inducible (iNOS), and endothelial (eNOS) [242]. iNOS and nNOS are soluble and found predominantly in the cytosol, whereas eNOS is membrane associated. eNOS and nNOS take part in the calcium/calmodulin-controlled NO signaling. The inducible isoform iNOS is involved in the immune response and it produces large amounts of NO as a defense mechanism, this being a direct cause of septic shock and playing a role in many autoimmune diseases. It must be noted here that the paramagnetic oxidonitrogen radical is not the only NO form involved in the reactions with diverse biological targets; more and more data bring evidence of the important regulatory functions played by the diamagnetic, one-electron oxidized NO^+ nitrosonium cation, and the one-electron reduced NO^-, the nitroxyl anion (or its protonated form, HNO) [167, 204, 243].

In biological systems nitrogen monoxide is incorporated into its transducers, such as transition metal complexes, in prevailing number of cases, being those of iron and nitrosothiols [61, 244, 245]. The fact that NO is a radical, inspired the idea that incorporation into these biological conveyors determines its existence in biofluids. This supposition was based on the notion that radicals are highly reactive, whereas it was shown that NO can be transported to a distance several-fold exceeding cell sizes [246]. Indeed, free NO can be readily caught by various endogenous cellular scavengers, for example, by exposed heme centers and inorganic ferrous clusters in proteins, oxygen species, and other free radicals [247], the two most obvious NO targets being hemoglobin and superoxide anion. Yet, the NO chemistry is limited.

From the chemical point of view, nitric oxide is a stable free radical, it acts as a σ-donor and π-acceptor. Unpaired electron is localized at 2pπ* orbital, and its loss produces nitrosonium ion NO^+, much more stable than NO. This is seen in the shift of ν vibrations of N–O bond in infrared spectroscopy. ν(NO) of nitrosonium ion is 2,273 cm^{-1}, while these of NO and NO^- are 1,880 cm^{-1} and 1,366 cm^{-1}, respectively. In nitrosyl complexes this band occurs at 1,900–1,500 cm^{-1} [248].

NO is neither readily reduced nor oxidized. Fairly high reduction potential of NO^+ into NO has been estimated to be 1.50 V vs. SCE according to the results of Lee [249], and ca. +1.2 V vs. NHE by Stanbury [250], whereas the recently reexamined reduction of NO to triplet and singlet NO^- was determined to be -0.8 ± 0.2 V for $^3NO^-$ and -1.7 ± 0.2 V for $^1NO^-$, rendering NO inert to direct, one-electron reduction processes under physiological conditions [251]. Conversely, NO^- and its protonated form, HNO show a quite high reactivity towards a broad range of oxidants, reductants, nucleophiles, or metalloproteins, often leading to the formal oxidation of NO^- to NO. Nitrosonium cation, NO^+, is moderately stable in aqueous solutions but highly reactive with nucleophiles or other nitrogen oxides [252]. Both NO^+ and NO^- were shown to display biological activity, distinct from that of NO radical [167]. NO• is isoelectronic with the dioxygen monocation (O_2^+). NO^+ is isoelectronic with CO and CN^-, whereas NO^- is isoelectronic with O_2 and both can exist in singlet and triplet state. This last relationship accounts for the continuing interest in the study of certain types of metal nitrosyl complexes due to their structural and electronic similarity to biological oxygen activators. NO can be an effective probe for examination of spatial and electronic structures and function of metalloenzymes, where a spectroscopic examination of the resting or oxygenated enzyme is difficult or impossible because of instability. Due to that and also to their similar size and hydrophobicity, O_2, CO, and NO have an access to the same cellular iron pools, either labile or protein-bound. They could theoretically compete for binding sites on ferrous iron. Paramagnetic properties of both NO and O_2 result in the addition of unpaired electrons to the system which initiates further interconversions besides simple reversible binding. Nitric oxide affinity for heme-bound Fe^{2+} is much greater than that of CO and O_2. This results mainly from its lower dissociation rate (in some cases $<10^{-5}$ s^{-1}) as compared to O_2 (15 s^{-1} from R-hemoglobin, 1,900 s^{-1} for T-Hb) and CO (ca 10^{-2} s^{-1}; for on-, off-rates and stability constants of NO complexes with ferrous iron in different proteins see [253], see Table 2 for thermodynamics of iron binding to heme). Simultaneously, binding rates of NO are slightly (up to an order of magnitude) higher than those of O_2 and CO. This susceptibility to heme, together with its paramagnetic nature made NO an excellent tool for structural studies of a wide range of heme–iron proteins by means of EPR, as described in the following sections/chapters. Due to this special significance of nitrosyl-iron chemistry, investigations of Fe–NO complexes dominated the literature of nitric oxide interactions with transition metals, although NO complexes with almost all transition metals are known (see references in section 2). NO binding to ferric heme is less reversible than that to ferrous heme due to the possibility of reductive nitrosylation, which yields ferrous iron and nitrite anion [74].

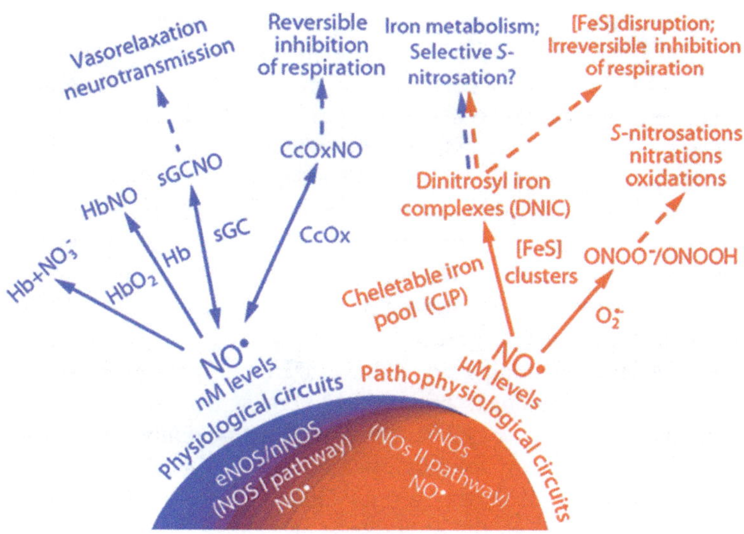

Fig. 2 Nitric oxide signaling relevant to metal nitrosyl chemistry. Reprinted from [254], with the permission of the American Chemical Society, copyright 2012

4.2 The Biological Targets of Nitrogen (II) Oxide

4.2.1 Heme Iron Proteins

Figure 2 illustrates an overview of biological NO signaling pathways relevant to metal nitrosyl chemistry. The two primary targets of NO-mediated regulatory action are: nitric oxide (NO)-sensitive guanylyl cyclase (NO-sensitive GC), the most important NO receptor which catalyzes cGMP formation, and cytochrome *c* oxidase (in the mitochondrial complex IV), which is responsible for mitochondrial O_2 consumption and is inhibited by NO in competition with O_2 [255]. Both these enzymes are hemoproteins and prosthetic heme group is responsible for the regulation of the enzymes by NO [256, 257]. In contrast to other hemoproteins such as hemoglobin or myoglobin, the heme of NO-sensitive GC does not bind oxygen [258]. In the case of cytochrome *c* oxidase inhibition is based upon competitive binding of NO to the O_2 binding site [255].

The soluble guanylate cyclase (sGC) serves as the biological NO sensor/receptor. This enzyme upon its interactions with NO mediates cardiovascular regulation [259–263]. In its active form, sGC contains a five-coordinate (5C) heme with proximal histidine coordination in the ferrous oxidation state [260]. Due to the strong σ-*trans* effect of NO on the axial His ligand, upon binding of NO the Fe(II)–His bond is broken, leading to the corresponding five-coordinated ferrous heme NO complex [28, 175, 264]. Formation of a stable NO–heme complex results in a low-activity species. This is believed to be accompanied by large structural changes that modulate the duration and intensity of the catalytic site of sGC enzyme activity, which in turn correlates with activation for the conversion of guanosine triphosphate (GTP) to cyclic guanosine monophosphate (cGMP). The latter serves as a secondary messenger molecule involved in relaxation of the vascular smooth muscles, this inducing vasodilation of the arteries, and hence, controlling the blood flow. In the absence of excess NO, GTP accelerates conversion of the transient six-coordinate intermediate to the five-coordinate final species, which is fully active. ATP blocks GTP from accelerating conversion of the intermediate to the final species. When ATP is present together with GTP, both the conversion to the final species and the dissociation of NO from the heme are slow. This species has low activity, and additional NO produces a transient, fully active enzyme [265].

NO binds to ferrous heme iron in a bent mode (i.e., angle between Fe–N–O is 130–150°), whereas in complexes with ferric iron a linear geometry is favored [9, 74]. The bent mode binding is assigned to NO being a one-electron donor, while the linear mode corresponds to NO being a three-electron donor [266]. Dissociation constants of ferric heme complexes fall within the range of 10^{-4}–10^{-7}; this makes them much less stable than the ferrous analogues. For numbers, see [74]. NO penetrates easily through the erythrocyte membrane, and under the action of oxyhemoglobin transforms into non-permeable nitrite ion, which accumulates inside the erythrocyte. Independently, a pool of NO converts to stable S-nitrosothiols with the SH groups of hemoglobin and other proteins. The erythrocyte cell converts into the "depot" of rather stable NO donors [267, 268]. Then, the nitrite ion is easily reduced to NO under the action of deoxyhemoglobin [269, 270]. Therefore, hemoglobin in erythrocytes could be one of the most important elements in the biological transformation of NO donors and NO transport inside the organism. Sanina et al. performed some investigations on the possible interaction of DNICs and hemoglobin and noticed that Hb cannot act as a reductant in the reaction of NO release from DNIC-type complexes [170, 271]. In the case of catalase NO binds to the ferric heme group and thereby inhibits the enzymatic conversion of hydrogen peroxide to water [254, 272]. In this respect, the binding and inhibition of catalase by NO provides a point of cross-talk between the H_2O_2 and NO-mediated redox signaling pathways. The primary site of interaction of NO with mitochondria is the cytochrome c oxidase in the complex IV of the mitochondrial respiratory chain. NO binds to the ferrous iron of heme a_3 in the binuclear site of the enzyme [273] with high affinity ($K_d = 1 \times 10^{-10}$ M) [274] and reaction is very fast with a rate constant of $1 \times 10^8\ M^{-1}\ s^{-1}$ [275].

Nitrite reacts both with oxy- and deoxyhemoglobin. The primary reaction, described in 1981 by Doyle and colleagues, leads to autocatalytic oxidation of HbO_2 to Hb^{3+}, accompanied by the formation of NO and H_2O, and excess nitrite binding weakly to $Hb(Fe^{3+})$ [276–278]. The affinity of NO to hemoglobin measured at half-saturation is $3 \times 10^{10}\ M^{-1}$, while those of CO and O_2 are 2×10^7 and $6 \times 10^4\ M^{-1}$, respectively. The main reason for such a high affinity is the very slow dissociation of NO from HbNO (mean $t_{1/2} \sim 3$ h) [159, 279] as compared to CO (mean $t_{1/2} \sim 35$ s) and O_2 (mean $t_{1/2} \sim 20$ ms) [280, 281]. It may seem surprising that although it has an exceptionally low off rate from hemoglobin, NO is a fast signal transducer. For example, NO-dependent inhibition of guanylate cyclase is restored in 50% after less than a minute. This is possible due to preventing bond formation and/or increasing the rate of NO dissociation [74]. In effect, even during endotoxemia in rats (the biological situation where NO production probably is at its maximum) only 2% of the total hemoglobin concentration is trapped as Hb-NO [282]. A complex reaction mechanism has been postulated to prevent the formation of a ferrous-nitrosyl bond during the reaction cycle [219]. It is claimed that heme-proteins' amino acid structure is able to modify the environment of NO–Fe site in a way that allows its much higher dissociation rates. NO dissociation from six-coordinate model heme occurs at least 1,000 times faster than from five-coordinate heme [160]. Another mechanism of NO removal from heme residues can be explained by the reaction of nitrosyl hemoglobin (or myoglobin) with oxygen which generates methemoglobin and nitrate. Different mechanisms have been proposed for this reaction [283]. Binding of NO to ferric heme in many cases leads to reductive nitrosylation. This slow reaction (minutes for hemoglobin in neutral pH) produces ferrous-nitrosyl complex and NO_2^- and can be reversed in the presence of oxygen [164, 284, 285].

4.2.2 Non-Heme Iron Proteins

As described above, the biological properties of NO are generally attributed to its interaction with iron in the heme groups of enzymes. However, NO also interacts with a wide range of other cellular components, many of which do not contain heme [286]. In the case of NO, the ferrous iron frequently remains high spin ($S = 2$) resulting in an EPR signal due to an $S = 3/2$ spin system (when spin coupled to the $S = 1/2$ NO molecule). Similar to iron enzymes, iron storage proteins can react with NO. A complex series of ferritin interactions with NO includes signals from $S = 1/2$ and $S = 3/2$ spin states [287, 288]. It is possible that these interactions are responsible for the ability of NO to release iron from the ferritin stores [287]. Metallothioneins (MTs) are small, sulfur-rich metal-binding polypeptides produced in response to a variety of physiological and environmental stresses. In the presence of iron and NO, both apo-metallothionein and Zn-metallothionein form EPR spectra similar to DNIC [289].

Enzymes that react with oxygen (e.g., monooxygenases, dioxygenases) have the potential to make nitrosyl complexes as shown in the case of lipoxygenase [290, 291]. Alike iron cluster-binding enzymes, iron storage proteins upon the reaction with NO give EPR spectra similar to those of DNICs [288, 292]. One should bear in mind that the EPR-observable DNICs formed with the cellular proteins may not be the only species responsible for nitrosative modification of proteins. Only recently Tinberg et al. [43] reported that in the reaction of NO with Rieske-type [2Fe–2S] clusters dinuclear RRE species instead of mononuclear DNICs are formed. This finding implies that in certain cases RREs may be the primary iron dinitrosyl species responsible for the pathological and physiological effects of nitric oxide.

The electron-transferring iron–sulfur centers in the mitochondrial electron transfer chain are inaccessible to oxygen or NO for steric reasons. However, in iron–sulfur enzymes with non-redox roles [293, 294] it is frequently necessary for LMW compounds (e.g., citrate) to interact with the cluster and therefore, also NO access could not be prevented. These are generally [4Fe–4S] clusters in which three of the iron atoms have cysteine coordination, while the fourth is coordinated by inorganic sulfur. The substrate reacts at the non-cysteine coordinated iron. Most notable of these are mitochondrial and cytoplasmic aconitases.

Mitochondrial aconitase catalyzes isomerization of citrate to isocitrate in the citric acid cycle. It contains a [4Fe–4S] center that is essential for complexing with citrate. $ONOO^-$ has been shown to inhibit the activity of isolated aconitase by a mechanism that is reversed by the addition of iron and thiols [295–297]. Intriguingly, substrate binding seems to prevent this damage [116], suggesting that peroxynitrite cannot access the iron cluster in the presence of citrate in the active site. High concentrations of NO (>100 μM) can also inhibit aconitase activity [295–297] but in contrast to the situation with peroxynitrite, substrate binding did not prevent this inhibition [116].

Cytoplasmic aconitase is identical to the iron regulatory protein 1 (IRP1). When the cellular iron level is low, the enzyme loses its [4Fe–4S] cluster and the apoprotein acquires the ability to bind specific RNA structures called iron responsive elements (IREs) [298–300]. The IRE sequences are located in the mRNA for ferritin (iron-storage protein), the transferrin receptor (cellular iron import protein), and erythroid 5-aminolevulinate synthase (heme biosynthesis catalyzing enzyme) [58, 301]. IRP binding represses the translation of mRNA for ferritin and erythroid 5-aminolevulinate synthase, and protects mRNA for the transferrin receptor against degradation. Described turning on post-transcriptional gene regulation assists the cell in accumulating iron, while preventing high levels of labile iron pool being diverted into storage or heme biosynthesis. The cluster-containing form of the protein has aconitase activity, but does not bind to the IRE, in contradiction to apoprotein. The [3Fe–4S] form of the cluster has neither aconitase activity nor does it bind to the IRE [302]. An interesting issue is interconversion of [4Fe–4S] cluster of this enzyme during formation of DNIC. It is reported that during formation of DNIC in aconitase, cluster is completely or partially removed from the active site. It has been demonstrated that specific NO inactivation of aconitase proceeds by attack at the non-cysteine coordinated iron site and protein-bound iron-sulfur dinitrosyl complexes, giving $g = 2.04$ EPR signals, are formed [100]. Inhibition of mitochondrial aconitase is pH-dependent. No inhibition can be seen in pH 7.5, while a slow inhibition ($\tau_{1/2} = 1$ h) was observed at pH 6.5 for the enzyme at rest [100, 296, 297, 303]. It has been shown that NO may play a role in activating the IRP [58, 304, 305]. When high (μM) levels of NO are produced in activated rodent macrophages and cell lines, the $g = 2.04$ DNIC EPR signal can be detected simultaneously with inhibition of cytoplasmic aconitase activity and activation of the IRP [306, 307]. The mitochondrial aconitase is a constituent of the Krebs cycle; so its inactivation by NO decreases cellular energy metabolism. This inactivation of aconitase may have a protective effect against additional oxidative stress by acting as a reversible "circuit breaker" [308]. Inactivation results in reduced electron flow through the mitochondrial electron transport chain, and thereby decreases the generation of reactive oxygen species (ROS), the natural by-products of respiration [309, 310]. In fact, it was also suggested, that NO-derived species, like peroxynitrate rather than $NO^{\bullet}$ *per se*, have been proposed to mediate the inactivation of aconitases. Indeed, peroxynitrite reacts vastly with the [4Fe–4S] cluster of aconitase yielding the inactive [3Fe–4S] enzyme [296, 297, 311], yet a number of evidence is also presented for NO-mediated inactivation of iron-sulfur proteins [309–311], the above question being resolved by the group of Drapier [305].

Nitrile hydratase (NHase, EC 4.2.1.84) is a microbial enzyme that catalyzes the hydration of diverse nitriles to their corresponding amides. NHase is a metalloenzyme which contains non-heme Fe(III) or non-corrinoid Co(III) in its catalytic center [312, 313]. Crystallographic studies on Fe-containing NHase from *Rhodococcus* sp. N-771 revealed that the metal center is composed of a cysteine cluster ($C^{109}SLCSC^{114}$) and sulfur atoms from the three cysteine residues Cys109, Cys112, Cys114 and two amide nitrogen atoms of the peptide backbone from Ser113 and Cys114 are coordinated to the iron [314–317]. Interestingly, two cysteine ligands, Cys112 and Cys114, are oxidized to a cysteine sulfinic acid ($Cys\text{-}SO_2H$) and cysteine sulfenic acid (Cys-SOH), respectively. It has been shown that the post-translational oxidation of the Cys-S residues is essential for the catalytic activity of the enzyme [318]. The activity of Fe-NHase is regulated by visible light and dependent on formation of the nitrosyl–iron complex. The active form of the enzyme ($NHase_{light}$) reacts with endogenous nitric oxide to form an $\{Fe\text{–}NO\}^6$ complex and yield the inactive form ($NHase_{dark}$). X-ray crystallographic studies of $NHase_{dark}$ reveal an exceptional $N_2S_3Fe\text{–}NO$ core with two amido nitrogen donors and three cysteine-derived sulfurs. In the inactive state, an NO molecule occupies the sixth coordination site of the Fe(III) center. Exposure to light causes a rapid loss of NO, and the subsequent binding of water activates the enzyme [319, 320].

Helicases are enzymes that unwind the DNA double helix. They are motor proteins that move directionally along a nucleic acid phosphodiester backbone, separating two annealed nucleic acid strands using the energy derived from ATP hydrolysis. There are many helicases adapted to the great variety of processes in which strand separation must be catalyzed. DinG helicase from *E. coli* plays an important role in recombinational DNA repair and the resumption of replication after DNA damage [321]. The protein contains a redox-active [4Fe–4S] cluster with a midpoint redox potential (E_m) of -390 ± 23 mV (pH 8.0). Reduction of the cluster reversibly switches off helicase activity. The DinG [4Fe–4S] cluster is very stable and the enzyme remains active after exposure to 100-fold excess of H_2O_2, but it can be modified by NO; this results in the formation of the dinitrosyl iron complex and the loss of helicase activity both in vitro and in vivo. It was proposed that modification of the iron–sulfur clusters in DinG and possibly other DNA repair enzymes such as human XPD helicase might contribute to the NO-mediated genomic instability [322].

Gene transcription is a complex process regulated by specific regulatory proteins that bind to the promoter regions of the gene. For example, the [2Fe–2S] centers in the regulatory protein, SoxR from *E. coli*, are essential for the activation of transcription of a regulon known as SoxS which controls the expression of several genes involved in the defense against activated macrophage-induced oxidative and nitrosative damage [323]. Interestingly, SoxR activation by NO occurs through direct modification of the [2Fe–2S] centers to form protein-bound dinitrosyl-iron-dithiol adducts, which have been observed both in intact bacterial cells and in purified SoxR protein after NO treatment. Another example of a bacterial transcription factor regulated by NO is the NorR protein, a regulatory protein in enteric bacteria known to serve exclusively as an NO-responsive transcription factor. In *E. coli*, NorR activates the transcription of the genes encoding flavorubredoxin and an associated flavoprotein, respectively, which together have NADH-dependent NO reductase activity. The wild-type NorR monomer consists of the three domains: (1) The regulatory, iron-containing N-terminal domain (GAF domain), which was shown to contain a mononuclear non-heme iron center, (2) the central catalytic AAA^+ domain, and (3) carboxy-terminal HTH DNA binding domain. The function of the GAF domain is to sense the signal and inhibit the ATPase activity of the central AAA^+ domain. When exposed to NO, iron center of GAF domain is modified by a single NO molecule to generate a mononitrosyl $\{Fe(NO)\}^7$ ($S = 3/2$) species. Formation of mononitrosyl iron complex triggers a conformational change in the protein that leads to the initiation of transcription: The interaction between the GAF domain and the AAA^+ domain is released, allowing ATP hydrolysis that is further coupled to transcriptional activation [324]. These two examples of functional activation through nitrosylation of iron–sulfur centers contrast with the inactivation typically caused by this modification [325, 326]. In contrast to most dinitrosyl–iron complexes, NO binding is reversible in mononitrosyl systems.

Early papers suggested that DNICs formation with iron–sulfur clusters occurs via transnitrosylation. It was concluded that external iron ions and LMW thiol ligands are necessary in the process of iron–sulfur cluster disruption by NO (see, e.g., [3, 5, 65, 252, 327, 328]). Nevertheless, recent studies on nitrosylation of the biomimetic compounds imply that the most straightforward and facile pathway of the formation of the $\{Fe(NO)_2\}^9$ type DNICs with iron–sulfur proteins is the direct nitrosylation of $[Fe_m(SR)_n]^{x-}$ clusters [118].

4.2.3 Cobalamins

NO is known to react with the cobalt of cobalamins (Cbl). The structure of cobalamin is presented in Scheme 7. Cobalamins containing various axial ligands *trans* to dimethylbenzimidazole moiety are known as vitamin B_{12} and are important cofactors for 5-methyltetrahydrofolate-homocysteine methyltransferase and methylmalonyl-coA mutase playing a key role in the normal functioning of the brain and nervous system and in red blood cell formation [330, 331].

Depending on the axial ligand *trans* to dimethylbenzimidazole, cobalamins differ with their affinity for NO: Aquacobalamin (H_2O-Cbl) and hydroxocobalamin (OH-Cbl) readily react with nitric oxide, unlike cyanocobalamin [CN-Cbl],

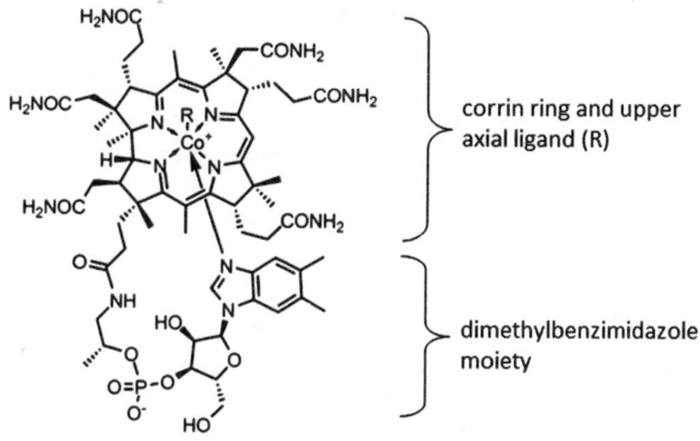

Scheme 7 The structure of cobalamin (α-(5,6-dimethylbenzimidazolyl)cobamidcyanide); R = CN; the axial group R can alternatively be methyl, OH, deoxyadenosyl group [329]

methylcobalamin [Me-Cbl], or adenosylcobalamin [Ado-Cbl] [330]. The reduced form of aquacobalamin binds nitric oxide very effectively to yield a nitrosyl adduct, Cbl(II)−NO [198]. Spectroscopic data suggest that the reaction product under physiological conditions is a six-coordinate complex with a weakly bound dimethylbenzimidazole moiety and a bent nitrosyl coordinated to cobalt at the β-site of the corrin ring. The nitrosyl adduct can formally be described as Cbl(III)−NO^- [198]. The kinetic studies on the binding of NO to Cbl suggested the operation of a dissociative interchange substitution mechanism at the Co(II) center [198]. The laser flash photolysis of cobalamin nitrosyl indicated the formation of water-bound intermediates in the laser flash experiments further supporting the dissociative interchange mechanism.

NO has a high affinity for cobalamin in the 2+ oxidation state [Cbl(II)] but does not react with Cbl(III). NO coordinates to Cbl(II) at all pH values and Cbl(III) does not react with NO at neutral pH. At low pH, however, a two-step process was observed that included the reduction of Cbl(III) to Cbl(II) and Cbl-NO adduct formation [332]. Reduced Cbl (II) with an odd number of outer shell electrons of the central ion is a free radical, and thus EPR spectroscopy can be used to investigate the possible NO interactions with reduced cobalamins [330].

Endogenous cobalamins and cobinamides (Cbi) may play important roles regulating NOS activity in normal and pathological conditions [329]. Regulatory function of cobalamins towards nitric oxide synthases was brought up and discussed by Weinberg et al. [329, 333] and also by Wheatley et al. [334, 335]. Recently, Weinberg et al. [329] reported that OH-Cbl, Cbi, and $(CN)_2$Cbi can potently inhibit the enzymatic function of NOS and thus block the biological formation of NO. It is to be reminded that (as stated above) OH-Cbl can bind and scavenge NO; the same was observed for OH-Cbi; thus, these two agents are able to regulate the NOS/NO system both by decreasing NOS activity and by quenching the existing NO pool. CN_2Cbi, the principal vit. B_{12} form, did not act as NO scavenger (due to the higher affinity of CN for the cobalt center), but it still directly interacted with NOS.

Hydroxocobalamin (OH-Cbl), cobinamide (Cbi), and dicyanocobinamide [$(CN)_2$-Cbi] were shown to potently inhibit ^{14}C-labeled L-arginine to L-citrulline conversion by NOS1, NOS2, and NOS3 [cyanocobalamin, methylcobalamin, and adenosylcobalamin had much less effect, but could be photoactivated (see further in the text)]. OH-Cbl and CN_2-Cbi directly bound to the reduced NOS1 and NOS2 oxygenase domain (as indicated by spectral perturbation analysis) and prevented binding of the oxygen analog carbon monoxide (CO) to heme. NOS inhibition by corrins was rapid and could not be reversed by dialysis with L-arginine (substrate) and tetrahydrobiopterin (a required NOS cofactor). Molecular modeling and UV–vis spectroscopy indicated that corrins could access the unusually large heme and substrate-binding pocket of NOS. The greater ability to inhibit NOS1 than NOS2 relates to the fact that the NOS1 active site pocket is larger than the NOS2 pocket. Best fits were obtained for the "base-off" conformations of the lower axial dimethylbenzimidazole ligand, $(CN)_2$Cbi being the most potent inhibitor.

Application of corrins would allow to regulate both NO and superoxide generation by NOS in circumstances when NO acts in a deleterious fashion (e.g., inflammatory diseases). These agents are well tolerated in high doses, as applied in cyanide poisoning [336, 337]. As high as millimolar blood levels were shown to be accompanied with minimal side effects (red urine and mild, reversible hypertension).

Various cobalamins are known to be light sensitive. Upon irradiation Cbl dissociates the upper (Co[β]) ligand converting to OH-Cbl, a form active as an NOS inhibitor [329]. It was thus proposed that this property of light-activation for NOS inhibition could be useful for site- and time-selective delivery of NOS inhibitors or NO quenchers. As an instance, Me-Cbl and Ado-Cbl are the two corrin derivatives of a very poor inhibitory activity towards NOS. During blue light irradiation of the methylated and adenosylated cobalamin the methyl and adenosyl groups dissociate which results in the formation of OH-Cbl.

Recently it has been reported that cobinamides can act as co-activators of nitric oxide receptor of soluble guanylyl cyclase by direct binding to the catalytic domain of sGC [338]. This offers new possibilities for its therapeutic applications in augmenting the effect of other sGC-targeting drugs.

4.2.4 The Pool of Labile Iron

As mentioned earlier, nitrosyl complexes have been ubiquitously detected in cells and tissues exposed to NO, but the exact origin of the ligands as well as the central ions remains obscure. These complexes are considered to be formed concurrently with low molecular cellular thiol–glutathione and with proteins, but the latter still remain undefined [76, 77, 339, 340]. Dinitrosyl–diglutathionyl–iron complex (DNDGIC) is readily formed in vitro starting from GSH, GSNO, and trace amounts of ferrous ions [341]. The reaction probably occurs after release of nitric oxide through an iron-mediated reaction between GSNO and GSH. DNDGIC may be monitored through the typical EPR spectrum of this complex which shows a characteristic shape with a maximum at $g = 2.03$. The detection limit of this procedure is about 1 μM concentration for EPR analysis performed at 25°C [59].

A considerable part of labile iron pool (LIP)-formed DNICs originates in lysosomes, since the inhibition of protein disassembly in lysosomes caused up to a 50% decrease in DNIC formation. The process of the DNIC's formation is strictly dependent on lysosomal proteolysis due to the presence of either high levels of labile iron or the free reduced thiols liberated in this process [342]. In addition, the primary iron source for cellular DNICs seems to be LIP and not a prosthetic iron such as iron–sulfur clusters [342, 343]. Nevertheless prosthetic group-bound DNICs were observed and are regarded as modifications having a regulatory role in cellular pathways. Toledo and coworkers found that no DNICs are detectable when cells are treated with very high (10 mM) concentrations of nitrite, ruling out this species as a source of NO ligands in DNIC formation. Comparison of the levels of formed paramagnetic DNICs to the simultaneous consumption of LIP proved that upon addition of close-to-physiological levels of the common NO donors most of the DNICs formed in the cell are paramagnetic [343]. DNICs were shown to emerge under physiologic conditions from low concentrations of •NO (~50 nM) and they formed the largest intracellular fraction of all •NO-derived adducts. At the physiological •NO dose, DNIC concentrations reached slowly (4 h in the conditions of the experiment) a stabile maximum level corresponding to the levels of the labile iron pool (LIP). However, upon exposure to higher steady-state •NO concentrations, DNICs accumulated at levels in two- to threefold excess of the LIP. At cytotoxic concentrations of •NO the LIP no longer was the limiting source of iron for DNIC assembly. It was speculated that, in addition to other, well-known toxic effects of the nitrosyl radical, one consequence of pathologic •NO concentrations may be the accumulation of non-LIP iron in the form of DNICs. Other potential sources of cellular iron, in addition to the LIP, are iron-storage proteins, iron–sulfur clusters, heme, and non-heme iron proteins [344]. These results are consistent with other research, showing high stability in solution and no effect of iron chelators on DNICs decomposition, pointing to an inert character of these complexes [59, 60, 102, 343]. This is confirmed by the analysis of temperature dependence of the EPR signal shape for biological DNICs. The spectra obtained by Watts et al. [97] in •NO-exposed mammary carcinoma cells at 150 K were essentially identical to those obtained at 265 K, implying a large molecular mass of DNICs. Moreover, it was shown that the magnitude of DNIC formation and disappearance in response to changes in O_2 concentrations was much less affected than the overall magnitude of •NO synthesis or •NO degradation in response to changes in O_2. For these reasons the formation of DNIC may function by buffering the dramatic effects of O_2 on •NO synthesis and degradation [344].

4.3 Biological Consequences of DNIC Formation

It is postulated that dinitrosyl iron complexes play an important role in signal transduction pathways regulating the cellular functions [60, 345, 346]. Both low

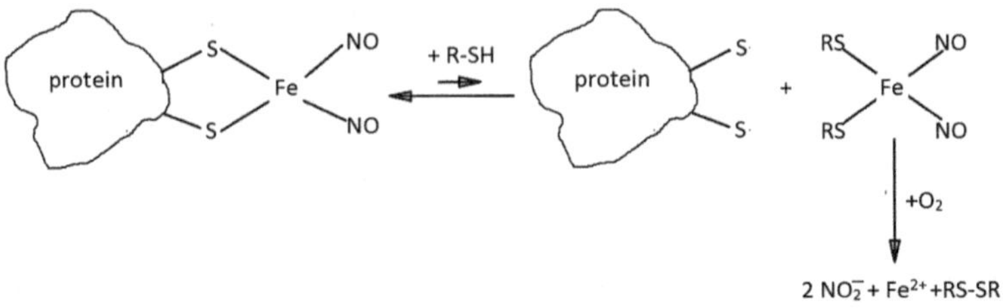

Scheme 8 A proposed model for the LMW-thiol mediated decomposition of protein-bound DNICs. LMW-thiol (e.g., L-Cysteine, glutathione) extrudes the DNIC from the protein-bound DNIC to form LMW-thiol-bound DNIC via thiol ligand exchange. Under anaerobic conditions, both the LMW-thiol-bound DNICs and the protein-bound DNICs are stable, and no DNIC is decomposed. Under aerobic conditions, the LMW-thiol-bound DNIC is rapidly disrupted by oxygen, resulting in eventual decomposition of the protein-bound DNIC [147, 286]

and high molecular DNICs are more stable than the nitrogen (II) oxide and are considered to be its transporters [61, 99]. DNICs also inhibit platelet aggregation, lower blood pressure, dilate blood vessels, induce the formation of proteins in the cellular response to oxidative stress, and modulate the activity of ion channels. Their appearance in cells exposed to NO is closely synchronized with the enzyme activity inhibition of proteins containing iron sulfur centers in the complexes I and II of mitochondrial respiratory chain and of mitochondrial aconitase. DNICs irreversibly inhibit the activity of glutathione reductase, and also reversibly of glutathione transferases. Extremely strong inactivation of these enzymes by DNICs indicates their important role in the metabolism of glutathione [60, 85, 345, 346].

The endogenous production of DNICs has been explained by the binding of NO to iron–sulfur cluster-containing proteins or enzymes in mitochondria and thiol-rich proteins in the presence of free iron [51, 90, 147, 347, 348]. It has been demonstrated that under physiological conditions, DNICs are generated in two forms having LMW thiols or cysteine residues of protein as ligands [68, 69, 90]. Although it has been suggested that an LMW-DNIC with cysteine [DNIC-$(Cys)_2$] possesses activities such as an endothelium-derived relaxing factor (EDRF) [69, 124] and exhibits S-nitrosating activity toward cysteine residues of serum albumin in vitro [61, 68], its physiological role has not been definitely elucidated. One of the reasons may be that in vivo detection of DNICs is very difficult because it can be formed in only a limited quantity under physiological conditions. Investigations of the in vivo distribution and behavior of LMW-DNICs have been conducted by means of EPR spectral measurements on the abdomen of mice treated by the DNIC-$(GS)_2$ [349]. The EPR spectra attributable to DNIC-$(GS)_2$ were detected in the blood, liver, kidney, and spleen; and it was shown that this complex has a relatively high affinity for the liver and kidney [68]. Ueno et al. demonstrated that when an LMW-DNIC-$(GS)_2$ was injected into living mice, different EPR spectral line-shapes were observed emanating from their isolated organs [349]. The appearance of such spectral conversions was explained mainly as the result of chemical modifications by the reaction of LMW-DNICs with thiol-containing compounds such as serum albumin and iron proteins (Scheme 8). It is probable that NO donation from

DNICs to a variety of in vivo targets and their resultant chemical modifications are closely associated with the physiological activities of DNICs.

DNICs have been reported to react with low molecular thiols or protein thiols to yield S-nitrosothiols. Further, DNIC may nitrosate endogenous secondary amines with the formation of nitrosamines [68]. In biological systems, therefore, transnitrosylation by DNIC would be in competition with nitrosation. This is consistent with findings of Ueno et al. [65] that the efficiency of transnitrosylation of Fe-DTC by DNICs is lower in the liver, a thiol-rich organ, than in the kidney.

Apart from their natural biological functions various metal nitrosyls (including organometallic dinitrosyl complexes of Cr and Mo) proved to be very convenient carriers supplying nitric oxide to various biological targets and are presently widely introduced into the therapy as vasodilators and in cancer treatment [350–352], see the chapter by Morris et al. in this volume of S&B.

4.3.1 Interactions of Glutathione Transferases and Dinitrosyl Iron Complexes

Glutathione transferases (GSTs) (EC 2.5.1.18) are a family of enzymes involved in the cellular defense against toxic compounds [353]. GSTs could be involved in the DNIC binding, storage, and detoxification as they can trap these complexes with one subunit, while maintaining their well-known detoxifying *effect* toward dangerous compounds with the other subunit [341]. The human cytosolic GSTs are dimeric proteins grouped into eight classes (Alpha, Kappa, Mu, Omega, Pi, Sigma, Theta, and Zeta) on the basis of amino acid sequence, three-dimensional structure, substrate and inhibitor specificity, and immunological properties [354]. Each subunit contains a binding site for the GSH (G-site) and a second one for the hydrophobic co-substrate (H-site). The most important reaction catalyzed by GSTs is the conjugation of the sulfur atom of GSH to an electrophilic center of many toxic organic compounds; this increases their solubility and enables excretion [353]. Other physiological roles of GSTs include chemical sequestration [355], regulation of Jun kinase[3] [356], inhibition of the proapoptotic action of Bax[4] protein [357], and modulation of calcium channels [358].

Dinitrosyl–diglutathionyl–iron complex binds with extraordinary affinity to the active site of all these dimeric enzymes, showing K_d values of 10^{-10}–10^{-9} M [59, 60, 341]. One of the glutathiones in the iron complex binds to the enzyme G-site, whereas the other GSH molecule is lost and replaced by a tyrosine phenolate in the coordination of the ferrous ion [60] Thus the bound complex is a

[3] c-Jun N-terminal kinases – mitogen-activated protein kinases which are responsive to stress stimuli, such as cytokines, ultraviolet irradiation, heat shock, and osmotic shock, and are involved in T cell differentiation and apoptosis. Kinase (phosphotransferase) – a type of enzyme that transfers phosphate groups from high-energy donor molecules, such as ATP, to specific substrates.

[4] Bcl-2–associated X protein (Bax) is a pro-apoptotic Bcl-2 type protein.

monoglutathionyl species. The binding of DNIC to the first subunit of the dimeric Alpha, Pi, and Mu GSTs triggers intersubunit communication, which lowers the affinity for DNIC of the second subunit that can still bind GSH and maintains the original enzyme activity [341]. In this way, GSTs may protect proteins against the toxic effect of DNIC and simultaneously prevent the extrusion of the free DNIC which would cause iron depletion. Both inactivation and reactivation of GSTs by DNGIC are relatively slow processes, the fastest for GSTA1-1 and the slowest for GSTT2-2. The reactivation is rate-limited by the release of the inactivator from the G-site. K_i values range from 10^{-10} M for GSTA1-1, 10^{-9} M for GSTM2-2, and GSTP1-1 to 10^{-7} M for GSTT2-2 [60]. DNDGIC binds to GSTA1-1, GSTM2-2, and GSTP1-1 with k_{on} values similar to those found for GSH (10^6–10^7 M^{-1} s^{-1}) [359], suggesting that the G-site is open enough to accommodate DNDGIC without gross structural changes. This perfectly agrees with the molecular modeling data that show DNDGIC partially exposed to the solvent and well stabilized in the G-sites of these GSTs. The observed strong affinity is likely due to the coordination of the iron atom to the phenolate group of the conserved Tyr residue of the active site (Tyr-7 in GSTP1-1, Tyr-6 in GSTM2-2, and Tyr-9 in GSTA1-1) which also causes a very slow extrusion of the complex from the G-site (low k_{off} value). A lower k_{on} value for DNDGIC binding appears to be the kinetic determinant of the decreased affinity of the second G-site in the half-saturated GSTs. Thus, negative cooperativity is not caused by a non-optimized geometry of iron ligands in the second G-site (which would cause an increased k_{off} value) but probably by an increased rigidity or shielding of the second G-site triggered by DNDGIC binding to the first subunit. Interestingly, the spectrum of the GST-bound dinitrosyl-iron complexes changes appreciably according to the GST isoform used. The A1-1 and M2-2 GSTs give essentially axial spectra, whereas P1-1 and T2-2 give strongly rhombic spectra; these differences concern the geometry of the ligand arrangement around the iron or the nature of the ligands themselves but not the number of ligands (four), thus indicating a four-coordinated iron with the ligand set [N, N, S^-, O^-] in the bound complex. Because the A1-1, M2-2, and P1-1 show similar K_d values for the dinitrosyl diglutathionyl iron complex, the actual conformation of the bound complex is not important for stability, as long as the iron is coordinated efficiently to both the glutathione and the tyrosine residue.

De Maria et al. noticed that from an evolutionary point of view it appears that the GST superfamily is under selective pressure in the direction of the optimization of the DNIC binding process as more recently evolved classes (Alpha, Mu, Pi) show a higher affinity for DNIC than the Theta class, which is close to the ancestral precursor of all GSTs [60].

4.3.2 The Role of Multidrug Resistance-Associated Protein 1 in the Efflux of DNIC-Bound Iron from Cells

It was shown by Watts et al. that GSH-bound nitrosyl iron takes part in NO-mediated iron mobilization from cells [96]. According to their results, NO-donors induced an increase in iron release from cells and a decrease in intracellular iron–ferritin levels. GSNO additionally decreased the mobilization of iron from cells and of the intracellular ferritin-iron levels. This decrease was a function of both time and increasing concentrations of nitrosoglutathione. The depletion of GSH prevented an NO-mediated iron efflux from cells, and the decrease in intracellular ferritin-iron levels seen in the presence of GSNO. Iron depletion was caused by a variety of permeable iron chelators. Extracellular high affinity Fe-binding proteins and the extracellular Fe chelators did not enhance NO-mediated iron efflux. GSNO, in contrast to desferrioxamine, did not mobilize iron from cytosolic lysates, while being more effective than DFO at mobilizing iron from intact cells. Finally, Fe mobilization was increased by incubating cells with D-glucose due to the subsequent generation of GSH [96, 97]. Taken together these results allowed the authors to conclude that the NO-mediated Fe efflux occurred as a GS–Fe–NO complex, which was exported from cells by the energy-dependent transport. Furthermore, the authors gave strong evidence that the iron and GSH efflux was the result of metabolic processes, by showing that these processes were dependent on the temperature and ATP. This inspired the authors to investigate the dependence of iron mobilization on the activity of a GSH transporter, multidrug resistance-associated protein 1 (MRP1). And so, the NO-mediated iron efflux was found to be greater in MRP1-hyperexpressing cell lines, while MRP1 inhibitors decreased NO-mediated iron efflux and induced DNIC accumulation in cells. The resulting outline of the interdependence of Fe, NO, GSH, and MRP1 is presented in the report by Watts et al. [97].

4.4 Biochemical Applications of Metal Nitrosyls

4.4.1 Nitric Oxide Donors

The rate of NO release and stability of metal nitrosyls depends mainly on the electron density of the metal centers and the nature of the ancillary ligands. Transition metal ions bind to ligands via interactions that are often strong and selective and the charge of the ion can be manipulated depending on the coordination environment. The thermodynamic and kinetic properties of metal–ligand interactions influence ligand exchange reactions. A partially filled d-shell confers unique electronic and magnetic properties, allows a range of coordination geometries (depending on the resulting electronic structure), and enables 1-electron oxidation and reduction reactions. The potential for designing transition metal nitrosylates with specific thermodynamic parameters makes them an interesting class of NO donors for biochemical and medical applications [360–362]. Some exemplary NO donors are depicted in Scheme 9. Attempts to design agents that deliver NO to a desired target in a controlled manner promise the prospect of obtaining new, NO-based

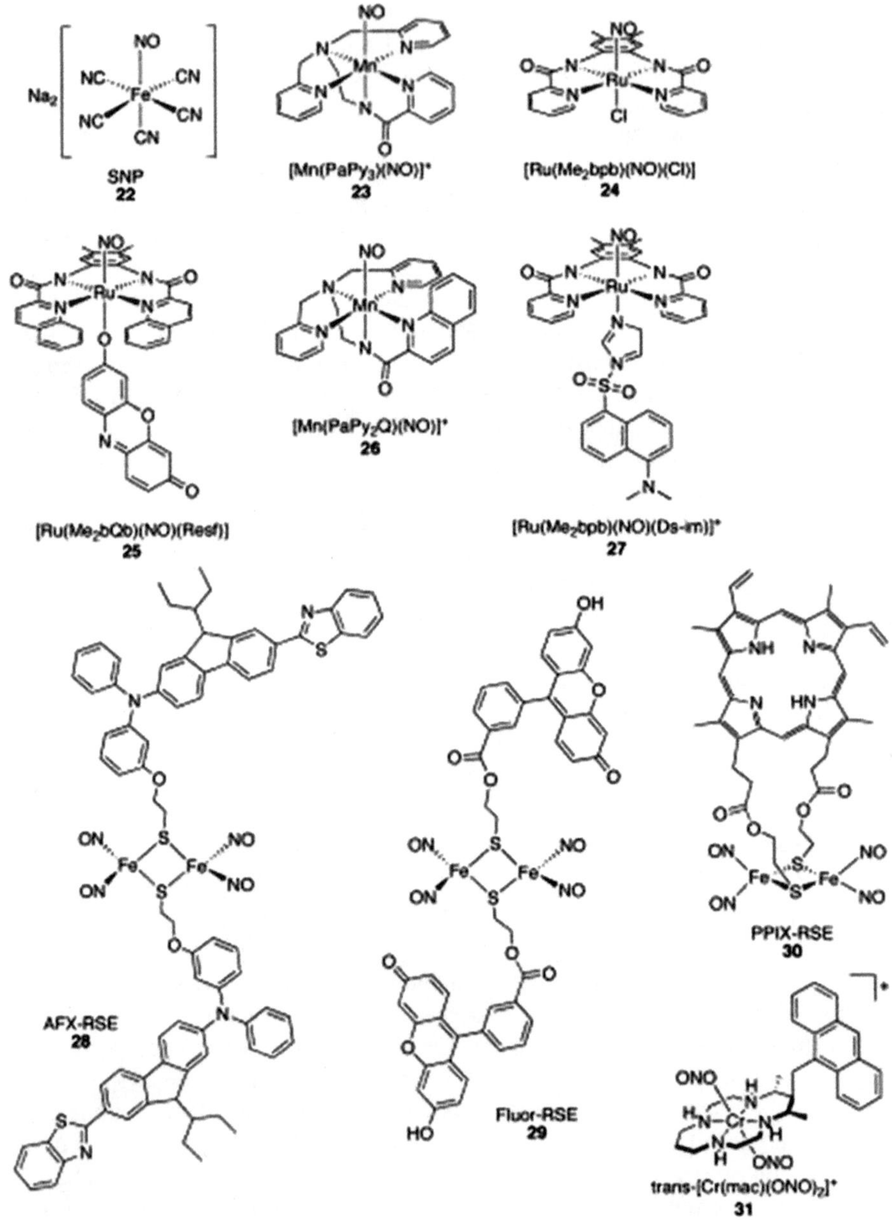

Scheme 9 The chosen NO-releasing metal complexes. Reprinted from [363], with permission of the American Chemical Society, copyright 2009

anti-infectious agents and γ-radiation sensitizers for photodynamic therapy. The known photosensitivity of many metal nitrosyl compounds makes light an attractive stimulus to release NO from a metal center.

The nitroprusside anion, $[Fe(CN)_5NO]^{2-}$, a seemingly simple octahedral coordination complex of Fe(II) with one NO and five cyanide ligands was the first metal-containing NO-donor identified and until recently used as a slow release agent for NO^+. Recently also DNICs have been considered as a potential new class of NO-donating drugs [364]. Depending on the ligand's electron properties DNICs' NO-donating ability can be easily modulated [86] which enables the controlling of DNICs' NO storage and transporting action in biological systems. Syntheses of such preparations are being conducted and initial toxicity tests have already been carried out [86, 89, 150, 365, 366]. Biological levels of NO are quantified by chelators containing iron, such as the iron complex of diethyldithiocarbamate (DETC) [73]. DETC administered to living animals can be visualized by EPR imaging in vivo and ex vivo for MNIC occurrence and thus NO distribution [65, 99, 367]. Another complex that can be used in NO imaging is the NO–Fe complex of N-(dithiocarboxy) sarcosine [368]. Pharmacokinetic methods with the use of EPR X-band imaging were elaborated [369]. The study allowed to evaluate the stability of NO–Fe(II)–DTCS in biological systems and the pharmacokinetic parameters were calculated on the basis of the two-compartment and hepatobiliary transport models. It was revealed that the compound is widely distributed in the peripheral organs and partially excreted into the bile. Also the organotransition–metal mononitrosyl complexes, $CpCr(NO)_2Cl$ and $CpMo(NO)_2Cl$, were examined and proposed as a new class of nitric oxide donors [351].

Photoactive Nitrosyl Donors

Several photosensitive metal nitrosyls including SNP, Roussin's salts, and simple Ru complexes like $[Ru(NO)Cl_5]^{2-}$ have been used as controlled sources of NO to elucidate the biological and neurophysiological roles of NO in cells and tissues. The disadvantages connected with the use of these simple salts include: production of toxic by-products like cyanide, the uncontrolled NO decay (due to pH and thermal instability, spontaneous disproportionation, or oxidation to NO_x) or the undesired reactions of the released metal with biomolecules as in the case of Ru forming DNA adducts [145, 200]. Some effort has been put into designing alternative multidentate ligands that support photoactive Mn and Ru nitrosyl adducts [145, 200, 363], as well as water-soluble Fe complexes [370, 371]. The photoactive metal nitrosylates can be divided into several groups, as described below. The applications of the various metal nitrosyls in therapy are discussed in the chapter by Morris in this volume of Structure and Bonding.

Two-Photon-Excitation-Sensitive NO Donors

The concept of two-photon excitation is based on the idea that two photons of comparably approximately half the energy necessary to excite the molecule also excite a fluorophore in one quantum event. NO donors coupled with the two-photon excitation sensitive fluorophores should be thermally stable but reactive under excitation at visible (vis) or near-infrared wavelengths where tissue transmission is optimal [372]. In order to enhance the light-gathering capability of light-sensitized NO donors, such as Roussin's red salt esters, Ford and coworkers have attached chromophores with high single- or two-photon absorption cross-sections to several photochemical NO precursors, including aminofluorene chromophore (AFX-RSE, RS = 2-thioethyl ester of N-phenyl-N-(3-(2-ethoxy)phenyl)-7-(benzothiazol-2-yl)-9,9-diethyl-fluoren-2-yl-amine) [373], fluorescein (Fluor-RSE) [374, 375], and protoporphyrin IX (PPIX-RSE) [376]. About 85% of the fluorescence intensity is quenched in these compounds, indicating energy transfer from the antenna to the Fe–NO cluster, ultimately inducing photochemical release of NO. The most commonly used fluorophores have excitation spectra in the 400–500 nm range, whereas the laser used to excite the two-photon fluorescence lies in the ~700–1,000 nm (infrared) range [360]. The probability of the near-simultaneous absorption of two photons is extremely low. Therefore a high flux of excitation photons is typically required, usually a femtosecond laser. Two-photon excitation (TPE) is of special interest, since the use of focused laser pulses to activate release could provide three-dimensional spatial control in therapeutic applications [377].

NO Donors Sensitive to Visible and Near-Infrared Light

Metalloporphyrins, being prone to photolysis-induced release of NO, were promptly recognized as bioregulatory compounds that could deliver NO to a target in a selective manner. Nevertheless, although having intense long-wavelength absorptions, the porphyrinate metal nitrosyls proved to be thermally unstable and oxygen sensitive. Ru–NO porphyrins, being relatively stable and convenient NO donors sensitive to light, suffer from complicated back and side reactions, such as rapid recombination of the photoproducts [360, 378–380]. In contrast, the nitrosyls with non-porphyrin ligands, such as amines, Schiff bases, thiolates, and ligands with carboxamide groups, readily release NO upon illumination and generate controlable photoproducts. For instance, complexes of Fe, Ru, and Mn with pentadentate ligand $PaPy_3H$ (*N,N-bis* (2-pyridylmethyl)amine-*N*-ethyl-2-pyridine-2-carboxamide) providing four nitrogen atom chelation around the equatorial plane of the metal and an additional chelating carboxamide group *trans* to NO proved promising, again the most stable and controllable being complexes of Mn and Ru [379–382]. The strong σ-donor character of the negatively charged carboxamide being *trans* to NO is a key feature in the photolability of these complexes, with the nature of the metal center dictating the wavelength of light required to achieve photo-release. Apart from being light-activated NO donors, these complexes allow the study of fast reactions of NO with heme proteins [383]. The Mn complex is promising with regard to therapeutic applications, as it is activated by visible light (500–650 nm) to release NO and the resulting photoreaction gives the Mn(II) complex, $[Mn(PaPy_3)(H_2O)]^+$, H_2O-substituted in place of NO [384]. Both the Mn and the Ru compounds stimulate soluble guanylate cyclase activity in vitro in a light- and concentration-dependent manner. They also elicit a concentration-dependent increase in cGMP in the vascular smooth muscle cells, demonstrating that the complexes release NO intracellularly under the control of light. Furthermore, the compounds showed light-dependent vasorelaxant activity in a rat thoracic aortic ring [385].

The photoactivity of metal nitrosyls requires promotion of an electron from a metal-based molecular orbital to a π*(NO) antibonding orbital. Therefore the sensitivity to light depends on the energy of the M→π*(NO) electronic transition [259, 382]. The therapeutic application of these complexes would be facilitated if it proved possible to downshift the activation absorption towards infrared because it would be able to reach deeper into the tissue (light penetration through mammalian tissue is mostly restricted to the 700–1,100 nm region). The bathochromic shift the M-NO bands can be achieved either by changing the metal or the field-strength of the ligand or by attaching light-harvesting chromophores to the complex. A good example is the series of modifications to the bipiridyl ligand, *N,N′-bis*(bipyridine-2-carboxamido)-1,2-diaminobenzene (H_2bpb); formation of a ruthenium nitrosyl complex with its dimethyl derivative yields a complex, which is only sensitive to low-intensity UV light. However, replacing the pyridal arms with quinolines red shifts the Ru–NO photoband from 380 nm to 455 nm, thereby accessing visible light photoactivation [386]. As the four-coordinate H_2bpb allows an additional coordination *trans* to NO, it is possible to attach a light-harvesting dye directly to the metal, and thus additional sensitization is achieved [387, 388]. The effects of the quinoline arms on the ligand and the dye attached to the metal are additive.

Trackable NO Donors

In addition to acting as light harvesters to increase the photosensitization of NO-donors, the incorporation of fluorescent dyes into NO-releasing compounds can provide a tracking signal to monitor the cellular distribution of the donor. Resorufin or dansyl-containing diamagnetic ruthenium nitrosyls [389] are species that retain appreciable fluorescence intensity of the coordinated dansyl or resorufin dyes. Photoinduced loss of NO gives paramagnetic Ru(III) species that quench fluorescence of the coordinated dyes to an extent related to the amount of NO released [387]. Cell culture studies in human breast cancer cells demonstrated the ability to track the visible light-triggered NO release from these compounds [387, 389].

4.4.2 Nitrosyl Formation-Based NO Sensors

Over the last decade, a number of metal complexes have appeared that give a fluorescence change in response to NO [390]. These include systems based on iron (II) [391, 392], cobalt (II) [393], ruthenium (II) [394, 395], rhodium (II) [396, 397], and copper (II) [398–401].

Metal-based NO probes, in contrast to their organic-based counterparts, take advantage of either direct NO reactivity at the metal center or reactivity at chelated ligand atoms. These probes provide an opportunity to explore direct and reversible sensing, because metals can interact reversibly with nitric oxide. Typically, metal-ligand constructs are assembled where the fluorophore is part of the ligand. In the NO-probe complex the fluorophore emission is quenched by one of the several PeT mechanisms [54]. There are three main strategies for eliciting a fluorescence response from these quenched systems in response to NO: In the first possible instance, resumption of fluorophore emission is accomplished by fluorophore displacement from the quenching site either by releasing the ligand entirely or by removing a chelating arm to a sufficient distance from the metal. In both cases, ligand displacement accompanies NO binding to form a metal nitrosyl. Many metal-based NO-probes utilize this approach, including those containing Co(II) [393], Fe(II) [391, 392], Ru(II) [394], or Rh(II) [396, 397].

Reduction of the metal to form a diamagnetic species can alleviate quenching caused by the paramagnetic metal ions [402]. Typically, this mechanism operates in protic solvents (ROH) and results in transfer of NO to the solvent to form an alkyl nitrite (RONO). This strategy has been employed primarily with Cu(II) probes, which are readily reduced by NO but fail to form stable Cu(I)–NO species [399, 401]. Copper(II) has also been incorporated into conjugated polymers to fashion NO-sensitive films with the use of the intrinsic fluorescence of the polymer as the emitter [400, 403, 404]. Certain modifications, such as incorporating water-soluble functional groups to facilitate biological compatibility, can be applied according to the requirements of the experiment [405].

A third mechanism for restoring fluorescence emission relies both on displacement of the fluorophore and on reductive nitrosylation of the paramagnetic metal center. In this process, NO can either coordinate to the metal [396, 405], as in case of paramagnetic cobalt (II) complex of dansyl fluorophore appended aminotroponiminate [406], or nitrosate either the ligand or solvent, when the latter is protic. In the first case, reaction with NO forms Co(I)–dinitrosyl adducts simultaneously inducing dissociation of a fluorescent ligand, thereby providing an ~8-fold emission enhancement. Secondly, a method recently gaining popularity is determination of nitric oxide using fluorescein-based copper (II) complex (CuFL) that allows the visualization of NO in living cells [407]. The FL ligand coordinates Cu(II) with a K_d of 1.5 μM at pH 7.0 to give a non-fluorescent complex that reacts rapidly and selectively with NO to give an 11-fold increase in fluorescence [408]. It was shown that fluorescence derives from the *N*-nitrosated product FL-NO and also is induced by reacting of FL with S-nitrosothiols. FL-NO no longer binds to Cu(II) or Cu(I) and has a 7.5-fold higher quantum yield than FL itself, indicating that it is the species responsible for the observed fluorescence signal. Neither removal of Cu(II) from the CuFL complex nor addition of Cu(I) to FL gives the dramatic fluorescence enhancement, providing further assurance that the response is NO-dependent and doesn't result of metal release or simple reduction of the probe compound. The advantage of this method of NO determination is also that CuFL easily permeates cell membranes [408]. Another Cu(II)-based NO sensor has also appeared recently

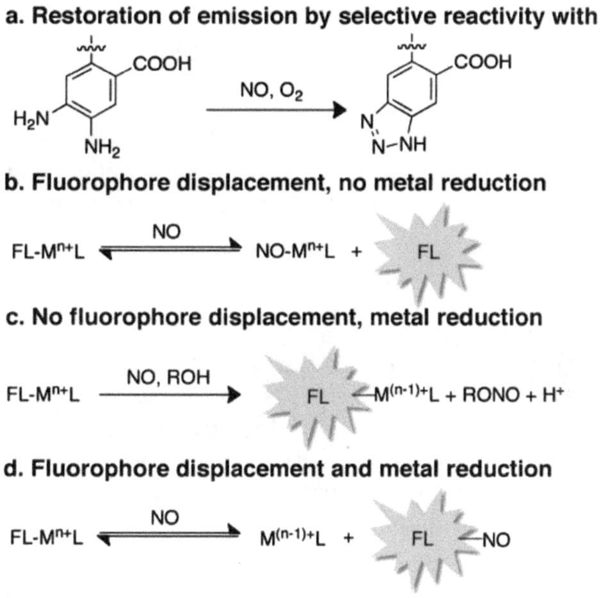

Fig. 3 Possible mechanisms of action for fluorescent NO-sensing probes. Reprinted from [54], with the permission of the American Chemical Society (ACS), copyright 2010

along with the in vivo imaging data. The fluorescent compound, 4-methoxy-2-(1*H*-naptho[2,3-*d*]imidazol-2-yl)phenol (MNIP), binds Cu(II) with 1:1 stoichiometry to give non-fluorescent Cu-MNIP with an apparent K_d in pH 7.4 water/DMSO of 0.6 μM. MNIP-Cu reacts rapidly and specifically with NO to generate a product with blue fluorescence that can be used in vitro and in vivo. The fluorescence enhancement was determined to be a reductive nitrosylation process, where Cu(I) is released from the complex and one of the nitrogen ligands is nitrosylated [409]. The described mechanisms are schematically illustrated in Fig. 3.

5 Conclusions

Nitric oxide (II) is a common free radical produced in living cells through the enzymatic degradation of L-arginine. It displays a number of regulatory functions: it is involved in the processes of neural conduction, regulation of cardiac function and immune defense, it triggers the pathways leading to the controlled cell death – apoptosis. One of the most important and earliest discovered function of NO was to stimulate vasodilation. At the same time, being a free radical, NO is deleterious to the components of living cells.

NO in the unbound form has a very short lifetime in the cell but can be stabilized by the formation of complexes, i.e. metal–porphyrin nitrosyls, dinitrosyl–iron complexes and S-nitrosothiols, which are considered to be its biological transporters. Nitric oxide has a very high affinity for iron contained in the active sites of proteins [74].

Basically one can distinguish three main binding sites of $Fe(NO)_2$ complexes in biological systems (1) the sulfhydryl groups of amino acids and inorganic sulfur atoms, (2) the iron–sulfur centers, (3) the imidazole rings present among others in histidine and purines, (4) the pyrole rings in heme type prosthetic groups. While thiols are responsible for iron binding in non-heme proteins, histidine binds globin to the heme iron of hemoglobin. Proteins containing iron–sulfur centers are the first molecules that undergo nitrosylation under conditions of increased NO production in biological systems [57]. The major protein targets of NO are heme-containing proteins, non-heme iron–sulfur proteins, proteins containing free thiol groups and protein radicals. Among DNICs forming non-protein ligands in living organisms are: the tripeptide glutathione and the amino acids cysteine and homocysteine. Nitrosylation of proteins occurs via direct binding of NO to thiol groups of the peptide chain or by binding of the nitrosylated iron to protein sulfhydryl and imidazole residues. Heme or iron–sulfur centers in proteins exhibit high affinity for nitric oxide. Much of the data indicate that this phenomenon has a significant regulatory function [57]. Nitric oxide and its complexes with biological conveyors are involved in neural conduction, cardiac function, and regulation of immune defense, as well as in directing pathways to the controlled cell death – apoptosis. Therefore, it appears worthwhile to collect spectral data that could clarify the question of metal nitrosyls' electronic structures and affinity of the MNO core to the various ligands of biological importance.

It is postulated that dinitrosyl complexes of iron play an important role in signal transduction pathways by regulating the cell homeostasis [410]. Yet, there is a lack of evidence which of the available thiol ligands in the cell form physiologically active iron nitrosyls; recent reports, however, indicate that a significant portion of these complexes is formed by proteins of high molecular weight [55, 70]. DNICs, both of low and high molecular weight are more stable than the nitrogen (II) oxide and are considered to be its transporters [410]. DNICs also inhibit platelet aggregation, lower the blood pressure, dilate the blood vessels, induce the formation of proteins in the cell response to oxidative stress, and modulate the activity of ion channels. Their appearance in cells exposed to NO is closely synchronized with the enzyme activity inhibition of iron sulfur centers in the complex I and complex II of the mitochondrial respiratory chain and of mitochondrial aconitase. DNICs irreversibly inhibit the activity of glutathione reductase, and reversibly that of glutathione transferases. The extremely strong inactivation of these enzymes by DNICs indicates their important role in the metabolism of glutathione [345, 411]. One- and bi-nucleated dinitrosyl iron complexes induce activation of the oxidative shock response (soxA) gene and cell division inhibiting gene sfiA in *E. coli*. Activation of these genes occurs as a result of DNICs action and is not caused by iron ions or nitric oxide alone [412]. Another proposed function of DNICs is to regulate the amount of labile iron in cells [97, 105]. Labile iron pool (LIP) was defined by Cabantchik et al. as the iron present in the cell in the form of highly dissociable complexes with low molecular weight ligands, and accessible in redox reactions [103]. In addition to these revealed and partially explored functions metal nitrosyls have within the organisms, there may yet be other potentially to be discovered. A putative new function of the $\{Fe(NO)_2\}^{10}$

type DNIC has recently been proposed by Tran et al. [413], who has reported the dioxygen-related reactivity of a bidentate N-ligand bound $\{Fe(NO)_2\}^{10}$ nitrosyl. In the presence of O_2, an unstable complex was formed being presumably a peroxynitrite [Fe(L)(NO)(ONOO)]. The complex exhibited nitrating properties towards 2,4,-di-tert-butylphenol, which suggests that DNIC-type d^{10} complexes might be involved in the processes of nitration of biological phenols, among which protein tyrosine nitration is an important posttranslational modification associated with various pathological conditions including inflammatory, neurodegenerative, and cardiovascular diseases.

The molecular and electronic structure of DNICs continues to be actively studied and discussed [77]. Most of the classification and theoretical structure predictions for nitrosyl complexes of iron and other transition metals are based on diagrams developed by Enemark and Feltham [25]. The construction of nitrosyl complexes proposed by these authors assumed $M(NO)_x$ to be a group of covalently bound atoms, whose structure is modified by the coordination of additional ligands. The energies of transition metal d orbitals are close to the π^* orbital energy of nitric oxide. Thus, the relative charge distribution between the metal and NO group at the M and N binding orbitals may be significantly different between the two isoelectronic complexes, and relatively small changes in the properties of metal or ligand can cause a significant change in the properties of the complex.

Formation of iron nitrosyl complexes is a major biological mechanism that allows direct detection of nitric oxide in cells [414]. Nitrosyl complexes of iron, due to their paramagnetic properties, were the first identified nitric oxide compounds in living organisms long before their numerous regulatory functions were discovered [87]. Currently, research is being conducted aimed at the synthesis of new dinitrosyl iron complexes, which would be applicable in biochemistry and medicine, as donors of nitrogen (II) oxide, mimicking the action of natural NO conveyors (see, e.g., [54, 119]).

Acknowledgments The author wishes to thank Professor I. Szumiel from the Institute of Nuclear Chemistry and Technology for her invaluable meritorical and editorial remarks.

References

1. Greenwood NN, Earnshaw A, Earnshaw A (1997) Chemistry of the elements. Butterworth-Heinemann, Oxford
2. Laane J, Ohlsen JR (1980) Characterization of nitrogen oxides by vibrational spectroscopy. In: Progress in inorganic chemistry. Wiley, New York
3. Ford PC, Lorkovic IM (2002) Mechanistic aspects of the reactions of nitric oxide with transition-metal complexes. Chem Rev 102:993–1018
4. Richter-Addo GB, Legzdins P (1992) Metal nitrosyls. Oxford University Press, Oxford
5. McCleverty JA (2004) Chemistry of nitric oxide relevant to biology. Chem Rev 104:403–418
6. De La Cruz C, Sheppard N (2011) A structure-based analysis of the vibrational spectra of nitrosyl ligands in transition-metal coordination complexes and clusters. Spectrochim Acta A Mol Biomol Spectrosc 78:7–28

7. McNaught AD, Wilkinson A (2012) IUPAC. Compendium of chemical terminology, Version 2.3.2 (the "Gold Book"). Blackwell Scientific, Oxford
8. Scheidt WR, Barabanschikov A et al (2010) Electronic structure and dynamics of nitrosyl porphyrins. Inorg Chem 49:6240–6252
9. Zavarine IS, Kini AD et al (1998) Photochemistry of nitrosyl metalloporphyrins: mechanisms of the photoinduced release and recombination of NO. J Phys Chem B 102:7287–7292
10. Brockway LO, Anderson JS (1937) The molecular structures of iron nitrosocarbonyl Fe $(NO)_2$ $(CO)_2$ and cobalt nitrosocarbonyl $Co(NO)(CO)_3$. Trans Faraday Soc 33:1233–1239
11. Dai RJ, Ke SC (2007) Detection and determination of the Fe(NO)2 core vibrational features in dinitrosyl-iron complexes from experiment, normal coordinate analysis, and density functional theory: an avenue for probing the nitric oxide oxidation state. J Phys Chem B 111:2335–2346
12. Brock CP, Collman JP et al (1973) Bent vs. linear nitrosyl paradox. Infrared and X-ray photoelectron spectra of dichloronitrosylbis (L)cobalt (II) and crystal structure with L = diphenylmethylphosphine. Inorg Chem 12:1304–1313
13. Haymore BL, Ibers JA (1975) Linear vs. bent nitrosyl ligands in four-coordinate transition metal complexes. Structure of dinitrosylbis (triphenylphosphine)osmium (−II) hemibenzene, Os $(NO)_2$ (P $(C_6H_5)_3)_2 \cdot 1/2C_6H_6$. Inorg Chem 14:2610–2617
14. Lin ZS, Chiou TW et al (2012) A dinitrosyliron complex within the homoleptic $Fe(NO)_4$ anion: NO as nitroxyl and nitrosyl ligands within a single structure. Inorg Chem 51:10092–10094
15. Peterson ES, Larsen RD, Abbott EH (1988) Crystal and molecular structures of potassium aquatetrakis (nitrito)nitrosylplatinate (IV), a blue, mononuclear platinum complex with a bent nitrosyl group, and of potassium trichlorobis (nitrito)nitrosylplatinate (IV). Inorg Chem 27:3514–3518
16. Calderon JL, Fontana S et al (1976) The crystal structure of tris-cyclopentadienyldimanganese trisnitrosyl; a compound containing unsymmetrically bonded bridging nitrosyl groups. Inorg Chim Acta 17:L31–L32
17. Calderon JL, Cotton FA et al (1971) Molecular structure of an unusual binuclear manganese complex with highly unsymmetrical nitrosyl bridges. J Chem Soc D 1476–1477
18. Norton JR, Collman JP et al (1972) Preparation and structure of ruthenium and osmium nitrosyl carbonyl clusters containing double-nitrosyl bridges. Inorg Chem 11:382–388
19. Yan B, Xie Y et al (2009) (Cyclopentadienyl)nitrosylmanganese compounds: the original molecules containing bridging nitrosyl groups. Eur J Inorg Chem 2009:3982–3992
20. Müller J, Manzoni de Oliveira G, Sonn I (1988) Reaktionen von Nitrosylkomplexen: VIII. Ein alternativer Weg zu $Cp_3Co_3(\mu3NO)_2$ sowie Synthese des isoelektronischen clusters $Cp_3Co_2Fe(\mu_3NH)(\mu_3NO)$. J Organomet Chem 340:C15–C18
21. Terrile C, Nascimento OR et al (1990) On the electronic structure of metastable nitroprusside ion in $Na_2[Fe(CN)_5NO]\cdot 2H_2O$. A comparative single crystal ESR study. Solid State Commun 73:481–486
22. Maier G, Reisenauer HP, De Marco M (2000) Isomerizations between nitrosyl halides X-N=O and isonitrosyl halides X-O-N: a matrix-spectroscopic study. Chem Eur J 6:800–808
23. Connelly NG, Hartshorn RM et al (2005) Nomenclature of inorganic chemistry. In: IUPAC recommendations, 1990. N. G. Con. Royal Society of Chemistry, London
24. Reginato N, McCrory CTC et al (1999) Synthesis, X-ray crystal structure, and solution behavior of Fe $(NO)_2$ $(1\text{-MeIm})_2$: implications for nitrosyl non-heme-iron complexes with g = 2.03. J Am Chem Soc 121:10217–10218
25. Enemark JH, Feltham RD (1974) Principles of structure, bonding, and reactivity for metal nitrosyl complexes. Coord Chem Rev 13:339–406
26. Ozawa S, Fujii H, Morishima I (1992) NMR studies of iron (II) nitrosyl Pi-cation radicals of octaethylchlorin and octaethylisobacteriochlorin as models for reaction intermediate of nitrite reductase. J Am Chem Soc 114:1548–1554

27. Afshar RK, Patra AK et al (2004) Syntheses, structures, and reactivities of $(Fe\text{-}NO)_6$ nitrosyls derived from polypyridine-carboxamide ligands: photoactive NO-donors and reagents for S-nitrosylation of alkyl thiols. Inorg Chem 43:5736–5743
28. Praneeth VKK, Nather C et al (2006) Spectroscopic properties and electronic structure of five- and six-coordinate iron (II) porphyrin NO complexes: effect of the axial N-donor ligand. Inorg Chem 45:2795–2811
29. Eroy-Reveles AA, Hoffman-Luca CG, Mascharak PK (2007) Formation of a triply bridged Mu-Oxo Diiron (III) core stabilized by two deprotonated carboxamide groups upon photorelease of NO from a $[Fe\text{-}NO]_6$ iron nitrosyl. Dalton Trans 5268–5274
30. Hoffman-Luca CG, Eroy-Reveles AA et al (2009) Syntheses, structures, and photochemistry of manganese nitrosyls derived from designed Schiff base ligands: potential NO donors that can be activated by near-infrared light. Inorg Chem 48:9104–9111
31. Horsken A, Zheng G et al (1998) Iron dinitrosyl complexes of TCNE: a synthetic, X-ray crystallographic, high field NMR and electrochemical study. J Organomet Chem 558:1–9
32. Rose MJ, Mascharak PK (2009) Photosensitization of Ruthenium nitrosyls to red light with an isoelectronic series of heavy-atom chromophores: experimental and density functional theory studies on the effects of O-, S- and Se-substituted coordinated dyes. Inorg Chem 48:6904–6917
33. Sandala GM, Hopmann KH et al (2011) Calibration of DFT functionals for the prediction of Fe Mossbauer spectral parameters in iron-nitrosyl and iron-sulfur complexes: accurate geometries prove essential. J Chem Theory Comput 7:3232–3247
34. Conradie J, Hopmann KH, Ghosh A (2010) Understanding the unusually straight: a search for MO insights into linear $\{FeNO\}^7$ units. J Phys Chem B 114:8517–8524
35. Hopmann KH, Ghosh A, Noodleman L (2009) Density functional theory calculations on mossbauer parameters of nonheme iron nitrosyls. Inorg Chem 48:9155–9165
36. Ghosh P, Bill E et al (2007) The molecular and electronic structure of the electron transfer series $[Fe_2(NO)_2(S_2C_2R_2)_3]$ (z) (z = 0, -1, -2; R = phenyl, p-tolyl, p-tert-butylphenyl). Inorg Chem 46:2612–2618
37. Conradie J, Ghosh A (2006) Iron (III)-nitro porphyrins: theoretical exploration of a unique class of reactive molecules. Inorg Chem 45:4902–4909
38. Wondimagegn T, Ghosh A (2001) A quantum chemical survey of metalloporphyrin-nitrosyl linkage isomers: insights into the observation of multiple FeNO conformations in a recent crystallographic determination of nitrophorin 4. J Am Chem Soc 123:5680–5683
39. Sizova OV, Sokolov AY et al (2007) Quantum chemical study of the bond orders in the ruthenium, diruthenium and dirhodium nitrosyl complexes. Polyhedron 26:4680–4690
40. Feltham RD, Nyholm RS (1965) Metal nitrosyls. VI. Some new six-coordinate mononitrosyl complexes of cobalt. Inorg Chem 4:1334–1339
41. Herberhold M, Razavi A (1972) Tetranitrosylchromium$[Cr(NO)_4]$. Angew Chem Int Ed Engl 11:1092–1094
42. Roussin ML (1858) Recherches sur les Nitrosulfures Doubles de Fer. (Nouvelle classe de sels.). Ann Chim Phys 52:285
43. Tinberg CE, Tonzetich ZJ et al (2010) Characterization of iron dinitrosyl species formed in the reaction of nitric oxide with a biological Rieske center. J Am Chem Soc 132: 18168–18176
44. Wang R, Camacho-Fernandez MA et al (2009) Neutral and reduced Roussin's red salt ester $[Fe_2\ \mu\text{-}RS)_2(NO)_4]$ (R = N-Pr, T-Bu, 6-Methyl-2-Pyridyl and 4,6-Dimethyl-2-Pyrimidyl): synthesis, X-ray crystal structures, spectroscopic, electrochemical and density functional theoretical investigations. Dalton Trans 777–786
45. Liu JG, Li MH (1989) Roussin red methyl ester, a tumor promoter isolated from pickled vegetables. Carcinogenesis 10:617–620
46. Janczyk A, Wolnicka-Glubisz A et al (2004) NO-dependent phototoxicity of Roussin's black salt against cancer cells. Nitric Oxide 10:42–50

47. Mitchell JB, Wink DA et al (1993) Hypoxic mammalian cell radiosensitization by nitric oxide. Cancer Res 53:5845–5848
48. Bourassa J, DeGraff W et al (1997) Photochemistry of Roussin's red salt, $Na_2[Fe_2S_2(NO)_4]$, and of Roussin's black salt, $NH_4[Fe_4S_3(NO)_7]$. in situ nitric oxide generation to sensitize γ-radiation induced cell death. J Am Chem Soc 119:2853–2860
49. Butler AR, Glidewell C et al (1985) Formation of paramagnetic mononuclear iron nitrosyl complexes from diamagnetic di- and tetranuclear iron-sulphur nitrosyls: characterisation by EPR spectroscopy and study of thiolate and nitrosyl ligand exchange reactions. Polyhedron 4:797–809
50. Wanat A, Schneppensieper T et al (2002) Kinetics, mechanism, and spectroscopy of the reversible binding of nitric oxide to aquated iron (II). An undergraduate text book reaction revisited. Inorg Chem 41:4–10
51. Butler AR, Glidewell C, Li MH (1988) Nitrosyl complexes of iron-sulfur clusters. In: Sykes AG (ed) Advances in inorganic chemistry, vol 32. Academic, London
52. Connelly NG, Gardner C (1976) Simple halogenonitrosyl anions of iron. J Chem Soc Dalton Trans 1525–1527
53. Lu TT, Chiou SJ et al (2006) Mononitrosyl tris(thiolate) iron complex $[Fe(NO)(SPh)_3]^-$ and dinitrosyl iron complex $[(EtS)_2Fe(NO)_2]^-$: formation pathway of dinitrosyl iron complexes (DNICs) from nitrosylation of biomimetic rubredoxin $[Fe(SR)_4]^{2-/1-}$ (R = Ph, Et). Inorg Chem 45:8799–8806
54. Tonzetich ZJ, McQuade LE, Lippard SJ (2010) Detecting and understanding the roles of nitric oxide in biology. Inorg Chem 49:6338–6348
55. Henry Y, Ducrocq C et al (1991) Nitric oxide, a biological effector. Electron paramagnetic resonance detection of nitrosyl-iron-protein complexes in whole cells. Eur Biophys J 20:1–15
56. Singel DJ, Lancaster JR Jr (1996) Electron paramagnetic resonance spectroscopy and nitric oxide biology. In: Feelisch M, Stamler JS (eds) Methods in nitric oxide research. Wiley, Chichester
57. van Faassen E, Vanin AF (2007) Radicals for life: the various forms of nitric oxide. Elsevier, Amsterdam
58. Bouton C, Drapier JC (2003) Iron regulatory proteins as NO signal transducers. Sci STKE 182:pe17
59. Turella P, Pedersen JZ et al (2003) Glutathione transferase superfamily behaves like storage proteins for dinitrosyl-diglutathionyl-iron complex in heterogeneous systems. J Biol Chem 278:42294–42299
60. De Maria F, Pedersen JZ et al (2003) The specific interaction of dinitrosyl-diglutathionyl-iron complex, a natural NO carrier, with the glutathione transferase superfamily: suggestion for an evolutionary pressure in the direction of the storage of nitric oxide. J Biol Chem 278:42283–42293
61. Vanin AF (1998) Dinitrosyl iron complexes and S-nitrosothiols are two possible forms for stabilization and transport of nitric oxide in biological systems. Biochemistry 63:782–793
62. Vanin AF, Malenkova IV, Serezhenkov VA (1997) Iron catalyzes both decomposition and synthesis of S-nitrosothiols: optical and electron paramagnetic resonance studies. Nitric Oxide 1:191–203
63. Stubauer G, Giuffre A, Sarti P (1999) Mechanism of S-nitrosothiol formation and degradation mediated by copper ions. J Biol Chem 274:28128–28133
64. Lim MD, Capps KB et al (2005) Further evidence supporting an inner sphere mechanism in the NO reduction of the copper (II) complex Cu $(dmp)_2^{2+}$ (dmp= 2,9-dimethyl-1,10-phenanthroline). Nitric Oxide 12:244–251
65. Ueno T, Suzuki Y et al (2002) In vivo nitric oxide transfer of a physiological NO carrier, dinitrosyl dithiolato iron complex, to target complex. Biochem Pharmacol 63:485–493
66. Wiegant FA, Malyshev IY et al (1999) Dinitrosyl iron complexes with thiol-containing ligands and S-nitroso-D, L-penicillamine as inductors of heat shock protein synthesis in H35 hepatoma cells. FEBS Lett 455:179–182

67. Manukhina EB, Smirin BV et al (2002) [Nitric oxide storage in the cardiovascular system]. Izv Akad Nauk Ser Biol 585–596
68. Boese M, Mordvintcev PI et al (1995) S-nitrosation of serum albumin by dinitrosyl-iron complex. J Biol Chem 270:29244–29249
69. Mulsch A, Mordvintcev P et al (1991) The potent vasodilating and guanylyl cyclase activating dinitrosyl-iron (II) complex is stored in a protein-bound form in vascular tissue and is released by thiols. FEBS Lett 294:252–256
70. Henry Y, Lepoivre M et al (1993) EPR characterization of molecular targets for NO in mammalian cells and organelles. FASEB J 7:1124–1134
71. Alencar JL, Chalupsky K et al (2003) Inhibition of arterial contraction by dinitrosyl-iron complexes: critical role of the thiol ligand in determining rate of nitric oxide (NO) release and formation of releasable NO stores by S-nitrosation. Biochem Pharmacol 66:2365–2374
72. Mulsch A (1994) Nitrogen monoxide transport mechanisms. Arzneimittelforschung 44:408–411
73. Vanin AF, Mordvintcev PI et al (1993) The relationship between L-arginine-dependent nitric oxide synthesis, nitrite release and dinitrosyl-iron complex formation by activated macrophages. Biochim Biophys Acta 1177:37–42
74. Cooper CE (1999) Nitric oxide and iron proteins. Biochim Biophys Acta 1411:290–309
75. Richardson DR, Lok HC (2008) The nitric oxide-iron interplay in mammalian cells: transport and storage of dinitrosyl iron complexes. Biochim Biophys Acta 1780:638–651
76. Lewandowska H, Stepkowski TM et al (2012) Coordination of iron ions in the form of histidinyl dinitrosyl complexes does not prevent their genotoxicity. Bioorg Med Chem 20:6732–6738
77. Lewandowska H, Kalinowska M et al (2011) Nitrosyl iron complexes-synthesis, structure and biology. Dalton Trans 40:8273–8289
78. Hieber W, Nast R (1940) Uber Stickoxydverbindungen von Nickel[I]- und Eisen [I] halogeniden. Z Anorg Allg Chem 244:23–47
79. Cotton FA (2013) Progress in inorganic chemistry. Wiley, New York
80. Herberhold M, Kremnitz W et al (1982) Solution and matrix photochemistry of the isoelectronic series of a 'half -sandwich' carbonyl (η_5-Cyclopentadienyl)nitrosyl complexes of manganese, chromium, and vanadium, [M (η_5-C_5H_5)$(CO)_3$(NO)] (n= 0, M = Mn; n= 1, M = Cr; n= 2, M = V). J Chem Soc Dalton Trans 1261–1273
81. Haller KJ, Enemark JH (1978) Structural chemistry of the $\{CoNO\}^8$ Group. III. The structure of N, N′-Ethylenebis (salicylideneiminato)nitrosylcobalt (II), Co(NO)(salen). Acta Crystallogr B 34:102–109
82. Collman JP, Farnham P, Dolcetti G (1971) Intramolecular redox equilibriums of cobalt-nitrosyl complexes. J Am Chem Soc 93:1788–1790
83. Pierpont CG, Van Derveer DG et al (1970) Ruthenium complex having both linear and bent nitrosyl groups. J Am Chem Soc 92:4760–4762
84. Martin RL, Taylor D (1976) Bending of linear nitric oxide ligands in four-coordinate transition metal complexes. Crystal and molecular structure of dinitrosyldithioacetylacetonatocobalt(−I), $Co(NO)_2$. Inorg Chem 15:2970–2976
85. Henry Y, Giussani A, Ducastel B (1997) Nitric oxide research from chemistry to biology: EPR spectroscopy of nitrosylated compounds. Springer, Berlin
86. Tsai FT, Chiou SJ et al (2005) Dinitrosyl iron complexes (DNICs) $[L_2Fe(NO)_2]^-$ (L = Thiolate): interconversion among $\{Fe(NO)_2\}^9$ DNICs, $\{Fe(NO)_2\}^{10}$ DNICs, and [2Fe-2S] clusters, and the critical role of the thiolate ligands in regulating NO release of DNICs. Inorg Chem 44:5872–5881
87. Vithayathil AJ, Ternberg JL, Commoner B (1965) Changes in electron spin resonance signals of rat liver during chemical carcinogenesis. Nature 207:1246–1249
88. Cleare MJ, Fritz HP, Griffith WP (1972) Vibrational spectra of nitrosyl and carbonyl complexes: nitrosylpentahalogeno complexes of osmium, ruthenium and iridium. Spectrochim Acta A Mol Spectrosc 28:2013–2018

89. Vanin AF, Sanina NA et al (2007) Dinitrosyl-iron complexes with thiol-containing ligands: spatial and electronic structures. Nitric Oxide 16:82–93
90. Woolum JC, Tiezzi E, Commoner B (1968) Electron spin resonance of iron-nitric oxide complexes with amino acids, peptides and proteins. Biochim Biophys Acta 160:311–320
91. Bostanci MO, Bagirici F (2007) Neuroprotection by 7-nitroindazole against iron-induced hippocampal neurotoxicity. Cell Mol Neurobiol 27:933–941
92. Bostanci MO, Bagirici F (2008) Nitric oxide synthesis inhibition attenuates iron-induced neurotoxicity: a stereological study. Neurotoxicology 29:130–135
93. Gupta A, Sharma S, Chopra K (2008) Reversal of iron-induced nephrotoxicity in rats by molsidomine, a nitric oxide donor. Food Chem Toxicol 46:537–543
94. Kadkhodaee M, Gol A (2004) The role of nitric oxide in iron-induced rat renal injury. Hum Exp Toxicol 23:533–536
95. Feger F, Ferry-Dumazet H et al (2001) Role of iron in tumor cell protection from the pro-apoptotic effect of nitric oxide. Cancer Res 61:5289–5294
96. Watts RN, Richardson DR (2002) The mechanism of nitrogen monoxide (NO)-mediated iron mobilization from cells. NO intercepts iron before incorporation into ferritin and indirectly mobilizes iron from ferritin in a glutathione-dependent manner. Eur J Biochem 269:3383–3392
97. Watts RN, Hawkins C et al (2006) Nitrogen monoxide (NO)-mediated iron release from cells is linked to NO-induced glutathione efflux via multidrug resistance-associated protein 1. Proc Natl Acad Sci USA 103:7670–7675
98. Fass U, Panickar K et al (2004) The role of glutathione in nitric oxide donor toxicity to SN56 cholinergic neuron-like cells. Brain Res 1005:90–100
99. Ueno T, Yoshimura T (2000) The physiological activity and in vivo distribution of dinitrosyl dithiolato iron complex. Jpn J Pharmacol 82:95–101
100. Boese M, Keese MA et al (1997) Inhibition of glutathione reductase by dinitrosyl-iron-dithiolate complex. J Biol Chem 272:21767–21773
101. Becker K, Savvides SN et al (1998) Enzyme inactivation through sulfhydryl oxidation by physiologic NO-carriers. Nat Struct Biol 5:267–271
102. Vlasova MA, Vanin AF et al (2003) Detection and description of various stores of nitric oxide store in vascular wall. Bull Exp Biol Med 136:226–230
103. Kakhlon O, Cabantchik ZI (2002) The labile iron pool: characterization, measurement, and participation in cellular processes (1). Free Radic Biol Med 33:1037–1046
104. Lipinski P, Drapier JC et al (2000) Intracellular iron status as a hallmark of mammalian cell susceptibility to oxidative stress: a study of L5178Y mouse lymphoma cell lines differentially sensitive to H_2O_2. Blood 95:2960–2966
105. Denninger JW, Marletta MA (1999) Guanylate cyclase and the NO/cGMP signaling pathway. Biochim Biophys Acta 1411:334–350
106. Kubrina LN, Mikoyan VD et al (1993) Iron potentiates bacterial lipopolysaccharide-induced nitric oxide formation in animal organs. Biochim Biophys Acta 1176:240–244
107. Vanin AF, Men'shikov GB et al (1992) The source of non-heme iron that binds nitric oxide in cultivated macrophages. Biochim Biophys Acta 1135:275–279
108. Riley RF, Ho L (1962) Evidence for a pentacyanonitrosyl complex of molybdenum. J Inorg Nuclear Chem 24:1121–1127
109. Szaciłowski K, Oszajca J et al (2001) Photochemistry of the $[Fe(CN)_5N(O)SR]^{3-}$ complex: a mechanistic study. J Photochem Photobiol A 143:93–262
110. Costanzo S, Ménage S et al (2001) Re-examination of the formation of dinitrosyl–iron complexes during reaction of S-nitrosothiols with Fe (II). Inorg Chim Acta 318:1–7
111. Baltusis LM, Karlin KD et al (1980) Synthesis and structure of $Fe(L'H)(NO)_2$, a tetracoordinate complex having a twelve-membered chelate ring, and its conversion to pentacoordinate FeL′ (NO) through formal loss of "HNO" (L′ = SCH2CH2NMeCH2CH2CH2NMeCH2CH2S-). Inorg Chem 19:2627–2632

112. Szaci+éowski K, Macyk W et al (2000) Ligand and medium controlled photochemistry of iron and ruthenium mixed-ligand complexes: prospecting for versatile systems. Coord Chem Rev 208:277–297
113. Crayston JA, Glidewell C, Lambert RJ (1990) Redox properties of iron-sulphur-nitrosyl complexes: exploratory electrochemistry of mononuclear, dinuclear and tetranuclear complexes. Polyhedron 9:1741–1746
114. Wang PG, Xian M et al (2002) Nitric oxide donors: chemical activities and biological applications. Chem Rev 102:1091–1134
115. Jaworska M, Stasicka Z (2004) Structure and UV–vis spectroscopy of nitrosylthiolatoferrate mononuclear complexes. J Organomet Chem 689:1702–1713
116. Kennedy MC, Antholine WE, Beinert H (1997) An EPR investigation of the products of the reaction of cytosolic and mitochondrial aconitases with nitric oxide. J Biol Chem 272:20340–20347
117. Bryar TR, Eaton DR (1992) Electronic configuration and structure of paramagnetic iron dinitrosyl complexes. Can J Chem 70:1917–1926
118. Harrop TC, Tonzetich ZJ et al (2008) Reactions of synthetic [2Fe-2S] and [4Fe-4S] clusters with nitric oxide and nitrosothiols. J Am Chem Soc 130:15602
119. Sanina NA, Rakova OA et al (2004) Structure of the neutral mononuclear dinitrosyl iron complex with 1,2,4-triazole-3-thione$[Fe(SC_2H_3N_3)(SC_2H_2N_3)(NO)_2]\cdot 0.5\ H_2O$. Mendeleev Commun 14:7–8
120. Sanina NA, Aldoshin SM (2004) Functional models of [Fe–S] nitrosyl proteins. Russ Chem Bull Int Ed 53:2428
121. Davies SC, Evans DJ et al (2002) J Chem Soc Dalton Trans 2473
122. Tsou CC, Lu TT, Liaw WF (2007) EPR, UV–vis, IR, and X-ray demonstration of the anionic dimeric dinitrosyl iron complex$[(NO)_2Fe(micro\text{-}S(t)Bu)_2Fe(NO)_2]^-$: relevance to the products of nitrosylation of cytosolic and mitochondrial aconitases, and high-potential iron proteins. J Am Chem Soc 129:12626–12627
123. Vanin AF, Stukan RA, Manukhina EB (1996) Physical properties of dinitrosyl iron complexes with thiol-containing ligands in relation with their vasodilator activity. Biochim Biophys Acta 1295:5–12
124. Vanin AF (1991) Endothelium-derived relaxing factor is a nitrosyl iron complex with thiol ligands. FEBS Lett 289:1–3
125. Michael D, Mingos P, Sherman DJ (1989) Transition metal nitrosyl complexes. In: Sykes AG (ed) Advances in inorganic chemistry, vol 34. Academic, London
126. Lu TT, Tsou CC et al (2008) Anionic Roussin's Red Esters (RREs) syn-/anti-$[Fe(\mu\text{-}SEt)(NO)_2]^{2-}$: the critical role of thiolate ligands in regulating the transformation of RREs into dinitrosyl iron complexes and the anionic RREs. Inorg Chem 47:6040–6050
127. Hung MC, Tsai MC et al (2006) Transformation and structural discrimination between the neutral $\{Fe(NO)_2\}^{10}$ dinitrosyliron complexes (DNICs) and the anionic/cationic $\{Fe(NO)_2\}^9$ DNICs. Inorg Chem 45:6041–6047
128. Yeh SW, Lin CW et al (2012) Insight into the dinuclear $\{Fe(NO)_2\}^{10}\{Fe\ (NO)_2\}^{10}$ and mononuclear $\{Fe(NO)_2\}^{10}$ dinitrosyliron complexes. Inorg Chem 51:4076–4087
129. Tsou CC, Liaw WF (2011) Transformation of the {Fe (NO)2}9 dinitrosyl iron complexes (DNICs) into S-nitrosothiols (RSNOs) triggered by acid–base pairs. Chemistry 17:13358–13366
130. Lin ZS, Lo FC et al (2011) Peptide-bound dinitrosyliron complexes (DNICs) and neutral/reduced-form Roussin's Red Esters (RREs/rRREs): understanding nitrosylation of [Fe-S] clusters leading to the formation of DNICs and RREs using a de novo design strategy. Inorg Chem 50:10417–10431
131. Lu TT, Lai SH et al (2011) Discrimination of mononuclear and dinuclear dinitrosyl iron complexes (DNICs) by S K-edge X-ray absorption spectroscopy: insight into the electronic structure and reactivity of DNICs. Inorg Chem 50:5396–5406

132. Tsai FT, Chen PL, Liaw WF (2010) Roles of the distinct electronic structures of the $\{Fe(NO)_2\}^9$ and $\{Fe(NO)_2\}^{10}$ dinitrosyliron complexes in modulating nitrite binding modes and nitrite activation pathways. J Am Chem Soc 132:5290–5299
133. Tsai FT, Kuo TS, Liaw WF (2009) Dinitrosyl iron complexes (DNICs) bearing O-bound nitrito ligand: reversible transformation between the six-coordinate $\{Fe\ (NO)_2\}^9[(1\text{-MeIm})_2(\text{eta}(2)\text{-}ONO)Fe(NO)_2]$ (g = 2.013) and four-coordinate $\{Fe\ (NO)_2\}^9[\ (1\text{-MeIm})(ONO)Fe(NO)_2]$ (g = 2.03). J Am Chem Soc 131:3426–3427
134. Harrop TC, Song D, Lippard SJ (2006) Interaction of nitric oxide with tetrathiolato iron (II) complexes: relevance to the reaction pathways of iron nitrosyls in sulfur-rich biological coordination environments. J Am Chem Soc 128:3528–3529
135. Thomas JT, Robertson JH, Cox EG (1958) The crystal structure of Roussin's red ethyl ester. Acta Crystallogr 11:599–604
136. Tsai ML, Hsieh CH, Liaw WF (2007) Dinitrosyl iron complexes (DNICs) containing S/N/O ligation: transformation of Roussin's red ester into the neutral $\{Fe(NO)_2\}^{10}$ DNICs. Inorg Chem 46:5110–5117
137. Lu TT, Huang HW, Liaw WF (2009) Anionic mixed thiolate-sulfide-bridged Roussin's red esters $[(NO)_2Fe(\mu\text{-SR})(\mu\text{-S})Fe(NO)_2]^-$ (R = Et, Me, Ph): a key intermediate for transformation of dinitrosyl iron complexes (DNICs) to [2Fe-2S] clusters. Inorg Chem 48:9027–9035
138. Sanina NA, Filipenko OS et al (2000) Influence of the cation on the properties of binuclear iron nitrosyl complexes. Synthesis and crystal structure of $[Pr_4nN]_2[Fe_2S_2\ (NO)_4]$. Russ Chem Bull 49:1109–1112
139. Collins DJ, Zhou HC (2011) Iron-sulfur models of protein active sites. In: Encyclopedia of inorganic and bioinorganic chemistry. Wiley, New York
140. King RB, Bitterwolf TE (2000) Metal carbonyl analogues of iron–sulfur clusters found in metalloenzyme chemistry. Coord Chem Rev 206–207:563–579
141. Noodleman L, Pique ME et al (2007) Iron-sulfur clusters: properties and functions. In: Wiley encyclopedia of chemical biology. Wiley, New York
142. Kurihara T, Mihara H et al (2003) Assembly of iron-sulfur clusters mediated by cysteine desulfurases, IscS, CsdB and CSD, from Escherichia coli. Biochim Biophys Acta Proteins Proteomics 1647:303–309
143. Ballmann J, Albers A et al (2008) A synthetic analogue of Rieske-type [2Fe-2S] clusters. Angew Chem Int Ed 47:9537–9541
144. Sharma VS, Traylor TG et al (1987) Reaction of nitric oxide with heme proteins and model compounds of hemoglobin. Biochemistry 26:3837–3843
145. Pavlos CM, Xu H, Toscano JP (2005) Photosensitive precursors to nitric oxide. Curr Top Med Chem 5:637–647
146. Tonzetich ZJ, Do LH, Lippard SJ (2009) Dinitrosyl iron complexes relevant to Rieske cluster nitrosylation. J Am Chem Soc 131:7964–7965
147. Foster MW, Cowan JA (1999) Chemistry of nitric oxide with protein-bound iron sulfur centers. Insights on physiological reactivity. J Am Chem Soc 121:4093–4100
148. Mitra D, Pelmenschikov V et al (2011) Dynamics of the [4Fe-4S] cluster in pyrococcus furiosus D14C ferredoxin via nuclear resonance vibrational and resonance Raman spectroscopies, force field simulations, and density functional theory calculations. Biochemistry 50:5220–5235
149. Huang HW, Tsou CC et al (2008) New members of a class of dinitrosyliron complexes (DNICs): interconversion and spectroscopic discrimination of the anionic $\{Fe(NO)_2\}^9[(NO)_2Fe(C_3H_3N_2)_2]^-$ and$[(NO)_2Fe(C_3H_3N_2)(SR)]^-$ ($C_3H_3N_2$ = deprotonated imidazole; R = tBu, Et, Ph). Inorg Chem 47:2196–2204
150. Tsai ML, Liaw WF (2006) Neutral $\{Fe(NO)_2\}^9$ dinitrosyliron complex (DNIC)$[\ (SC_6H_4\text{-o-}NHCOPh)(Im)Fe(NO)_2]$ (Im = Imidazole): interconversion among the anionic/neutral $\{Fe(NO)_2\}^9$ DNICs and Roussin's red ester. Inorg Chem 45:6583–6585
151. Chen TN, Lo FC et al (2006) Dinitrosyl iron complexes $[E_5Fe(NO)_2]^-$ (E = S, Se): a precursor of Roussin's black salt $[Fe_4E_3(NO)_7]^-$. Inorg Chim Acta 359:2525–2533

152. Chen HW, Lin CW et al (2005) Homodinuclear iron thiolate nitrosyl compounds [(ON)Fe (S, S-C_6H_4)$_2$Fe(NO)$_2$]- and [(ON)Fe(SO_2, S-C_6H_4)(S, S-C_6H_4)Fe(NO)$_2$]$^-$ with {Fe(NO)}^{7}-{Fe (NO)$^{2-9}$ electronic coupling: new members of a class of dinitrosyl iron complexes. Inorg Chem 44:3226–3232
153. Rogers PA, Eide L et al (2003) Reversible inactivation of E. coli endonuclease III via modification of its [4Fe-4S] cluster by nitric oxide. DNA Repair 2:809–817
154. Crack JC, Smith LJ et al (2010) Mechanistic insight into the nitrosylation of the [4Fe-4S] cluster of WhiB-like proteins. J Am Chem Soc 133:1112–1121
155. Crack JC, den Hengst CD et al (2009) Characterization of [4Fe-4S]-containing and cluster-free forms of Streptomyces WhiD. Biochemistry 48:12252–12264
156. Smith LJ, Stapleton MR et al (2010) Mycobacterium tuberculosis WhiB1 is an essential DNA-binding protein with a nitric oxide-sensitive iron-sulfur cluster. Biochem J 432:417–427
157. Alderton WK, Cooper CE, Knowles RG (2001) Nitric oxide synthases: structure, function and inhibition. Biochem J 357:593–615
158. Cassoly R, Gibson QH (1975) Conformation, cooperativity and ligand binding in human hemoglobin. J Mol Biol 91:301–313
159. Sharma VS, Ranney HM (1978) The dissociation of NO from nitrosylhemoglobin. J Biol Chem 253:6467–6472
160. Kharitonov VG, Sharma VS et al (1997) Kinetics of nitric oxide dissociation from five- and six-coordinate nitrosyl hemes and heme proteins, including soluble guanylate cyclase. Biochemistry 36:6814–6818
161. Moore EG, Gibson QH (1976) Cooperativity in the dissociation of nitric oxide from hemoglobin. J Biol Chem 251:2788–2794
162. Hoshino M, Ozawa K et al (1993) Photochemistry of nitric oxide adducts of water-soluble iron (III) porphyrin and ferrihemoproteins studied by nanosecond laser photolysis. J Am Chem Soc 115:9568–9575
163. Ascenzi P, Coletta M et al (1994) Nitric oxide binding to ferrous native horse heart cytochrome c and to its carboxymethylated derivative: A spectroscopic and thermodynamic study. J Inorg Biochem 53:273–280
164. Hoshino M, Maeda M et al (1996) Studies on the reaction mechanism for reductive nitrosylation of ferrihemoproteins in buffer solutions. J Am Chem Soc 118:5702–5707
165. Sharma VS, Isaacson RA et al (1983) Reaction of nitric oxide with heme proteins: studies on metmyoglobin, opossum methemoglobin, and microperoxidase. Biochemistry 22:3897–3902
166. Sharpe MA, Cooper CE (1998) Reactions of nitric oxide with mitochondrial cytochrome C: a novel mechanism for the formation of nitroxyl anion and peroxynitrite. Biochem J 332 (Pt 1):9–19
167. Osipov AN, Borisenko GG, Vladimirov YA (2007) Biological activity of hemoprotein nitrosyl complexes. Biochemistry 72:1491–1504
168. Khrapova NV, Malenkova IV, Vanin AF (1995) S-nitrosothiols and dinitrosothiol iron complexes as a source of nitric oxide in animals. Biofizika 40:117–121
169. Shumaev KB, Kosmachevskaya OV et al (2008) Dinitrosyl iron complexes bind with hemoglobin as markers of oxidative stress. Methods Enzymol 436:445–461
170. Timoshin AA, Vanin AF et al (2007) Protein-bound dinitrosyl-iron complexes appearing in blood of rabbit added with a low-molecular dinitrosyl-iron complex: EPR studies. Nitric Oxide 16:286–293
171. Kadish KM, Smith KM, Guillard R (2000) Binding and activation of nitric oxide by metalloporphyrins and heme. In: Kadish KM, Smith KM, Guillard R (eds) The porphyrin handbook. Academic, New York
172. Wyllie GR, Scheidt WR (2002) Solid-state structures of metalloporphyrin NO_x compounds. Chem Rev 102:1067–1090
173. Goodrich LE, Paulat F et al (2010) Electronic structure of heme-nitrosyls and its significance for nitric oxide reactivity, sensing, transport, and toxicity in biological systems. Inorg Chem 49:6293–6316

174. Sulok CD, Bauer JL et al (2012) A detailed investigation into the electronic structures of macrocyclic iron (II)-nitrosyl compounds and their similarities to ferrous heme-nitrosyls. Inorg Chim Acta 380:148–160
175. Praneeth VK, Neese F, Lehnert N (2005) Spin density distribution in five- and six-coordinate iron (II)-porphyrin NO complexes evidenced by magnetic circular dichroism spectroscopy. Inorg Chem 44:2570–2572
176. Larkworthy LF, Sengupta SK (1991) Mononitrosyl derivatives of iron and cobalt complexes of quadridentate ligands from 2-hydroxy-1-naphthaldehyde and ethylenediamine, O-phenylenediamine, and 4-methyl-O-phenylenediamine. Inorg Chim Acta 179:157–160
177. Groombridge CJ, Larkworthy LF et al (1992) Synthesis and ^{15}N nuclear magnetic resonance shift tensors of bent nitrosyl complexes with N-substituted salicylideneiminate coligands; the shift tensor as a criterion of MNO bond angle. J Chem Soc Dalton Trans 3125–3131
178. Zeng W, Silvernail NJ et al (2005) Direct probe of iron vibrations elucidates NO activation of heme proteins. J Am Chem Soc 127:11200–11201
179. Sage JT et al (2001) Nuclear resonance vibrational spectroscopy of a protein active-site mimic. J Phys Condens Matter 13:7707
180. Wolfgang S (2004) Nuclear resonant spectroscopy. J Phys Condens Matter 16:S497
181. Lehnert N, Galinato MG et al (2010) Nuclear resonance vibrational spectroscopy applied to [Fe (OEP) (NO)]: the vibrational assignments of five-coordinate ferrous heme-nitrosyls and implications for electronic structure. Inorg Chem 49:4133–4148
182. Ellison MK, Schulz CE, Scheidt WR (2000) Structural and electronic characterization of nitrosyl (octaethylporphinato)iron (III) perchlorate derivatives. Inorg Chem 39:5102–5110
183. Ellison MK, Schulz CE, Scheidt WR (2002) Nitrosyliron (III) porphyrinates: porphyrin core conformation and FeNO geometry. Any correlation? J Am Chem Soc 124:13833–13841
184. Ellison MK, Scheidt WR (1998) Tilt/asymmetry in nitrosyl metalloporphyrin complexes the cobalt case. Inorg Chem 37:382–383
185. Richter-Addo GB, Wheeler RA et al (2001) Unexpected nitrosyl-group bending in six-coordinate$[M(NO)]^6$ sigma-bonded aryl (iron) and -(ruthenium) porphyrins. J Am Chem Soc 123:6314–6326
186. Guilard R, Lagrange G et al (1985) Reactions of sigma-bonded alkyl- and aryliron porphyrins with nitric oxide. Synthesis and electrochemical characterization of six-coordinate nitrosyl sigma-bonded alkyl- and aryliron porphyrins. Inorg Chem 24:3649–3656
187. Nasri H, Ellison MK et al (1997) Sharing the π-bonding. An iron porphyrin derivative with trans, π-accepting axial ligands. Synthesis, EPR and Mössbauer spectra, and molecular structure of two forms of the complex nitronitrosyl(α, α, α, α-tetrakis(o-pivalamidophenyl)-porphinato)ferrate(II). J Am Chem Soc 119:6274–6283
188. Wyllie GR, Schulz CE, Scheidt WR (2003) Five- to six-coordination in (nitrosyl)iron (II) porphyrinates: effects of binding the sixth ligand. Inorg Chem 42:5722–5734
189. Mason J, Larkworthy LF, Moore EA (2002) Nitrogen NMR spectroscopy of metal nitrosyls and related compounds. Chem Rev 102:913–934
190. Kadish KM, Caemelbecke EV (2007) Electrochemistry of metalloporphyrins in nonaqueous media. In: Encyclopedia of electrochemistry. Wiley-VCH, New York
191. Li CG, Rand MJ (1993) Effects of hydroxocobalamin and haemoglobin on NO-mediated relaxations in the rat anococcygeus muscle. Clin Exp Pharmacol Physiol 20:633–640
192. Rajanayagam MAS, Li CG, Rand MJ (1993) Differential effects of hydroxocobalamin on NO-mediated relaxations in rat aorta and anococcygeus muscle. Br J Pharmacol 108:3–5
193. Rand MJ, Li CG (1993) Differential effects of hydroxocobalamin on relaxations induced by nitrosothiols in rat aorta and anococcygeus muscle. Eur J Pharmacol 241:249–254
194. Jenkinson KM, Reid JJ, Rand MJ (1995) Hydroxocobalamin and haemoglobin differentiate between exogenous and neuronal nitric oxide in the rat gastric fundus. Eur J Pharmacol 275:145–152

195. De Man JG, Boeckxstaens GE et al (1995) Comparison of the pharmacological profile of S-nitrosothiols, nitric oxide and the nitrergic neurotransmitter in the canine ileocolonic junction. Br J Pharmacol 114:1179–1184
196. Akinori A, Yutaka T et al (1993) Protective effects of a vitamin B12 analog, methylcobalamin, against glutamate cytotoxicity in cultured cortical neurons. Eur J Pharmacol 241:1–6
197. Greenberg SS, Xie J et al (1995) Hydroxocobalamin (vitamin B_{12a}) prevents and reverses endotoxin-induced hypotension and mortality in rodents: role of nitric oxide. J Pharmacol Exp Ther 273:257–265
198. Wolak M, Zahl A et al (2001) Kinetics and mechanism of the reversible binding of nitric oxide to reduced cobalamin B_{12r} (Cob(II)alamin). J Am Chem Soc 123:9780–9791
199. Ignarro L (2000) Nitric oxide: biology and pathobiology. Academic, San Diego
200. Rose MJ, Mascharak PK (2008) Photoactive ruthenium nitrosyls: effects of light and potential application as NO donors. Coord Chem Rev 252:2093–2114
201. McGarvey BR, Ferro AA et al (2000) Detection of the EPR spectra of NO in ruthenium (II) complexes. Inorg Chem 39:3577–3581
202. Callahan RW, Meyer TJ (1977) Reversible electron transfer in ruthenium nitrosyl complexes. Inorg Chem 16:574–581
203. Videla M, Jacinto JS et al (2006) New ruthenium nitrosyl complexes with tris(1-pyrazolyl) methane (tpm) and 2,2′-bipyridine (bpy) coligands. Structure, spectroscopy, and electrophilic and nucleophilic reactivities of bound nitrosyl. Inorg Chem 45:8608–8617
204. Roncaroli F, Videla M et al (2007) New features in the redox coordination chemistry of metal nitrosyls {$M\text{-}NO^+$; M-NO; $M\text{-}NO^-$ (HNO)}. Coord Chem Rev 251:1903–1930
205. Roncaroli F, Ruggiero ME et al (2002) Kinetic, mechanistic, and DFT study of the electrophilic reactions of nitrosyl complexes with hydroxide. Inorg Chem 41:5760–5769
206. Patra AK, Mascharak PK (2003) A ruthenium nitrosyl that rapidly delivers NO to proteins in aqueous solution upon short exposure to UV light. Inorg Chem 42:7363–7365
207. Serres RG, Grapperhaus CA et al (2004) Structural, spectroscopic, and computational study of an octahedral, non-heme $[Fe\text{-}NO]^{6-8}$ series: $[Fe(NO)\ (cyclam\text{-}ac)]^{2+/+/0}$. J Am Chem Soc 126:5138–5153
208. Coppens P, Novozhilova I, Kovalevsky A (2002) Photoinduced linkage isomers of transition-metal nitrosyl compounds and related complexes. Chem Rev 102:861–884
209. Cheng L, Novozhilova I et al (2000) First observation of photoinduced nitrosyl linkage isomers of iron nitrosyl porphyrins. J Am Chem Soc 122:7142–7143
210. Schmitt F, Govindaswamy P et al (2008) Ruthenium porphyrin compounds for photodynamic therapy of cancer. J Med Chem 51:1811–1816
211. Xu N, Yi J, Richter-Addo GB (2010) Linkage isomerization in heme-NO_x compounds: understanding NO, nitrite, and hyponitrite interactions with iron porphyrins. Inorg Chem 49:6253–6266
212. Fomitchev V, Coppens P et al (1999) Photo-induced metastable linkage isomers of ruthenium nitrosyl porphyrins. Chem Commun 2013–2014
213. Adman ET, Murphy ME (2001) Copper nitrite reductase. In: Encyclopedia of inorganic and bioinorganic chemistry. Wiley, New York
214. Kroneck PMH, Beuerle J, Schumacher (1992) Metal-dependent conversion of inorganic nitrogen and sulfur compounds. In: Sigel A, Sigel H (ed) Metal ions in biological systems: degradation of environmental pollutants by microorganisms and their metalloenzymes. Marcel Dekker, New York
215. Tocheva EI, Rosell FI et al (2004) Side-on copper-nitrosyl coordination by nitrite reductase. Science 304:867–870
216. Tocheva EI, Rosell FI et al (2007) Stable copper-nitrosyl formation by nitrite reductase in either oxidation state. Biochemistry 46:12366–12374
217. Paul PP, Tyeklar Z et al (1990) Isolation and X-ray structure of a dinuclear copper-nitrosyl complex. J Am Chem Soc 112:2430–2432

218. Gennari M, Marchio L (2009) Cu-complexes with scorpionate ligands as models for the binding sites of copper proteins. Curr Bioact Compounds 5:244–263
219. Averill BA (1996) Dissimilatory nitrite and nitric oxide reductases. Chem Rev 96:2951–2964
220. Hulse CL, Tiedje JM, Averill BA (1989) Evidence for a copper-nitrosyl intermediate in denitrification by the copper-containing nitrite reductase of Achromobacter cycloclastes. J Am Chem Soc 111:2322–2323
221. Adman ET, Godden JW, Turley S (1995) The structure of copper-nitrite reductase from achromobacter cycloclastes at five pH values, with NO-2 bound and with type II copper depleted. J Biol Chem 270:27458–27474
222. Murphy MEP, Turley S, Adman ET (1997) Structure of nitrite bound to copper-containing nitrite reductase from Alcaligenes faecalis: mechanistic implications. J Biol Chem 272:28455–28460
223. Kataoka K, Furusawa H et al (2000) Functional analysis of conserved aspartate and histidine residues located around the type 2 copper site of copper-containing nitri reductase. J Biochem 127:345–350
224. Suzuki S, Kataoka K, Yamaguchi K (2000) Metal coordination and mechanism of multicopper nitrite reductase. Acc Chem Res 33:728–735
225. Olesen K, Veselov A et al (1998) Spectroscopic, kinetic, and electrochemical characterization of heterologously expressed wild-type and mutant forms of copper-containing nitrite reductase from Rhodobacter sphaeroides 2.4.3. Biochemistry 37:6086–6094
226. Veselov A, Olesen K et al (1998) Electronic structural information from Q-band ENDOR on the type 1 and type 2 copper liganding environment in wild-type and mutant forms of copper-containing nitrite reductase. Biochemistry 37:6095–6105
227. Antonyuk SV, Strange RW et al (2005) Atomic resolution structures of resting-state, substrate- and product-complexed Cu-nitrite reductase provide insight into catalytic mechanism. Proc Natl Acad Sci USA 102:12041–12046
228. Ghosh S, Dey A et al (2007) Resolution of the spectroscopy versus crystallography issue for NO intermediates of nitrite reductase from Rhodobacter sphaeroides. J Am Chem Soc 129:10310–10311
229. Fujisawa K, Tateda A et al (2008) Structural and spectroscopic characterization of mononuclear copper (I) nitrosyl complexes: end-on versus side-on coordination of NO to copper (I). J Am Chem Soc 130:1205–1213
230. Woolum JC, Commoner B (1970) Isolation and identification of a paramagnetic complex from the livers of carcinogen-treated rats. Biochim Biophys Acta 201:131–140
231. Vanin AF, Nalbandian RM (1965) Free radicals of a new type in yeast cells. Biofizika 10:167–168
232. Vanin AF, Nalbandian RM (1966) Free radical states with localization of the unpaired electron on the sulfur atom in yeast cells. Biofizika 11:178–179
233. Maruyama T, Kataoka N et al (1971) Identification of three-line electron spin resonance signal and its relationship to ascites tumors. Cancer Res 31:179–184
234. Palmer RM, Ferrige AG, Moncada S (1987) Nitric oxide release accounts for the biological activity of endothelium-derived relaxing factor. Nature 327:524–526
235. Waldman SA, Rapoport RM et al (1986) Desensitization to nitroglycerin in vascular smooth muscle from rat and human. Biochem Pharmacol 35:3525–3531
236. Ignarro LJ (1999) Nitric oxide: a unique endogenous signaling molecule in vascular biology. Biosci Rep 19:51–71
237. Ignarro LJ (2002) Nitric oxide as a unique signaling molecule in the vascular system: a historical overview. J Physiol Pharmacol 53:503–514
238. Yamamoto T, Bing RJ (2000) Nitric oxide donors. Exp Biol Med 225:200–206
239. Garthwaite J, Boulton CL (1995) Nitric oxide signaling in the central nervous system. Annu Rev Physiol 57:683–706
240. Liu Q, Gross SS (1996) Binding sites of nitric oxide synthases. In: Lester P (ed) Methods in enzymology. Nitric oxide Part A: Sources and detection of NO; NO synthase, vol 268. Academic, London

241. del Río LA, Corp J, Barroso JB (2004) Nitric oxide and nitric oxide synthase activity in plants. Phytochemistry 65:783–792
242. Knowles RG, Moncada S (1994) Nitric oxide synthases in mammals. Biochem J 298 (Pt 2):249–258
243. Martinez MC, Andriantsitohaina R (2009) Reactive nitrogen species: molecular mechanisms and potential significance in health and disease. Antioxid Redox Signal 11:669–702
244. Myers PR, Minor RL et al (1990) Vasorelaxant properties of the endothelium-derived relaxing factor more closely resemble S-nitrosocysteine than nitric oxide. Nature 345:161–163
245. Myers PR, Guerra R, Harrison DG (1992) Release of multiple endothelium-derived relaxing factors from porcine coronary arteries. J Cardiovasc Pharmacol 20:392–400
246. Lancaster JR (1994) Simulation of the diffusion and reaction of endogenously produced nitric oxide. Proc Natl Acad Sci USA 91:8137–8141
247. Wink DA, Mitchell JB (1998) Chemical biology of nitric oxide: insights into regulatory, cytotoxic, and cytoprotective mechanisms of nitric oxide. Free Radic Biol Med 25:434–456
248. Nakamoto K (2009) Infrared and Raman spectra of inorganic and coordination compounds, applications in coordination, organometallic, and bioinorganic chemistry. Wiley, New York
249. Le XC, Xing JZ et al (1998) Inducible repair of thymine glycol detected by an ultrasensitive assay For DNA damage. Science 280:1066–1069
250. Stanbury DM (1989) Reduction potentials involving inorganic free radicals in aqueous solution. In: Sykes AG (ed) Advances in inorganic chemistry, vol 33. Academic, London
251. Bartberger MD, Liu W et al (2002) The reduction potential of nitric oxide (NO) and its importance to NO biochemistry. Proc Natl Acad Sci USA 99:10958–10963
252. Stamler JS, Singel DJ, Loscalzo J (1992) Biochemistry of nitric oxide and its redox-activated forms. Science 258:1898–1902
253. Henry Y, Singel DJ (1996) Metal-nitrosyl interactions in nitric oxide biology probed by electron paramagnetic resonance spectroscopy. In: Feelisch M, Stamler JS (eds) Methods in nitric oxide research. Wiley, Chichester
254. Brown GC (1995) Reversible binding and inhibition of catalase by nitric oxide. Eur J Biochem 232:188–191
255. Bellamy TC, Griffiths C, Garthwaite J (2002) Differential sensitivity of guanylyl cyclase and mitochondrial respiration to nitric oxide measured using clamped concentrations. J Biol Chem 277:31801–31807
256. Craven PA, DeRubertis FR (1978) Restoration of the responsiveness of purified guanylate cyclase to nitrosoguanidine, nitric oxide, and related activators by heme and hemeproteins. Evidence for involvement of the paramagnetic nitrosyl-heme complex in enzyme activation. J Biol Chem 253:8433–8443
257. Ignarro LJ (1990) Haem-dependent activation of guanylate cyclase and cyclic GMP formation by endogenous nitric oxide: a unique transduction mechanism for transcellular signaling. Pharmacol Toxicol 67:1–7
258. Ferrero R, Rodriguez-Pascual F (2000) Nitric oxide-sensitive guanylyl cyclase activity inhibition through cyclic GMP-dependent dephosphorylation. J Neurochem 75:2029–2039
259. Works CF, Jocher CJ et al (2002) Photochemical nitric oxide precursors: synthesis, photochemistry, and ligand substitution kinetics of ruthenium salen nitrosyl and ruthenium salophen nitrosyl complexes. Inorg Chem 41:3728–3739
260. Garbers DL, Lowe DG (1994) Guanylyl cyclase receptors. J Biol Chem 269:30741–30744
261. Karow DS, Pan D et al (2004) Spectroscopic characterization of the soluble guanylate cyclase-like heme domains from Vibrio cholerae and Thermoanaerobacter tengcongensis. Biochemistry 43:10203–10211
262. Gilles-Gonzalez MA, Gonzalez G (2005) Heme-based sensors: defining characteristics, recent developments, and regulatory hypotheses. J Inorg Biochem 99:1–22
263. Zhao Y, Hoganson C et al (1998) Structural changes in the heme proximal pocket induced by nitric oxide binding to soluble guanylate cyclase. Biochemistry 37:12458–12464

264. Paulat F, Berto TC et al (2008) Vibrational assignments of six-coordinate ferrous heme nitrosyls: new insight from nuclear resonance vibrational spectroscopy. Inorg Chem 47:11449–11451
265. Cary SPL, Winger JA et al (2006) Nitric oxide signaling: no longer simply on or off. Trends Biochem Sci 31:231–239
266. Caulton KG (1975) Synthetic methods in transition metal nitrosyl chemistry. Coord Chem Rev 14:317–355
267. Patel RP, Gladwin MT (2004) Physiologic, pathologic and therapeutic implications for hemoglobin interactions with nitric oxide. Free Radic Biol Med 36:399–401
268. Huang Z, Shiva S et al (2005) Enzymatic function of hemoglobin as a nitrite reductase that produces NO under allosteric control. J Clin Invest 115:2099–2107
269. Lancaster JR Jr, Gaston B (2004) NO and nitrosothiols: spatial confinement and free diffusion. Am J Physiol Lung Cell Mol Physiol 287:L465–L466
270. Robinson JM, Lancaster JR Jr (2005) Hemoglobin-mediated, hypoxia-induced vasodilation via nitric oxide: mechanism(s) and physiologic versus pathophysiologic relevance. Am J Respir Cell Mol Biol 32:257–261
271. Sanina NA, Syrtsova LA et al (2007) Reactions of sulfur-nitrosyl iron complexes of "g=2.03" family with hemoglobin (Hb): kinetics of Hb-NO formation in aqueous solutions. Nitric Oxide 16:181–188
272. Brunelli L, Yermilov V, Beckman JS (2001) Modulation of catalase peroxidatic and catalatic activity by nitric oxide. Free Radic Biol Med 30:709–714
273. Giuffre A, Barone MC et al (2000) Reaction of nitric oxide with the turnover intermediates of cytochrome C oxidase: reaction pathway and functional effects. Biochemistry 39:15446–15453
274. Cooper CE (2002) Nitric oxide and cytochrome oxidase: substrate, inhibitor or effector? Trends Biochem Sci 27:33–39
275. Blackmore RS, Greenwood C, Gibson QH (1991) Studies of the primary oxygen intermediate in the reaction of fully reduced cytochrome oxidase. J Biol Chem 266:19245–19249
276. Doyle MP, Herman JG, Dykstra RL (1985) Autocatalytic oxidation of hemoglobin induced by nitrite: activation and chemical inhibition. J Free Radic Biol Med 1:145–153
277. Spagnuolo C, Rinelli P et al (1987) Oxidation reaction of human oxyhemoglobin with nitrite: a reexamination. Biochim Biophys Acta 911:59–65
278. Doyle MP, Pickering RA et al (1981) Kinetics and mechanism of the oxidation of human deoxyhemoglobin by nitrites. J Biol Chem 256:12393–12398
279. Antonini E, Brunori M et al (1966) Preparation and kinetic properties of intermediates in the reaction of hemoglobin with ligands. J Biol Chem 241:3236–3238
280. Gray RD, Gibson QH (1971) The effect of inositol hexaphosphate on the kinetics of CO and O_2 binding by human hemoglobin. J Biol Chem 246:7168–7174
281. Gray RD, Gibson QH (1971) The binding of carbon monoxide to and chains in tetrameric mammalian hemoglobin. J Biol Chem 246:5176–5178
282. Wang QZ, Jacobs J et al (1991) Nitric oxide hemoglobin in mice and rats in endotoxic shock. Life Sci 49:L55–L60
283. Arnold EV, Bohle DS (1996) Isolation and oxygenation reactions of nitrosylmyoglobins. In: Lester P (ed) Methods in enzymology. Nitric oxide Part B: Physiological and pathological processes, vol 269. Academic, London
284. Addison AW, Stephanos JJ (1986) Nitrosyliron (III) hemoglobin: autoreduction and spectroscopy. Biochemistry 25:4104–4113
285. Yonetani T, Yamamoto H et al (1972) Electromagnetic properties of hemoproteins: V. Optical and electron paramagnetic resonance characteristics of nitric oxide derivatives of metalloporphyrin-apohemoprotein complexes. J Biol Chem 247:2447–2455
286. Salerno JC (1996) Nitric oxide complexes of metalloproteins: an introductory overview. In: Lancaster JR Jr (ed) Nitric oxide: principles and actions. Academic, San Diego

287. Le Brun NE, Andrews SC et al (1997) Interaction of nitric oxide with non-haem iron sites of Escherichia coli bacterioferritin: reduction of nitric oxide to nitrous oxide and oxidation of iron (II) to iron (III). Biochem J 326:173–179
288. Lee M, Arosio P et al (1994) Identification of the EPR-active iron-nitrosyl complexes in mammalian ferritins. Biochemistry 33:3679–3687
289. Kim HR, Kim TH et al (2012) Direct detection of tetrahydrobiopterin (BH4) and dopamine in rat brain using liquid chromatography coupled electrospray tandem mass spectrometry. Biochem Biophys Res Commun 419:632–637
290. Nelson MJ (1987) The nitric oxide complex of ferrous soybean lipoxygenase-1. Substrate, pH, and ethanol effects on the active-site iron. J Biol Chem 262:12137–12142
291. Rubbo H, Parthasarathy S et al (1995) Nitric oxide inhibition of lipoxygenase-dependent liposome and low-density lipoprotein oxidation: termination of radical chain propagation reactions and formation of nitrogen-containing oxidized lipid derivatives. Arch Biochem Biophys 324:15–25
292. Kennedy MC, Gan T et al (1993) Metallothionein reacts with Fe^{2+} and NO to form products with a $g = 2.039$ ESR signal. Biochem Biophys Res Commun 196:632–635
293. Flint DH, Allen RM (1996) Iron-sulfur proteins with nonredox functions. Chem Rev 96:2315–2334
294. Brzoska K, Meczynska S, Kruszewski M (2006) Iron-sulfur cluster proteins: electron transfer and beyond. Acta Biochim Pol 53:685–691
295. Cheung PY, Danial H et al (1998) Thiols protect the inhibition of myocardial aconitase by peroxynitrite. Arch Biochem Biophys 350:104–108
296. Hausladen A, Fridovich I (1994) Superoxide and peroxynitrite inactivate aconitases, but nitric oxide does not. J Biol Chem 269:29405–29408
297. Castro L, Rodriguez M, Radi R (1994) Aconitase is readily inactivated by peroxynitrite, but not by its precursor, nitric oxide. J Biol Chem 269:29409–29415
298. Haile DJ, Rouault TA et al (1992) Reciprocal control of RNA-binding and aconitase activity in the regulation of the iron-responsive element binding protein: role of the iron-sulfur cluster. Proc Natl Acad Sci USA 89:7536–7540
299. Rouault TA, Stout CD et al (1991) Structural relationship between an iron-regulated RNA-binding protein (IRE-BP) and aconitase: functional implications. Cell 64:881–883
300. Rouault TA, Haile DJ et al (1992) An iron-sulfur cluster plays a novel regulatory role in the iron-responsive element binding protein. Biometals 5:131–140
301. Drapier JC, Hibbs JB Jr (1996) Aconitases: a class of metalloproteins highly sensitive to nitric oxide synthesis. Methods Enzymol 269:26–36
302. Haile DJ, Rouault TA et al (1992) Cellular regulation of the iron-responsive element binding protein: disassembly of the cubane iron-sulfur cluster results in high-affinity RNA binding. Proc Natl Acad Sci USA 89:11735–11739
303. Gardner PR, Costantino G et al (1997) Nitric oxide sensitivity of the aconitases. J Biol Chem 272:25071–25076
304. Pantopoulos K, Weiss G, Hentze MW (1994) Nitric oxide and the post-transcriptional control of cellular iron traffic. Trends Cell Biol 4:82–86
305. Soum E, Brazzolotto X et al (2003) Peroxynitrite and nitric oxide differently target the iron-sulfur cluster and amino acid residues of human iron regulatory protein 1. Biochemistry 42:7648–7654
306. Weiss G, Goossen B et al (1993) Translational regulation via iron-responsive elements by the nitric oxide/NO-synthase pathway. EMBO J 12:3651–3657
307. Drapier JC, Hirling H et al (1993) Biosynthesis of nitric oxide activates iron regulatory factor in macrophages. EMBO J 12:3643–3649
308. Gardner PR, Fridovich I (1991) Superoxide sensitivity of the Escherichia coli aconitase. J Biol Chem 266:19328–19333
309. Navarre DA, Wendehenne D et al (2000) Nitric oxide modulates the activity of tobacco aconitase. Plant Physiol 122:573–582

310. Gupta KJ, Shah JK et al (2012) Inhibition of aconitase by nitric oxide leads to induction of the alternative oxidase and to a shift of metabolism towards biosynthesis of amino acids. J Exp Bot 63:1773–1784
311. Tortora V, Quijano C et al (2007) Mitochondrial aconitase reaction with nitric oxide, S-nitrosoglutathione, and peroxynitrite: mechanisms and relative contributions to aconitase inactivation. Free Radic Biol Med 42:1075–1088
312. Kobayashi M, Shimizu S (2000) Nitrile hydrolases. Curr Opin Chem Biol 4:95–102
313. Endo I, Nojiri M et al (2001) Fe-type nitrile hydratase. J Inorg Biochem 83:247–253
314. Nakasako M, Odaka M et al (1999) Tertiary and quaternary structures of photoreactive Fe-type nitrile hydratase from Rhodococcus sp. N-771: roles of hydration water molecules in stabilizing the structures and the structural origin of the substrate specificity of the enzyme. Biochemistry 38:9887–9898
315. Nagashima S, Nakasako M et al (1998) Novel non-heme iron center of nitrile hydratase with a claw setting of oxygen atoms. Nat Struct Biol 5:347–351
316. Huang W, Jia J et al (1997) Crystal structure of nitrile hydratase reveals a novel iron centre in a novel fold. Structure 5:691–699
317. Harrop TC, Olmstead MM, Mascharak PK (2004) Structural models of the bimetallic subunit at the A-cluster of acetyl coenzyme a synthase/CO dehydrogenase: binuclear sulfur-bridged Ni-Cu and Ni-Ni complexes and their reactions with CO. J Am Chem Soc 126:14714–14715
318. Murakami T, Nojiri M et al (2000) Post-translational modification is essential for catalytic activity of nitrile hydratase. Protein Sci 9:1024–1030
319. Odaka M, Fujii K et al (1997) Activity regulation of photoreactive nitrile hydratase by nitric oxide. J Am Chem Soc 119:3785–3791
320. Endo I, Odaka M, Yohda M (1999) An enzyme controlled by light: the molecular mechanism of photoreactivity in nitrile hydratase. Trends Biotechnol 17:244–248
321. Voloshin ON, Camerini-Otero RD (2007) The ding protein from Escherichia coli is a structure-specific helicase. J Biol Chem 282:18437–18447
322. Shestakov AF, Shul'ga YM et al (2009) Experimental and theoretical study of the arrangement, electronic structure and properties of neutral paramagnetic binuclear nitrosyl iron complexes with azaheterocyclic thyolyls having 'S-C-N type' coordination of bridging ligands. Inorg Chim Acta 362:2499–2504
323. Nunoshiba T, DeRojas-Walker T et al (1993) Activation by nitric oxide of an oxidative-stress response that defends Escherichia coli against activated macrophages. Proc Natl Acad Sci USA 90:9993–9997
324. D'Autreaux B, Tucker NP et al (2005) A non-haem iron centre in the transcription factor NorR senses nitric oxide. Nature 437:769–772
325. Hidalgo E, Bollinger JM Jr et al (1995) Binuclear[2Fe-2S] clusters in the Escherichia coli SoxR protein and role of the metal centers in transcription. J Biol Chem 270:20908–20914
326. Ding H, Demple B (2000) Direct nitric oxide signal transduction via nitrosylation of iron-sulfur centers in the SoxR transcription activator. Proc Natl Acad Sci USA 97:5146–5150
327. Stamler JS (1994) Cell 78:931
328. Butler AR, Megson IL (2002) Non-heme iron nitrosyls in biology. Chem Rev 102:1155–1166
329. Weinberg JB, Chen Y et al (2009) Inhibition of nitric oxide synthase by cobalamins and cobinamides. Free Radic Biol Med 46:1626–1632
330. Brouwer M, Chamulitrat W et al (1996) Nitric oxide interactions with cobalamins: biochemical and functional consequences. Blood 88:1857–1864
331. Bunn HF, Aster JC (2010) Pathophysiology of blood disorders. McGraw Hill, Boston
332. Sharma VS, Pilz RB et al (2003) Reactions of nitric oxide with vitamin B12 and its precursor, cobinamide. Biochemistry 42:8900–8908
333. Weinberg JB, Chen YC et al (2005) Cobalamins and cobinamides inhibit nitric oxide synthase enzymatic activity. Blood 106:2225

334. Wheatley C (2006) A scarlet pimpernel for the resolution of inflammation? The role of supra-therapeutic doses of cobalamin, in the treatment of systemic inflammatory response syndrome (SIRS), sepsis, severe sepsis, and septic or traumatic shock. Med Hypotheses 67:124–142
335. Wheatley C (2007) The return of the scarlet pimpernel: cobalamin in inflammation II-cobalamins can both selectively promote all three nitric oxide synthases (NOS), particularly iNOS and eNOS, and, as needed, selectively inhibit iNOS and nNOS. J Nutr Environ Med 16:181–211
336. Forsyth JC, Mueller PD et al (1993) Hydroxocobalamin as a cyanide antidote: safety, efficacy and pharmacokinetics in heavily smoking normal volunteers. Clin Toxicol 31:277–294
337. Houeto P, Borron SW et al (1996) Pharmacokinetics of hydroxocobalamin in smoke inhalation victims. Clin Toxicol 34:397–404
338. Sharina I, Sobolevsky M et al (2012) Cobinamides are novel coactivators of nitric oxide receptor that target soluble guanylyl cyclase catalytic domain. J Pharmacol Exp Ther 340:723–732
339. Lewandowska H, Brzoska K et al (2010) [Dinitrosyl iron complexes–structure and biological functions]. Postepy Biochem 56:298–304
340. Meczynska S, Lewandowska H et al (2008) Variable inhibitory effects on the formation of dinitrosyl iron complexes by deferoxamine and salicylaldehyde isonicotinoyl hydrazone in K562 cells. Hemoglobin 32:157–163
341. Lo BM, Nuccetelli M et al (2001) Human glutathione transferase P1-1 and nitric oxide carriers; a new role for an old enzyme. J Biol Chem 276:42138–42145
342. Lewandowska H, Meczynska S et al (2007) Crucial role of lysosomal iron in the formation of dinitrosyl iron complexes in vivo. J Biol Inorg Chem 12:345–352
343. Toledo JC Jr, Bosworth CA et al (2008) Nitric oxide-induced conversion of cellular chelatable iron into macromolecule-bound paramagnetic dinitrosyliron complexes. J Biol Chem 283:28926–28933
344. Hickok JR, Sahni S et al (2011) Dinitrosyliron complexes are the most abundant nitric oxide-derived cellular adduct: biological parameters of assembly and disappearance. Free Radic Biol Med 51:1558–1566
345. Pedersen JZ, De Maria F et al (2007) Glutathione transferases sequester toxic dinitrosyl-iron complexes in cells. A protection mechanism against excess nitric oxide. J Biol Chem 282:6364–6371
346. Hidalgo E, Ding H, Demple B (1997) Redox signal transduction via iron-sulfur clusters in the SoxR transcription activator. Trends Biochem Sci 22:207–210
347. Chiang RW, Woolum JC, Commoner B (1972) Biochim Biophys Acta 257:452
348. Pellat C, Henry Y, Drapier JC (1990) IFN-gamma-activated macrophages: detection by electron paramagnetic resonance of complexes between L-arginine-derived nitric oxide and non-heme iron proteins. Biochem Biophys Res Commun 166:119–125
349. Ueno T, Suzuki Y et al (1999) In vivo distribution and behavior of paramagnetic dinitrosyl dithiolato iron complex in the abdomen of mouse. Free Radic Res 31:525–534
350. Ignarro LJ, Ross G, Tillisch J (1991) Pharmacology of endothelium-derived nitric oxide and nitrovasodilators. West J Med 154:51–62
351. Wang YX, Legzdins P et al (2000) Vasodilator effects of organotransition-metal nitrosyl complexes, novel nitric oxide donors. J Cardiovasc Pharmacol 35:73–77
352. Saklayen MG, Mandal AK (1996) Nitric oxide: biological and clinical perspectives. Int J Artif Organs 19:630–632
353. Armstrong RN (1997) Structure, catalytic mechanism, and evolution of the glutathione transferases. Chem Res Toxicol 10:2–18
354. Mannervik B, Alin P et al (1985) Identification of three classes of cytosolic glutathione transferase common to several mammalian species: correlation between structural data and enzymatic properties. Proc Natl Acad Sci USA 82:7202–7206
355. Litwack G, Ketterer B, Arias IM (1971) Ligandin: a hepatic protein which binds steroids, bilirubin, carcinogens and a number of exogenous organic anions. Nature 234:466–467

356. Adler V, Yin Z et al (1999) Regulation of JNK signaling by GSTp. EMBO J 18:1321–1334
357. Kampranis SC, Damianova R et al (2000) A novel plant glutathione S-transferase/peroxidase suppresses Bax lethality in yeast. J Biol Chem 275:29207–29216
358. Dulhunty A, Gage P et al (2001) The glutathione transferase structural family includes a nuclear chloride channel and a ryanodine receptor calcium release channel modulator. J Biol Chem 276:3319–3323
359. Caccuri AM, Antonini G et al (1999) Proton release on binding of glutathione to alpha, mu and delta class glutathione transferases. Biochem J 344(Pt 2):419–425
360. Ford PC (2008) Polychromophoric metal complexes for generating the bioregulatory agent nitric oxide by single- and two-photon excitation. Acc Chem Res 41:190–200
361. Rose MJ, Mascharak PK (2008) Fiat Lux: selective delivery of high flux of nitric oxide (NO) to biological targets using photoactive metal nitrosyls. Curr Opin Chem Biol 12:238–244
362. Silva JJN, Osakabe AL et al (2007) In vitro and in vivo antiproliferative and trypanocidal activities of ruthenium NO donors. Br J Pharmacol 152:112–121
363. Haas KL, Franz KJ (2009) Application of metal coordination chemistry to explore and manipulate cell biology. Chem Rev 109:4921–4960
364. Vanin AF, Vanin AF (2008) P103. Dinitrosyl iron complexes with thiolate ligands as future medicines. Nitric Oxide 19:67
365. Vanin AF, Mokh VP et al (2007) Vasorelaxing activity of stable powder preparations of dinitrosyl iron complexes with cysteine or glutathione ligands. Nitric Oxide 16:322–330
366. Lakomkin VL, Vanin AF et al (2007) Long-lasting hypotensive action of stable preparations of dinitrosyl-iron complexes with thiol-containing ligands in conscious normotensive and hypertensive rats. Nitric Oxide 16:413–418
367. Fujii S, Yoshimura T (2000) Detection and imaging of endogenously produced nitric oxide with electron paramagnetic resonance spectroscopy. Antioxid Redox Signal 2:879–901
368. Yoshimura T, Yokoyama H et al (1996) In vivo EPR detection and imaging of endogenous nitric oxide in lipopolysaccharide-treated mice. Nat Biotechnol 14:992–994
369. Yasui H, Fujii S et al (2004) Spinnokinetic analyses of blood disposition and biliary excretion of nitric oxide (NO)-Fe (II)-N-(Dithiocarboxy)sarcosine complex in rats: BCM-ESR and BEM-ESR studies. Free Radic Res 38:1061–1072
370. Chen YJ, Ku WC et al (2008) Nitric oxide physiological responses and delivery mechanisms probed by water-soluble Roussin's red ester and $\{Fe(NO)_2\}^{10}$ DNIC. J Am Chem Soc 130:10929–10938
371. Dillinger SAT, Schmalle HW et al (2007) Developing iron nitrosyl complexes as NO donor prodrugs. Dalton Trans 3562–3571
372. Huerta S, Chilka S, Bonavida B (2008) Nitric oxide donors: novel cancer therapeutics. Int J Oncol 33:909–927
373. Wecksler SR, Mikhailovsky A et al (2006) Single- and two-photon properties of a dye-derivatized Roussin's red salt ester ($Fe_2(mu\text{-}RS)_2(NO)_4$) with a large TPA cross section. Inorg Chem 46:395–402
374. Wecksler SR, Mikhailovsky A et al (2006) A two-photon antenna for photochemical delivery of nitric oxide from a water-soluble, dye-derivatized iron nitrosyl complex using NIR light. J Am Chem Soc 128:3831–3837
375. Wecksler SR, Hutchinson J, Ford PC (2006) Toward development of water soluble dye derivatized nitrosyl compounds for photochemical delivery of NO. Inorg Chem 45:1192–1200
376. Wecksler S, Mikhailovsky A, Ford PC (2004) Photochemical production of nitric oxide via two-photon excitation with NIR light. J Am Chem Soc 126:13566–13567
377. Zheng Q, Bonoiu A et al (2008) Water-soluble two-photon absorbing nitrosyl complex for light-activated therapy through nitric oxide release. Mol Pharm 5:389–398
378. Fry NL, Rose MJ et al (2010) Ruthenium nitrosyls derived from tetradentate ligands containing carboxamido-N and phenolato-O donors: syntheses, structures, photolability, and time dependent density functional theory studies. Inorg Chem 49:1487–1495

379. Fry NL, Mascharak PK (2011) Photoactive ruthenium nitrosyls as NO donors: how to sensitize them toward visible light. Acc Chem Res 44:289–298
380. Fry NL, Mascharak PK (2012) Photolability of NO in designed metal nitrosyls with carboxamido-N donors: a theoretical attempt to unravel the mechanism. Dalton Trans 41:4726–4735
381. Afshar RK, Patra AK, Mascharak PK (2005) Light-induced inhibition of papain by a {Mn-NO}6 nitrosyl: identification of papain-SNO adduct by mass spectrometry. J Inorg Biochem 99:1458–1464
382. Eroy-Reveles AA, Leung Y et al (2008) Near-infrared light activated release of nitric oxide from designed photoactive manganese nitrosyls: strategy, design, and potential as NO donors. J Am Chem Soc 130:4447–4458
383. Szundi I, Rose MJ et al (2006) A new approach for studying fast biological reactions involving nitric oxide: generation of NO using photolabile ruthenium and manganese NO donors. Photochem Photobiol 82:1377–1384
384. Ghosh K, Eroy-Reveles AA et al (2004) Reactions of NO with Mn (II) and Mn (III) centers coordinated to carboxamido nitrogen: synthesis of a manganese nitrosyl with photolabile NO. Inorg Chem 43:2988–2997
385. Madhani M, Patra AK et al (2006) Biological activity of designed photolabile metal nitrosyls: light-dependent activation of soluble guanylate cyclase and vasorelaxant properties in rat aorta. J Med Chem 49:7325–7330
386. Patra AK, Rose MJ et al (2004) Photolabile ruthenium nitrosyls with planar dicarboxamide tetradentate N (4) ligands: effects of in-plane and axial ligand strength on NO release. Inorg Chem 43:4487–4495
387. Rose MJ, Fry NL et al (2008) Sensitization of ruthenium nitrosyls to visible light via direct coordination of the dye resorufin: trackable NO donors for light-triggered NO delivery to cellular targets. J Am Chem Soc 130:8834–8846
388. Rose MJ, Olmstead MM, Mascharak PK (2007) Photosensitization via dye coordination: a new strategy to synthesize metal nitrosyls that release NO under visible light. J Am Chem Soc 129:5342–5343
389. Rose MJ, Mascharak PK (2008) A photosensitive $\{Ru\text{-}NO\}^6$ nitrosyl bearing dansyl chromophore: novel NO donor with a fluorometric on/off switch. Chem Commun 3933–3935
390. Lim MH, Lippard SJ (2007) Metal-based turn-on fluorescent probes for sensing nitric oxide. Acc Chem Res 40:41–51
391. Soh N, Imato T et al (2002) Ratiometric direct detection of nitric oxide based on a novel signal-switching mechanism. Chem Commun 2650–2651
392. Hilderbrand SA, Lippard SJ (2004) Nitric oxide reactivity of fluorophore coordinated carboxylate-bridged diiron (II) and dicobalt (II) complexes. Inorg Chem 43:5294–5301
393. Hilderbrand SA, Lippard SJ (2004) Cobalt chemistry with mixed aminotroponiminate salicylaldiminate ligands: synthesis, characterization, and nitric oxide reactivity. Inorg Chem 43:4674–4682
394. Lim MH, Lippard SJ (2004) Fluorescence-based nitric oxide detection by ruthenium porphyrin fluorophore complexes. Inorg Chem 43:6366–6370
395. Lim MH, Kuang C, Lippard SJ (2006) Nitric oxide-induced fluorescence enhancement by displacement of dansylated ligands from cobalt. Chembiochem 7:1571–1576
396. Smith RC, Tennyson AG, Lippard SJ (2006) Polymer-bound dirhodium tetracarboxylate films for fluorescent detection of nitric oxide. Inorg Chem 45:6222–6226
397. Hilderbrand SA, Lim MH, Lippard SJ (2004) Dirhodium tetracarboxylate scaffolds as reversible fluorescence-based nitric oxide sensors. J Am Chem Soc 126:4972–4978
398. Lim MH (2007) Preparation of a copper-based fluorescent probe for nitric oxide and its use in mammalian cultured cells. Nat Protoc 2:408–415
399. Lim MH, Lippard SJ (2006) Fluorescent nitric oxide detection by copper complexes bearing anthracenyl and dansyl fluorophore ligands. Inorg Chem 45:8980–8989

400. Smith RC, Tennyson AG et al (2005) Conjugated polymer-based fluorescence turn-on sensor for nitric oxide. Org Lett 7:3573–3575
401. Lim MH, Lippard SJ (2005) Copper complexes for fluorescence-based NO detection in aqueous solution. J Am Chem Soc 127:12170–12171
402. Ford PC, Fernandez BO, Lim MD (2005) Mechanisms of reductive nitrosylation in iron and copper models relevant to biological systems. Chem Rev 105:2439–2455
403. Smith RC, Tennyson AG et al (2006) Conjugated metallopolymers for fluorescent turn-on detection of nitric oxide. Inorg Chem 45:9367–9373
404. Xing C, Yu M et al (2007) Fluorescence turn-on detection of nitric oxide in aqueous solution using cationic conjugated polyelectrolytes. Macromol Rapid Commun 28:241–245
405. Do L, Smith RC et al (2006) Luminescent properties of water-soluble conjugated metallopolymers and their application to fluorescent nitric oxide detection. Inorg Chem 45:8998–9005
406. Franz KJ, Singh N et al (2000) Aminotroponiminates as ligands for potential metal-based nitric oxide sensors. Inorg Chem 39:4081–4092
407. Lim MH, Wong BA et al (2006) Direct nitric oxide detection in aqueous solution by copper (II) fluorescein complexes. J Am Chem Soc 128:14364–14373
408. Lim MH, Xu D, Lippard SJ (2006) Visualization of nitric oxide in living cells by a copper-based fluorescent probe. Nat Chem Biol 2:375–380
409. Ouyang J, Hong H et al (2008) A novel fluorescent probe for the detection of nitric oxide in vitro and in vivo. Free Radic Biol Med 45:1426–1436
410. Vanin AF (2009) Dinitrosyl iron complexes with thiolate ligands: physico-chemistry, biochemistry and physiology. Nitric Oxide 21:1–13
411. Keese MA, Bose M et al (1997) Dinitrosyl-dithiol-iron complexes, nitric oxide (NO) carriers in vivo, as potent inhibitors of human glutathione reductase and glutathione-S-transferase. Biochem Pharmacol 54:1307–1313
412. Vasilieva SV, Moshkovskaya EY et al (2004) Genetic signal transduction by nitrosyl-iron complexes in Escherichia coli. Biochemistry 69:883–889
413. Tran NG, Kalyvas H et al (2011) Phenol nitration induced by an $\{Fe(NO)_2\}^{10}$ dinitrosyl iron complex. J Am Chem Soc 133:1184–1187
414. Bellota-Anton C, Munnoch J et al (2011) Spectroscopic analysis of protein Fe-NO complexes. Biochem Soc Trans 39:1293–1298

Struct Bond (2014) 153: 115–166
DOI: 10.1007/430_2013_109

Published online: 19 June 2013

Spectroscopic Characterization of Nitrosyl Complexes

Hanna Lewandowska

Abstract Structural and spectroscopic data when combined with theoretical calculations provide adequate descriptions of the electronic structures of the nitrosyl moiety in nitric oxide(II) complexes. This concise overview discusses the spectroscopic features of metal nitrosylates. The results of IR, Raman, UV–Vis, EPR, Mössbauer, magnetic circular dichroism, NRVS, X-ray absorption spectroscopy, and other methods are reviewed and spectroscopy-based conclusions concerning the structure and reactivities of nitrosyls are summarized.

Keywords Complex · EPR · Mössbauer · IR · Magnetic circular dichroism · Nitrogen (II) oxide · Nitrosyl · NRVS · Raman · Spectroscopy · Transition metal · UV–Vis · X-ray absorption spectroscopy · Iron

Contents

H. Lewandowska (✉)
Institute of Nuclear Chemistry and Technology, Centre for Radiobiology and Biological Dosimetry, 16 Dorodna Str., 03-195 Warsaw, Poland
e-mail: h.lewandowska@ichtj.waw.pl

Abbreviations

1-MeIm	1-Methylimidazole
5C	Five coordinate
6C	Six coordinate
B3LYP	Becke three-parameter, Lee–Yang–Parr exchange-correlation functional
BSA	Bovine serum albumin
Bu	Butyl
DEP	Diethyl pyrocarbonate
DFT	Density functional theory
DNIC	Dinitrosyl iron complex
DPIXDME	Deuteroporphyrin IX dimethyl ester
dppe	1,2-Bis(diphenylphosphino)ethane
dtci-Pr_2	*Di*-isopropyl dithiocarbamate
ENDOR	Electron nuclear double beam resonance
EPR	Electron paramagnetic resonance
ESEEM	Orientation-selected electron spin-echo envelope modulation
Et	Ethyl
EXAFS	X-ray absorption fine structure
GSH	Glutathione
GST	Glutathione transferase
Hb	Hemoglobin
HYSCORE	2D hyperfine sublevel correlation
Im	Imidazole
INDO/S	Intermediate neglect of differential overlap/screened approximation.
IPTG	Isopropyl β-D-1-thiogalactopyranoside
LCP	Left circular polarization
LMCT	Ligand-to-metal charge transfer
LMW	Low-molecular weight
Mb	Myoglobin
MB	Mössbauer
MCD	Magnetic circular dichroism
Me	Methyl
Me_3TACN	*N,N′,N″*-trimethyl-1,4,7-triazacyclononane
MLCT	Metal-to-ligand charge transfer
MNIC	Mononitrosyl iron complex
MPIXDME	Mesoporphyrin IX dimethyl ester
N-EtHPTB	*N,N,N'N'-tetrakis*(2-benzimidazolylmethyl)-2-hydroxy-1, 3-diaminopropane anion
NorR	Anaerobic nitric oxide reductase transcription regulator
NRVS	Nuclear resonance vibrational spectroscopy
OEP	Octaethylporphyrin
ORTEP	Oak Ridge thermal-ellipsoid plot program

Ph	Phenyl
PMB	*p*-(chloromercuri)-benzoate
PPIXDME	Protoporphyrin IX dimethyl ester
PPN	Bis-triphenylphosphineiminium cation
RCP	Right circular polarization
RRE	Roussin's red salt ester
SOD	Superoxide dismutase
SOMO	Single occupied molecular orbital
TBP	Trigonal bipyramidal
TD-DFT	Time-dependent density functional theory
THF	Tetrahydrofuran
T*o*-F_2PP	*Tetrakis*-5,10,15,20-(*o*-difluorophenyl)porphyrin
TP	Tetragonal pyramidal
TpivPP	α,α,α,α-*tetrakis*(*o*-pivalamidophenyl)-porphyrinato^{2-}
TPP	Tetraphenylporphyrin
TZVP	Triple zeta valence plus polarization functional
UV–Vis	Ultraviolet–Visible spectroscopy
XANES	X-ray absorption near-edge spectroscopy
XAS	X-ray absorption spectroscopy

1 Introduction

Structural and spectroscopic properties with theoretical calculations permit adequate descriptions of the MNO moiety's electronic structure. Nevertheless one must be careful in the interpretation of these results (i.e., in assigning limiting oxidation states to the metal and NO ligand), due to the delocalized nature of the electronic density. Among the spectroscopic methods, which are most helpful in the study of nitrosylates, are those traditionally used for structural research, i.e., IR, EPR, UV–Vis, resonance Raman, but also more specific ones, such as Mössbauer, XAS measurements, magnetic circular dichroism and EPR spectroscopy, magnetic susceptibility measurements, and magnetic resonance imaging for biological systems [1–4]. Advanced theoretical calculations are now essential when interpreting the results of spectroscopic measurements. In the following review a general overview is given of the results obtained to date by the most common methods for investigating metal nitrosyls. The issues, which can be resolved by these methods, are discussed.

2 Vibrational Spectra of Metal Nitrosylates

The stretching frequency for uncoordinated NO is 1,870 cm^{-1}. Upon complexation ν(NO) may increase or decrease, depending on the nature of the co-ligands in the complex, the electronic configuration, and the overall charge of the complex,

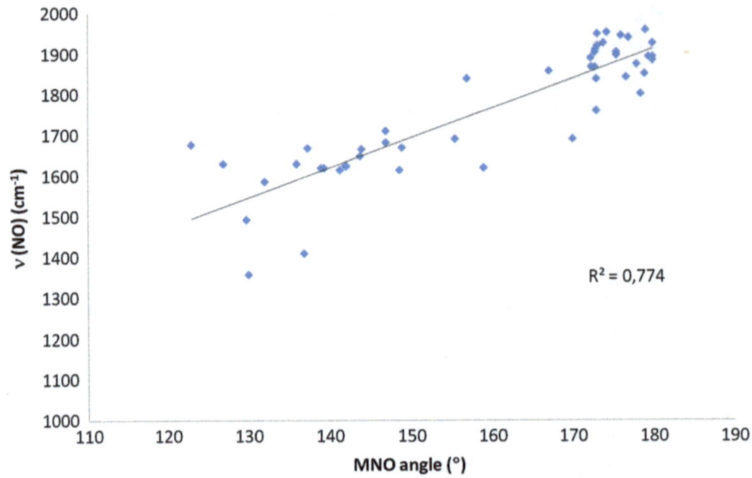

Fig. 1 Dependence between the MNO bond angle and ν(NO) wave number of 48 transition metal mononitrosyls, according to the data reported in the literature. The quoted data and references are listed in the supplementary Table 1

as well as the complex' geometry. In general, the IR stretching frequencies of the ν(NO) decrease with increasing charge, from 2,377 (NO^+) through 1,875 (NO) to 1,470 cm^{-1} (NO^-) [5].

Yet, drawing conclusions on the electronic structure of metal nitrosyls is not as easy as it might seem from the above description. For a long time it was considered that the dependencies between the structure and the infrared features for metal nitrosyls were too complex to draw up systemized regularities. The energies of ν(NO) vibrations cover a very wide range and are poorly correlated with geometrical parameters (see Fig. 1). Better correlations can be found, if chemically related series of complexes are considered: For the nitrosyls of 8 group metals (Fe, Ru, Os) cluster-type correlation can be observed for the structure-vibration dependencies for the $n = 6$ to 8 type nitrosyls (where n denotes the total number of electrons associated with the metal d and π^* (NO) orbitals), as seen in Fig. 2. The vibration bands for $n = 8$ occur in the region 1,270–1,500 cm^{-1}, for $n = 7$ in the region: 1,600–1,700 cm^{-1}, and for $n = 6$ in. 1,800–1,960 cm^{-1}. Unfortunately this rule cannot be applied to heme metal nitrosyls and metallo-organic compounds (see, e.g., [6]). Also, different ν(NO) ranges of nitrosyl complexes are found, when the metal is changed from one group to another. The electron charge distribution in the MNO fragment is very sensitive to the nature of the auxiliary ligands, the total charge of the complex, geometry of the coordination sphere, etc. This sensitivity may be a factor determining the biological significance of these nitrogen(II) oxide compounds, and is also responsible for the apparent lack of a clear relationship between the formal electron state of the nitrosyl group and its ν(NO) frequency [7].

If analyzed in terms of geometry, NO stretching frequencies for complexes containing linear M–N–O bonds occur in the range 1,950–1,450 cm^{-1}, those for

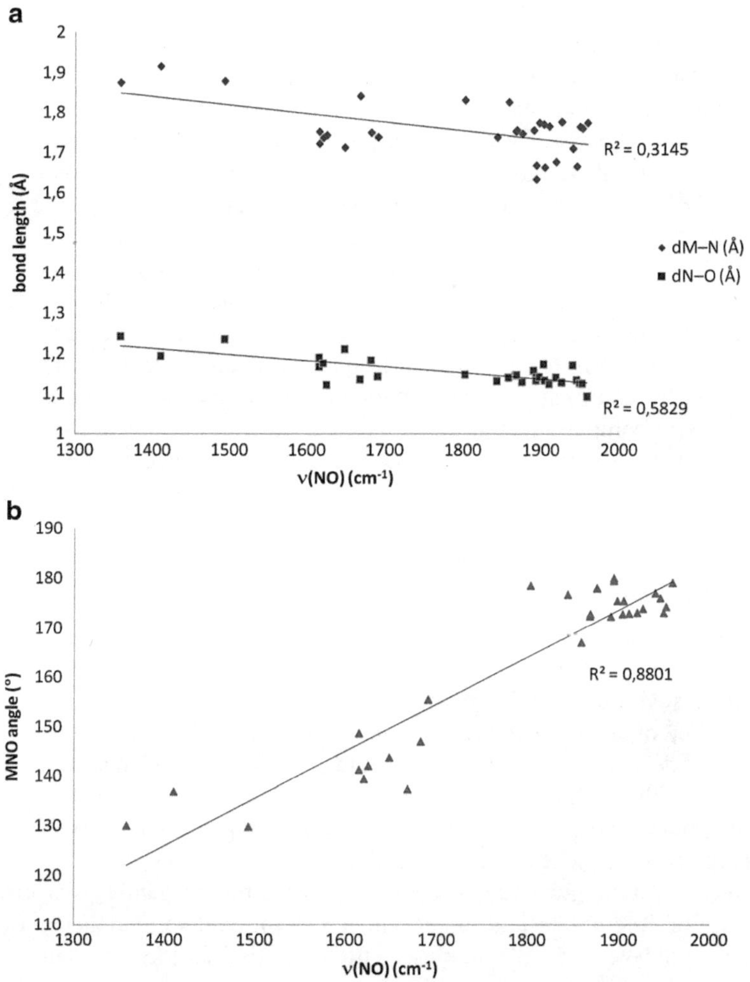

Fig. 2 (**a**) Charts of M–N and N–O bond lengths, and (**b**) MNO angle versus ν(NO) wave number in metal mononitrosyls of the 8 group transition metals (Fe, Ru, Os). The data and references are listed in the supplementary Table 1

complexes containing bent M–N–O bonds in the range 1,720–1,400 cm^{-1}, and when NO bridges two or more metal atoms, ν(NO) may occur anywhere between 1,650 and 1,300 cm^{-1}. It was noted that certain electronic factors (such as the position of the metal in the periodic table, coordination number of the metal, complex charge, and the type of the ancillary ligands) influence the ν(NO) value without changing the basic coordination geometry [8–12]. This inspired an early attempt of Haymore and Ibers [13] to formulate estimates of adjustments for the observed ν(NO) wave numbers in nitrosyls. Adjusted wave numbers fell in the two ranges, above and below 1,606–1,611 cm^{-1}, allowing the classification of a nitrosyl as linear or bent, respectively. These authors also noticed that the corrected ν(NO) values (found toward the end of the lower wave number range) could be used to distinguish between bridging and non-bridging NO groups.

The most recent and the most successful attempt of systemizing the ν(NO) frequencies on the basis of estimated adjustments was performed by De La Cruz and Sheppard [14]. A good insight into the relationship between spectral characteristics and the structure is provided by the analysis of the spectra recorded for simple complexes in the noble gas matrix (cryogenic matrices; see, e.g., [15–20]). The spectra of uncharged, closed-shell nitrosyl/carbonyl metal complexes turned out to be useful as references, because the presence of a carbonyl ligand was found to be neutral in its effect on the ν(NO) values. The effect that each particular ligand exerts on the nitrosyl vibration frequency (and electronic charge density) was studied by comparing spectral parameters for donation-saturated nitrosyl/carbonyl metal complexes with the analogical complexes substituted with some electron-withdrawing or electron-donating co-ligands bonded to the same metal atom and for the pairs of complexes in a different oxidation state of metal. Ligands disrupted the electron charge density of nitrosyl groups in varying degrees. Estimation of the ν(NO) wave number shifts led the above-mentioned authors [14] to draw the following conclusions:

1. The effect of a unit of positive charge was estimated to give an approximate increase of plus 100 cm^{-1} associated with a net unit positive charge on the metal complexes with linear NO ligands.
2. The effect of a unit negative charge was estimated to give approximately 145 cm^{-1} decrease of ν(NO) per unit of the negative charge on metal complexes with linear NO groups.
3. A value of ca. 180 cm^{-1} for the decrease of ν(NO) per unit negative charge was observed for complexes with bent nitrosyls.
4. The electron-withdrawing effect of halogens as terminal ligands gives an average increase of 30 cm^{-1} in ν(NO) per halogen atom. Bridging halogens effect was estimated to be ca. 15 cm^{-1} in plus. A slight positive shift (5 cm^{-1}) in the ν(NO) wave number is noticed, along with the increasing halogen electronegativity.
5. The substitution of one terminal NO by one CN leads to an increase of approximately 50 cm^{-1} in the ν(NO).
6. The trivalent phosphorus derivatives were shown to effect the nitrosyl stretching frequency accordingly to their electron-donating or electron-withdrawing properties: A decrease of about 70 cm^{-1} per trialkyl-phosphine ligand, a decrease of about 55 cm^{-1} per triphenylphosphine ligand, a lowering of ν(NO) by 40 cm^{-1} per trialkoxy and 30 cm^{-1} per triphenoxy ligands and an increase of about 10 cm^{-1} per PF_3 ligand were found. It is notable that the presence of increasingly electron-withdrawing substituents on phosphorus reduces or eliminates the effective basicity of the phosphine ligands.
7. A decrease in ν(NO) by ca. 60, 70, and 80 cm^{-1} was observed for the pentahapto-cyclopentadienyl group, η^5-C_5H_4Me and η^5-C_5Me_5, respectively.

Table 1 Spectroscopic amendments for factors on metal-nitrosyl complexes affecting the observed ν(NO) wave numbers elaborated by De La Cruz and Sheppard [14]

Co-ligands	$\Delta\nu$(NO) (cm^{-1})
Unit of positive charge	+100
Linear cyanide	+50
Linear halogens	+30
Bridging halogens	<+15
PF_3	≈+10
Linear or bridging CO ligand	≈0
$P(OPh)_3$	−30
$P(OR)_3$ (R=Me, Et, Bu)	−40
PPh_3	−55
PR_3 (R=Me, Et, Bu, C_6H_{11})	−70
Pentahapto C_5H_5	−60
Pentahapto C_5H_4Me	−70
Pentahapto C_5Me_5	−80
Unit of negative charge	−145

8. The ν(NO) wave numbers registered for nitrosyls in solution to the substantial degree depend on the dielectric properties of the solvent. One can conclude that the electronic nature of the polar NO group is such that it is particularly influenced by the actual or induced electric fields generated by its surroundings [21], which may have substantial consequences on the behavior and reactivities of nitrosyls as biological signal transducers.

The spectroscopic adjustments on metal nitrosyl complexes' ν(NO) wave numbers elaborated by De La Cruz and Sheppard [14] are listed in Table 1.

Although the effects of the electron-donating and electron-withdrawing co-ligands may not always be straightforwardly additive, the observed ν(NO) values for a wide range of nitrosyl complexes of mixed co-ligands can usually be predicted to within 20 or 30 cm^{-1}, by taking into account the individual approximate raising or lowering corrections for the active ligands as given above. With the increasing electron charge of the MNO moiety π backbonding contribution increases, along with the strengthening of the M–N bond and weakening of the N–O bond. When the charge decreases the bond order of M–N is decreased and N–O bonding increases up to bond order of 3, in the limiting case. The existence of a limit to the latter processes explains why the upward shift of the ν(NO) wave number per unit of positive charge is less than the downward shift for a unit of negative charge.

Adjusted NO stretching vibrations for the set of complexes (with linear NO group) for various central atoms indicate periodic regularities, which are expected to occur along with the change of the metal atom (see Fig. 3).

X-ray diffraction studies [22] indicated that $[M(NO)(L)_4]$ complexes of 8 and 9 group of the periodic table adopt either trigonal bipyramidal (TBP) or tetragonal pyramidal (TP) structures. Sometimes, however, the distinctions between these two structures are not unambiguous due to the distortions which are caused by the steric hindrance produced by the auxiliary ligands or by the presence of strong σ-donor in the base of the TP structure or in the axial and equatorial positions of the TBP

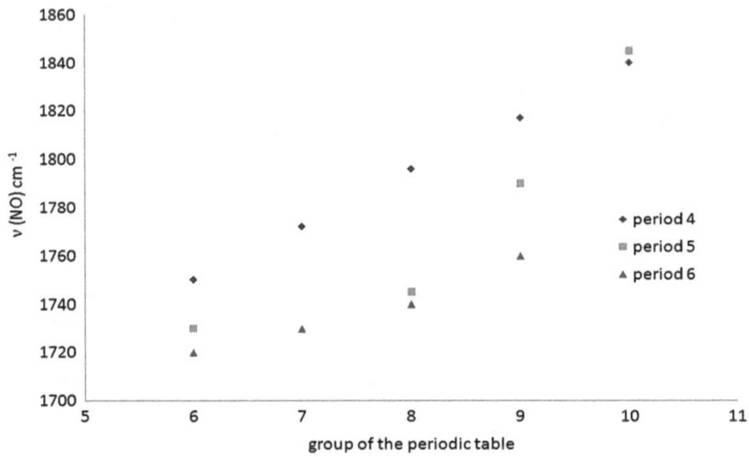

Fig. 3 The transition metal dependence of estimated ν(NO) values for linear nitrosyls with otherwise electronically neutral co-ligands, using data from complexes of metals of Periods 5, 6, and 7 of the periodic table based on the data taken from [15, 19, 21, 219–231] and according to [14]

structure [23, 24]. The linear and bent NO groups in a very wide range of compounds absorb infrared radiation over the rather wide ranges 1,862–1,690 cm^{-1} and 1,720–1,525 cm^{-1}, respectively, and as mentioned above, an overlap of these regions precludes the classification of these structures. Yet, another spectroscopic criterion can help to distinguish between the linear and bent NO groups, namely the $\nu(^{14}NO) \rightarrow \nu(^{15}NO)$ isotopic shifts. As was explicitly demonstrated for the linear nitrosyls, these extensive regions reflect the presence of a very wide range of co-ligands or ionic charges associated with the metal atom of the nitrosyl group. The isotopic band shift (IBS) which is defined as $\nu(^{14}NO)-\nu(^{15}NO)$ plotted against M–N–O bond angle for penta-, hexa-, and heptacoordinated mononitrosyls reveals that the IBS values are clustered between 45 and 30 cm^{-1} or between 37 and 25 cm^{-1} for linear or bent NO groups, respectively. Overall it is suggested that the bent nitrosyls absorb ca. 60–100 cm^{-1} below and have smaller co-ligand band shifts than their linear counterparts. The twofold bridged nitrosyls with a metal–metal bond order of one, or greater than one, absorb at ca. 1,610–1,490 cm^{-1}; the twofold bridged nitrosyl ligands with a longer non-bonding M . . . M distance absorb at ca. 1,520–1,490 cm^{-1}; the threefold bridged nitrosyls absorb at ca. 1,470–1,410 cm^{-1}; the isonitrosyls, from the few known examples, appear to absorb below ca. 1,100 cm^{-1}.

The vibrational spectra provide an insight into the structural features of the ON–M–NO moiety present in the dinitrosyl complexes with the general formula $[M(NO)_2(L)_y]$ ($y = 4, 3, 2$). The angle between the two nitrosyl groups (ON–M–NO) can take the value between 180° in the *trans* case and 90° in the *cis* case. The angle value and its changes may be estimated by vibrational spectroscopy: In the *trans* case

Table 2 Structural and vibrational MNO core parameters of representative metal nitrosyls, classified with respect to the character and symmetry of MNO bonding

Compound	ν(NO) (1)	ν(NO) (2)	$\Delta\nu$	$\angle$MNO (1)	$\angle$MNO (2)	$\Delta\angle$MNO	Reference
Linear plus bent MNO group							
$[Ru(NO)_2(PPh_3)_3(Cl)]^+$	1,850	1,687	163	179.5	136	43.5	[236]
$[Os(NO)_2(PPh_3)_2(OH)]^+$	1,842	1,632	210	177.6	133.6	44.0	[237, 238]
Linear MNO groups							
$[Fe(NO_2),Cl]_2,(\mu\text{-dppe})$	1,786	1,724	62	169.6	165.8	3.8	[239]
$Fe(NO)_2(dppe)$	1,707	1,657	50	178	176	2.0	[239]
$[PPN][S_5Fe(NO)_2]$	1,739	1,695	44	172.8	165.9	6.9	[32]
$[PPN][(SPh)_2Fe\text{-}(NO)_2]$	1,727	1,692	35	169.5	168.6	0.9	[32]
Linear/bent MNO isomers							
$[Co(NO)(PPhMe_2)_2Cl_2]/CH_2Cl_2$	1,760	1,650	110	–	–	–	[240]
$[Co(NO)(PMe_3)_2Br_2]/CH_2Cl_2$	1,750	1,670	80	–	–	–	[241]
$[Co(NO)(PMe_3)_2Cl_2]/CH_2Cl_2$	1,750	1,655	95	–	–	–	[241]

only the out-of-phase $\nu(NO)_{asym}$ mode, which gives rise to a vibrational dipole moment, is infrared active, while the [ν(NO)-sym] mode is only active in Raman. In the *cis* case both modes are infrared and Raman active. The intensity of the ν(NO)-sym band is therefore expected to increase as the angle between the two NO groups decreases from 180° to 90° [25, 26]. The ν(NO)-sym mode occurs at a higher wave number than the ν(NO)-asym mode, as is the case for analogous dicarbonyls, and as demonstrated for tetrahedral dinitrosyls with triphenylphosphine of Fe, Co, Rh, and Ir [27–30]. Wave number separations between the coupled modes are ca. 140–80 cm^{-1} for the ON–M–NO bond angles between ca. 80 and 100° (the approximate *cis* case) and ca. 70–20 cm^{-1} for the bond angles between 100 and 130° (the *trans*-case), respectively. If the interaction between the two nitrosyls was purely due to the mechanical vibrational coupling, then a small splitting would be expected for *cis* case (where the two nitrosyl groups are mutually orthogonal) and a large one for the *trans* case. The opposite direction with respect to the ON–M–NO angle implies that electronic rather than mechanical coupling is the predominant factor [14]. It is noteworthy that for dinitrosyls that comprise both linear and bent MNO groups, the two ν(NO) wave numbers are strongly decoupled and can be firmly assigned to the bond stretching of the linear (higher frequency) and bent (lower frequency) NO group. A substantial (60–100 cm^{-1}) difference in the two nitrosyl group stretching vibrations may also indicate the presence of linear versus bent MNO group isomerism, as it takes place in some cobalt nitrosyls which coexist as linear and bent nitrosyl isomers in equilibrium. A difference of ca. 60 cm^{-1} is found for nitrosyl stretches of geometrical isomers displaying either an apical or equatorial NO location dependence (see Table 2).

The metal–nitrogen bond stretching ν(MN) and the metal–nitrogen–oxygen angular deformation δ(MNO) modes of metal nitrosyls are observed with the relatively lower intensity below 800 cm^{-1}, as summarized by Durig and Wertz [31]. More recent data can be found in [32] and some complementary considerations are given in [33]. In brief, the analysis of isotopically edited Raman data, together with the normal coordinate analysis and DFT-computed frequencies, permitted the authors to assign the ν(NO) and ν(Fe–NO) stretching as well as δ (Fe–NO) bending modes in complexes containing triphenylphosphine along with some thiolate and/or selenate ligands. Correlation analysis of the ν(NO) and ν(Fe–NO) stretching frequencies in the dinitrosyl complexes of iron (DNICs) was performed. Analysis of the ratio of the ν(Fe–NO) to the ν(NO) stretching vibration energies was first carried out by Obayashi et al. [34] and Linder et al. [35, 36] for the ν(Fe–NO)/ν(NO) correlation in six-coordinate $\{FeNO\}^6$ systems. The authors discovered a direct correlation between the indicators of the bond strengths, ν(Fe–NO) and ν(NO) frequencies, and the Fe–N–O bond lengths. The electronic basis for this behavior was found to be centered around a high-energy molecular orbital which is σ-antibonding throughout the Fe–N–O triatomic unit. These findings were then developed by Dai and Ke [32]. A comparison of the ν(Fe–NO) to the ν(NO) stretching vibration energies allowed the authors to draw correlation plots characteristic for the d^6 and d^7 configurations of iron chelated in the five-coordinated (5C) and six-coordinated (6C) mononitrosyls of iron(II) and (III), respectively. The authors proved that in the nitrosyls of iron (II) or (III) the ratios of the ν(Fe–NO) to the ν(NO) vibration values are included in the defined ranges. It was also shown that among a group of these ratios for five- and six-coordinated complexes a good linear correlation can be found, which shows the effect of π* and σ* backbonding in mononitrosyls, depending on the degree of coordination and electron-donating properties of non-nitrosyl ligands. The correlation between the N–O and Fe–NO frequencies (and, hence, bond strengths) is informative on a character of changes in electronic structure of the FeNO moiety. Namely, the directly proportional change in the ν(Fe–NO) together with the ν(NO) indicates changes in σ bonding that is transatomic in character, whereas a change in the π backbond would lead to an inverse correlation. This is well illustrated by the effect of thiolate coordination to ferric heme nitrosyls: in summary, a thiolate ligand *trans* to NO lowers both the N–O and Fe–NO stretching frequencies, as compared to the ferric heme nitrosyls bound to imidazole as a sixth ligand. The axially coordinated thiolate ligand imposes a σ *trans* effect on the coordinated NO, which weakens both Fe–NO and N–O bonds and induces bending of the Fe–N–O unit [37]. The values of ν(Fe–NO)/ν(NO) for DNICs are located in a completely different range, which clearly indicates the different electronic nature of the iron ion in these complexes. With the use of the vibrational data available in the literature, Dai and Ke [32] demonstrated that the ν(Fe–NO)/ν(NO) correlation diagrams can be easily applied to determine the electron configurations and the effective charges on the central ion in metal nitrosyls.

3 Electron Paramagnetic Resonance Studies

Electron paramagnetic resonance (EPR) spectroscopy is widely used to study nitrosyl complexes of transition metals, in which coordination of one or more nitrosyl groups results in the presence of an unpaired electron in the coordination sphere. Unbound nitric oxide in frozen solutions gives rise to a signal with $g \approx 1.95$, detectable at very low temperatures (<20 K), and requires high concentrations of NO [38]. Paramagnetic complexes of iron with nitric oxide are more readily detectable and can be used to observe the molecular targets for NO in biological systems [39]. The DNICs of the type $[Fe(SR)_2(NO)_2]^-$ are low-spin ferrous-nitrosyl compounds ($S = 1/2$), broadly found in the biological material in the conditions associated with the presence of nitric oxide [40, 41]; these complexes give narrow and distinct EPR signals usually detectable at room temperature, with g values between 2.012 and 2.04 [42, 43]. Among the factors contributing to the strong EPR signals of paramagnetic metal nitrosylates are the large zero-field splitting (only the ground state is populated at lower temperatures), relatively small deviations from axial symmetry, small distributions in the E/D (zero-field splitting terms) ratio, and the near-coincidence of the g and hyperfine tensors. The nitrosyl groups in nitrosyl complexes may be described as NO^+ ($S = 0$), NO• ($S = 1/2$), NO^- ($S = 0$), NO^- ($S = 1$), and NO^{2-} ($S = 1/2$). The overall numbers of iron d electrons plus NO π^* electrons are $\{FeNO\}^{6-8}$ in most experimentally accessible model compounds. The EPR visible complexes are those comprising the $\{Fe–NO\}^7$ and $\{Fe–(NO)_2\}^9$ cores. Both $S = 3/2$ and $S = 1/2$ $\{Fe–NO\}^7$ complexes are found. Mononitrosyl $S = 1/2$ complexes with heme iron are well known and described, while the low-spin nonheme iron nitrosyls are scarce. The ground state $S = 1/2$ is characteristic for relatively strong ligand field systems. It is assumed that in view of the electronic structure the metal-nitrosyl core should be treated as a non-separable ensemble (see the chapter by Mingos and the chapter on the coordination issues by Lewandowska). This is well illustrated by the differences in conclusions that can be drawn according to the different spectroscopic methods: As discussed by Serres [44], the variations in the infrared NO stretching frequencies suggest that the redox chemistry of the $M(NO)_n$ moiety is ligand centered, while the Mössbauer isomer shifts indicate a significant contribution of the metal-associated orbitals. Serres et al. [44] performed a detailed experimental and theoretical study of the series $[Fe(cyclam\text{-}acetate)(NO)]^{2+,+,0}$ complexes (cyclam = 1,4,8,11-tetraazacyclotetradecane). An insight into the physical properties and the bonding was achieved by closely matching the experimental observables from EPR, UV–Vis, Mössbauer, and IR spectroscopies to the DFT calculations. The redox steps in the series $NO^+ \rightarrow NO \rightarrow NO^-$ ($S_{NO} = 0$) were found ligand centered, while the central ion was assigned in the low-spin Fe(II) state throughout. The main effect observed for the metal center was rationalized from the backbonding ability of the NO ligand, ranging from extremely strong (NO^+) to relatively weak (NO^-). The backbonding effect was found responsible for the pronounced changes of the Mössbauer parameters. On the other hand, the essentially

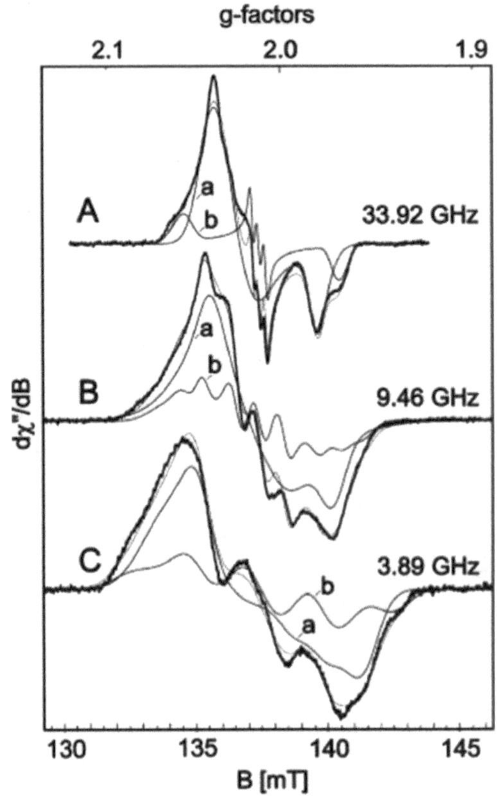

Fig. 4 EPR spectra of [Fe(NO)(cyclam-ac)](PF_6) in frozen butyronitrile solution recorded at different microwave frequencies: (*A*) 33.9202 GHz, 0.2 mW power, 1 mT/100 kHz modulation, temperature 40 K; (*B*) 9.4565 GHz, 2 mW power, 0.5 mT/100 kHz modulation, temperature 10 K; (*C*) 3.8835 GHz, 2 mW power, 0.8 mT/100 kHz modulation, temperature 13 K. Two forms, in a ratio of 70:30, were used in a simulation (*thin lines*). Reprinted from reference [44], with the permission of the American Chemical Society, copyright 2004

ligand-centered redox steps explain the trends in the vibrational data as well as the EPR results. The spectra derived by Serres (measured for three EPR frequencies) are given in Fig. 4, representing typical EPR signals for the d^7 low-spin iron nitrosyls.

The value of the g_x, g_y and g_z components of the EPR spectra can serve as a probe of the Fe–N–O covalency. Namely, the increasing covalency is accompanied by the general shift to higher g values, as explicitly illustrated by the comparison of the EPR data for Na^+NO adducts in zeolites via Cu^+NO complexes [45], to the Ru nitrosylates [46]. Also, the invariance of the EPR parameters along with the change of the ligand points to a higher degree of covalency. This is the case of the mentioned $\{RuNO\}^7$ nitrosylates as compared to the corresponding $\{FeNO\}^7$ complexes, the latter having more accessible valence tautomeric $Fe^{I}(NO^+)$ state, and possible interchanges between low-spin and high-spin states on the contrary to the dominant M^{II}(NO·) configuration for the Ru compounds. The small variation in g tensor splitting and hyperfine coupling for ruthenium nitrosyls suggests very similar electronic structures, described essentially by the Ru^{II}(NO·) formulation. On average, the g anisotropy is about twice as large for the ruthenium complexes as for the related Fe^{II}(NO) compounds [47]. This is an immediate consequence of the higher spin–orbit coupling constant for ruthenium [48]. In the case of iron

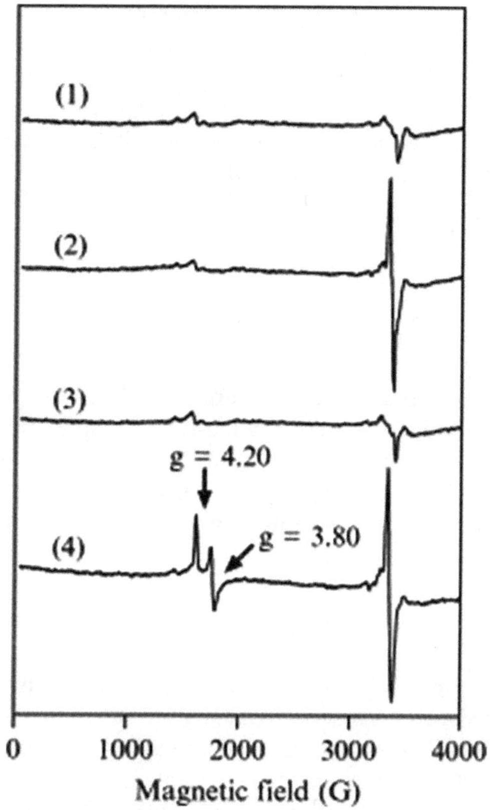

Fig. 5 In vivo EPR of NorR from *E. coli* cells. EPR spectra were recorded on intact *E. coli* BL21(DE3) cells expressing NorR. Cultures were given the following treatments: (1) none, (2) treatment with NO, (3) induced for protein expression with 50 μM isopropyl β-D-1-thiogalactopyranoside (IPTG), and (4) induced for protein expression with 50 μM IPTG and then treated with NO. Reprinted from reference [51], with the permission of Elsevier Science Ltd, copyright 2008

compounds, the 6C complexes approach the valence state of an $Fe^{II}(NO\cdot)$, whereas the 5C compounds have more noticeable $Fe^{I}(NO^{+})$ character [49, 50].

Nonheme ferrous centers in some metalloproteins react reversibly with NO forming nitrosyl complexes with $S = 3/2$ characterized by the g values of about 4.0 and 2.0 [51]. The EPR spectrum of the nitrosylated NorR (a bacterial NO-responsive transcription factor, the enhancer binding protein) is typical of a d^7 high-spin $Fe^{III}NO^-$, where the $S = 5/2$ iron is antiferromagnetically coupled to the NO^- (Fig. 5, [52]). This is confirmed by the X-ray, resonance Raman, MCD, Mössbauer spectroscopies, and DFT calculations. Similar structures were proposed for the classical complexes, [Fe(NO)(1-isopropyl-4,7-(4-*tert*-butyl-2mercaptobenzyl)-1,4, 7-triazacyclononane)], [53], Fe(EDTA)NO [54–56], the "brown-ring" compound, $Fe(H_2O)_5NO^{2+}$ [57], and for the Fe(*N*,*N*',*N*''-trimethyl-1,4,7-triazacyclononane) $(N_3)_2NO$ [54]. Interestingly, for the latter a spin equilibrium between the valence tautomers $S = 1/2$ and 3/2 in the solid state was observed.

Another EPR signal, at $g\perp = 1.999$ and $g\| = 1.927$, with resolved ^{14}N hyperfine splitting, is observed for an intermediate $Fe^{I}NO^{+}$ complex during the reduction of nitroprusside by thiols [58]. Reaction of nitroprusside with a thiol such as cysteine leads to a one-electron reduction of the complex and formation of

$[Fe(CN)_5NO]^{3-}$[59]. This compound is paramagnetic, with the unpaired electron centered largely on the NO group (EPR g values: $g_x = 1.9993$, $g_y = 1.9282$, and $g_z = 2.008$). Subsequent reactions involve internal electron transfer and the formation of an $Fe^{I}NO^{+}$ complex $[Fe(CN)_4NO]^{2-}$ ($g_x = 2.036, g_y = 2.0325, g_z = 2.0054$). Subsequently, this intermediate decomposes to a variety of products, including NO and, if an excess of thiol is present, it rearranges to mononucleated DNIC ($g = 2.035$) [58].

3.1 *Ferrous Heme-Nitrosyls*

NO, besides being a natural ligand in some porphyrinato complexes, is a probe for investigating Fe(II) heme-type complexes by means of paramagnetic electron spectroscopies: For instance, nitrosyl (NO)-ligated hemoglobin (HbNO) is considered to be a useful model to provide insight into the biochemistry of oxyhemoglobin (HbO_2) [60]. Ferrous heme nitrosyls are EPR active and exhibit low-spin ground states of $S = 1/2$ total spin. Binding of an N-donor ligand *trans* to NO has a profound effect on the Fe–N–O unit [61]. This strong influence can be clearly seen in the EPR spectra of nitrosylated hemoglobins and myoglobins [62–68], as well as of some model complexes [66, 68] as shown in Fig. 6. In the five-coordinate complex, the unpaired electron resides in the *a'* single occupied molecular orbital (SOMO), composed mainly of d_{z2} and π*(NO) [69]; a strong Fe–NO σ bond between π^*_h and d_{z2} leads to a large transfer of spin density from the NO ligand to Fe(II). The detailed DFT calculations based on the data from MCD, Mössbauer, NRVS and IR spectroscopies support the above assignment for Fe(TPP)NO, with a mixed electronic distribution, ca. 50% in Fe and NO for the 5C complexes [33, 49, 50, 70]. The six-coordinate Fe(TPP)(1-MeIm)NO shows a nitrosyl-centered SOMO (ca. 20% in Fe) and thus is described as $\{Fe^{II}NO\bullet\}$. The level of the described effect depends on the σ-donor strength of the axial N-donor ligand. The dependency of the resulting electron distribution on the ligand σ-donating ability of the sixth ligand has further biological implications, as described in detail in the chapter by Lehnert et al. As shown by the orientation-selected electron spin-echo envelope modulation (ESEEM) and 2D hyperfine sublevel correlation (HYSCORE) experiments, the variation of binding geometry in the NO-heme is controlled by the heme's protein surrounding and could provide an important contribution to the discussion on the physiological role of NO related to its interactions with protein metal centers [65].

Characteristic EPR spectra of 5C ferrous heme mononitrosyls exhibit average g values of about 2.10, 2.06, and 2.01 [71]. Single crystal EPR measurements on the model complex [Fe(OEP)(NO)] revealed that while the principal axis of the minimum g value, g_{min}, is closely aligned with the Fe–NO bond vector [72], perpendicular to the heme plane, the g_{max} (~2.1) vector is in plane with heme and perpendicular to the (bent) Fe–N–O plane, and g_{mid} (~2.06) is in the planes of Fe–N–O and heme (see Fig. 7). In addition, three strong hyperfine lines due to the presence of the ^{14}N nucleus of NO are typically resolved on g_{min}, giving rise to the

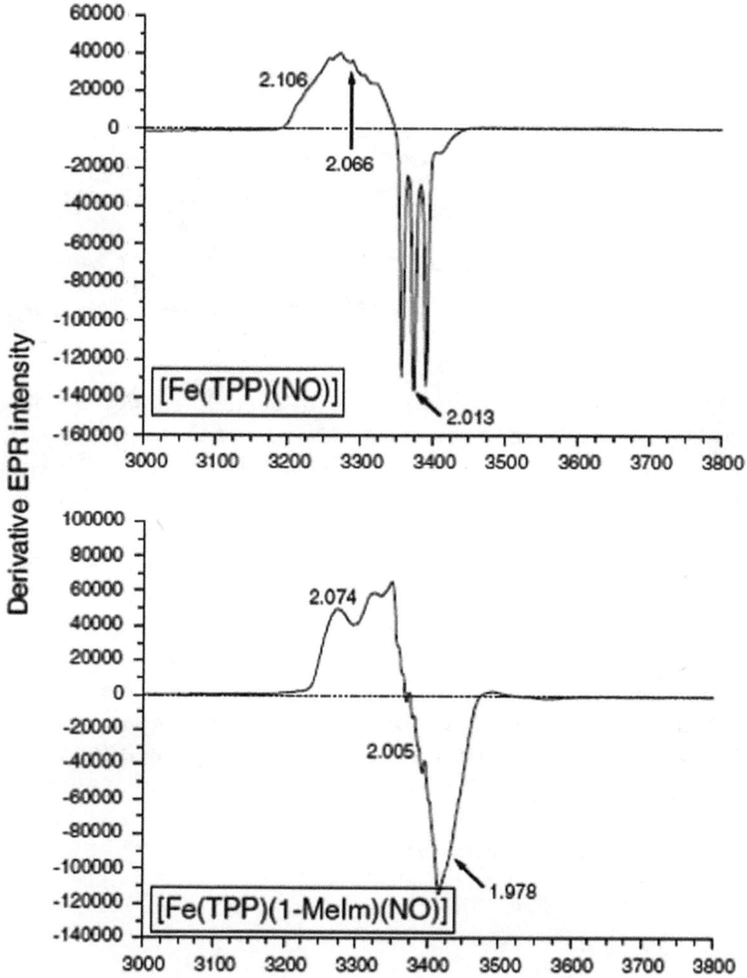

Fig. 6 *Up*: EPR spectrum of the 5C complex [Fe(TPP)(NO)]. *Down*: EPR spectrum of the 6C complex [Fe(TPP)(1-MeIm)(NO)]. Reprinted from [232], with the permission of Elsevier Science Ltd., copyright 2005

very characteristic EPR signal of 5C ferrous heme nitrosyls, as shown in Fig. 6, upper panel. The three-line hyperfine pattern originates from the nuclear spin $I = 1$ of ^{14}N. The ^{14}N hyperfine tensor components, available for a few model systems, are quite isotropic in the 5C case [71]. Typical g values for 6C complexes [71] are markedly smaller than those for their 5C analogues. This can be explained based on the results from magnetic circular dichroism (MCD) spectroscopy: MCD on 5C [Fe(TPP)(NO)] and its 6C analogue with 1-methylimidazole (1-MeIm). The sixth ligand [Fe(TPP)(1-MeIm)(NO)] induces a distinct decrease of the spin population on iron, as evident from the appearance of strong diamagnetic MCD signals for the

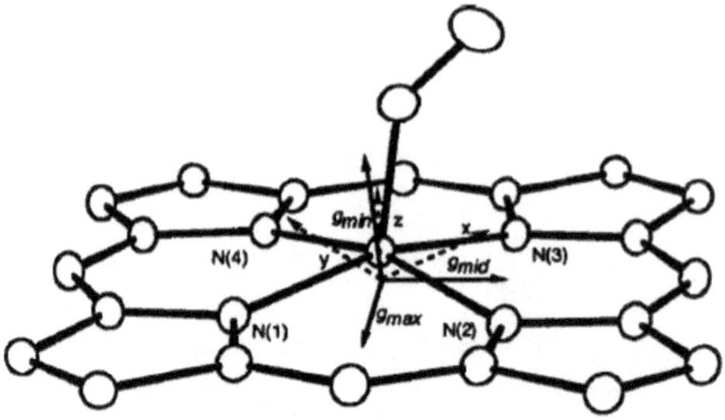

Fig. 7 ORTEP diagram of [Fe(OEP)(NO)] depicting the orientations of the principal axes of the *g* tensor with respect to the molecular coordinate system. The ethyl groups and hydrogen atoms are omitted for clarity. Reprinted from the reference [72], with the permission of the American Chemical Society, copyright 2000

6C complex shifting the spin density to the NO ligand [50]. The decrease in the spin density on Fe(II) generally reduces spin–orbit coupling matrix elements and, in this way, leads to the observed decrease in the g shifts [50]. For the ^{14}N-substituted complex, EPR data of the 6C complexes exhibit the well-resolved hyperfine nitrosyl ^{14}N lines on the medium *g* value, g_{mid}, as indicated in Fig. 6, lower panel, whereas the well-resolved hyperfine lines in the 5C case are observed on g_{min}. The three-line hyperfine pattern of ^{14}N(O) is further split by interaction with the ^{14}N of the proximal N-donor ligand in the 6C case, generating the nine-line hyperfine pattern (see Fig. 6, lower panel) [73]. This distinct difference in the appearance of the EPR spectra of the 5C and 6C complexes is due to a rotation of the *g* tensor in the 6C case, as evident from single-crystal EPR data on Mb–NO and HbNO [67, 74] and DFT calculations [75–77]. In the 6C complexes, the principal axis of g_{mid} is closest to the Fe–NO axis, whereas in the 5C case, the principal axis of g_{min} is aligned with the Fe–NO vector. Hence, the hyperfine lines are well resolved only for the g value which axis is aligned closest to the Fe–NO bond.

3.2 Dinitrosyl Complexes of Iron

As mentioned in the previous chapter by Lewandowska, the exact targeting of the nitrosyl compounds within the milieu of biological material is not clear and depends on the type of tissue, cellular compartment, the redox state of the cells, cellular regulation pathway response, as well as general physicochemical conditions. Simple EPR measurements are useful in the research intended to determine the biological targets of nitric oxide. The signal of the DNICs was observed in biological fluids treated with NO independent of their origin [1–3, 39, 60, 78–87]. The paramagnetic DNICs found in tissue are generally characterized by the EPR spectra with isotropic *g* values of around 2.03 [88, 89]. In the case of high-molecular (e.g., for protein-bound) DNICs an axial symmetry of the *g*-factor tensor is reported, e.g., $g_{\perp} = 2.04$, $g_{||} = 2.014$ [90, 91], and a rhombic signal is observed in the case of DNICs with low-molecular weight ligands (due to tumbling of small molecules) [92], if measured at room temperature. The *g* values vary slightly according to the nature of the ligand, yet the spectra for each particular complex are not distinctive in a manner that would permit treating each signal as the fingerprint of a given compound [92]. The intensity of the EPR signal of DNICs strongly depends on the pH of the solution. According to Vanin et al., the maximal intensity of the EPR signal for most of the water-soluble DNICs was observed at pH = 10–11 [93]. Decreasing the pH to 7.4 reduced the intensity of EPR signals 5–10-fold, and these complexes became yellowish due to the formation of the paramagnetic, dimeric forms of DNIC [94]. Low-molecular DNICs have been observed by EPR in extracts of rat liver following administration of certain chemical carcinogens [89] High-molecular DNICs signals were ascribed to complexes with ferritin (horse or human spleen) [95]. The characteristic 2.03 signal was also obtained in some biomimetic compounds that served as models of iron–sulfur clusters found in proteins ([2Fe-2S] or [4Fe-4S]) reacting with NO [96, 97]. An interesting research presenting the dependence of the EPR signal in a protein-bound nitrosyl iron complex from the kind of ligand taking part in complex formation can be found in the early work of Lee et al. [95]. The authors proved that the EPR signal of holoferritin is a combination of that coming from the complexes bound to either histidine or cysteine residues. Using *p*-(chloromercuri)-benzoate (PMB), the authors were able to modify the –SH groups so that after the nitrosation a complex with histidinyl residues was formed, while modification of histidines with diethyl pyrocarbonate (DEP) led to the formation of iron nitrosyls with cysteinyl residues of protein (see Fig. 8b). The two compounds gave slightly different EPR signals, centered around the same 2.03 value, but displaying different symmetries – rhombic for the histidinyl complex, and axial for the cysteinyl one. The symmetry of the EPR signal may also depend on the geometry of the ligand protein, as illustrated by the spectra of DNICs with various isoforms of GST (see Fig. 8c).

Nitric oxide fixation by non-transition metal complexes was reported by Ilyakina et al. [98]. Lead(II) and zinc(II) catecholato complexes readily react with nitric oxide to give the corresponding nitrosyl-containing mono-*o*-semiquinonates; the process of binding and transfer of NO was monitored by EPR spectroscopy [98].

4 Electronic Spectra

The UV–Vis spectrum of the ferrous iron 0.1 M acetate buffer at pH 5.0 and 23°C, upon saturation of the solution with NO, exhibits three characteristic absorption bands at 336 ($\varepsilon = 440$), 451 ($\varepsilon = 265$), and 585 nm ($\varepsilon = 85\ M^{-1}\ cm^{-1}$), when recorded at room temperature in water, anaerobically [57]. This was assigned to the formation of the mononitrosylated iron(II) complex. Yet, upon the bubbling with an inert gas, the spectrum was completely reversed to that characteristic for the aquated iron (II). This illustrates the great lability of the complex (Fig. 9).

Some calculations concerning the electronic structure of [Ru(salen)(NO)Cl] complex in CH_3CN were obtained by Bordini et al. [99], with the INDO/S

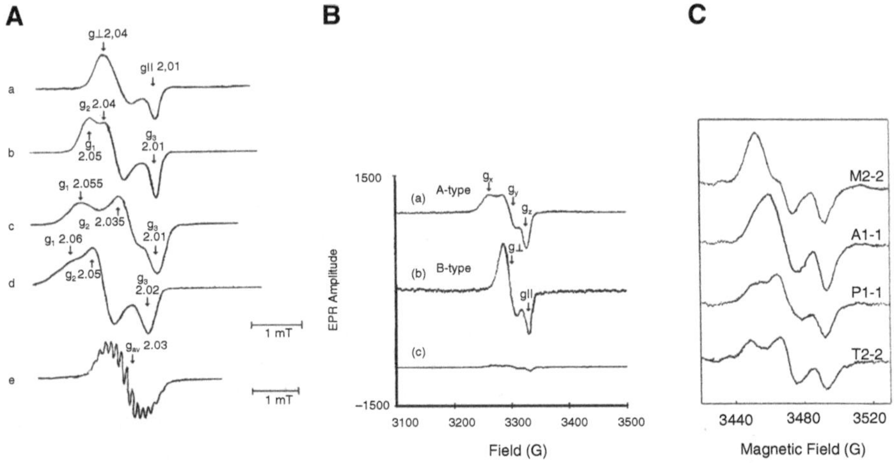

Fig. 8 Representative EPR spectra of nitrosyl-iron complexes. (**A**) Spectra of dinitrosyl-iron complexes at 77 K, X-band. The position of the g factors is indicated by arrows. Note that spectra *a–d* were recorded at four-fold higher magnetic field sweep than spectrum *e*, as indicated by the calibration bars (4 and 1 millitesla). *a*, DNIC-L-cysteine (25 μM); *b*, desalted DNIC–BSA (30 μM); *c*, BSA–DNIC (30 μM) from *N*-ethylmaleimide-treated BSA; *d*, DNIC (40 μM) in Griess reagent; *e*, DNIC-L-cysteine (200 μM) recorded at 295 K. (**B**) Spectra of nitrosylated holoferritin. Note that signals differ depending on the nitrosylated residue in protein (see text); A-type, complex with histidinyl residues (protein was modified with PMB in order to inactivate –SH groups); B-type, complex with cysteinyl residues (protein was modified with DEP in order to modify histidinyl groups). PMB- or DEP-modified proteins were flushed with NO for 5 min under anaerobic conditions at pH 7. (**C**) EPR spectra of different GST isoforms after reacting with the dinitrosyl diglutathionyl-iron complex. The Tyr residue conserved in the A1-1, M2-2, and P1-1 isoforms [233] is in the appropriate position to act as the fourth ligand of the iron atom (together with the GSH sulfur atom and the two NO molecules), probably displacing the second GSH. Reprinted from the reference [234], with the permission of the American Society for Biochemistry and Molecular Biology, copyright 2003

(the Intermediate Neglect Of Differential Overlap, Screened Approximation; semi-empirical) method of Zerner and co-workers [100, 101]. The low-energy region of the calculated spectrum is dominated by a sequence of very weak ligand (π–π*) transitions mixed with very little MLCT Ru to π*(salen) and Ru to π*(NO). The low-lying transitions calculated at 495, 437, 418, 383, and 379 nm corresponded to the very weak, broad band observed between 500 and 400 nm. The absorption observed at about 370 nm was found to derive from the ligand LMCT mixed excitations calculated at 354 and 346 nm. The broad absorption centered at 380 nm was assigned to the ligand excitations with some minor contribution from Ru d–d* excitations. The low-intensity (15–50 L mol^{-1} cm^{-1}) charge-transfer band Ru to π*(NO), expected between 400 and 450 nm [102] was calculated to be mixed with more intense transitions [103, 104]. Schreiner et al. [105] discussed the UV–Vis bands for the *trans*-$[Ru(NH_3)_4(NO)L]^{2+}$-type complexes. A broad band of a very low intensity at ca. 420–480 nm is characteristic of the interconfigurational spin-forbidden singlet–triplet transitions in 4d and 5d compounds and

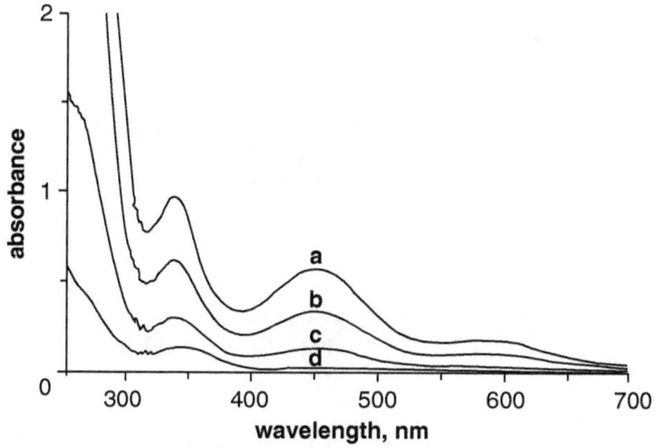

Fig. 9 Absorption spectral changes recorded for the reaction of 3 × 10^{-3} M $[Fe^{II}(H_2O)_6]^{2+}$ with NO. Experimental conditions: 0.2 M acetate buffer, pH = 5.0, 25°C; (*a*) $[Fe^{II}(H_2O)_6]^{2+}$ solution saturated with NO; (*b*) a + Ar brief (10 s); (*c*) b + Ar brief (10 s); (*d*) c + Ar brief (10 s). Reprinted from the reference [57], with the permission of the American Chemical Society, copyright 2002

also t_{2g}-+ π^*_{NO} transfers (see [102, 105] for discussion). The second band of medium intensity at 330 nm derives mostly from the spin-allowed d–d transitions. For some *trans*-ligands L, a third band, observed at above 300 nm, is very intense ($\varepsilon > 10^3$) and it is assigned to an electric dipole-allowed charge-transfer band pπL → ($x^2 - y^2$, z^2) Ru or πL → π*L. As expected on the basis of the spectrochemical series the d–d bands for *trans*-$[Ru(NH_3)_4L(NO)]^{3+}$-type complexes of nicotinamide, pyridine, imidazole, and L-histidine occur at higher energies than those for $[Ru(NH_3)_5(NO)]^{3+}$, which spin-allowed d–d transition is observed at 304 nm [105]. For the low-energy band, observed for all the complexes, an alternative interpretation to the mentioned Schreiner assignment was proposed by Borges et al. [102]. $^1A_1 \rightarrow {}^3\Gamma[{}^3\Gamma_1, {}^3\Gamma_2]$, π*(NO) could be considered, in which this absorption is assumed to be charge transfer in character. This is consistent with the red shift of this band along with the increase (toward less negative) of the formal reduction potential for the coordinated nitric oxide (for discussion, see [102, 106]).

The UV–Vis experimental spectra of the thiolate DNICs can be measured only in the presence of a large excess of the auxiliary ligand. Otherwise the monomer readily undergoes a reversible conversion to the RRE-type dinucleated form, $[Fe_2(SR)_2(NO)_4]$, displaying a much stronger absorption [79, 107–109]. A typical spectrum for a thiolate DNIC consists of a strong band at around 400 nm and two moderate bands at ca. 600 and 770 nm (see Table 3) [110]. The calculations revealed that all the bands are of the CT character. The lower energy bands are mostly of the LLCT ($S\pi \rightarrow \pi^*$ NO) and LMCT type ($S\pi \rightarrow$ d); some higher energy transitions are mostly the LMCT (with the prevailing d → π* NO) type. The MLCT transitions of the d → π* NO type occur in the calculated spectra at higher energy (below 270 nm) and are mixed with the Rydberg transitions to the 4p orbitals of iron. As shown by calculation for the $[(CH_3S)_2Fe(NO)_2]^-$ the transition at

Table 3 Characteristic UV–Vis features for the representatives of nitrosyl classes

Type of the complex	Complex formula	UV–Vis band(nm)/ε (M^{-1} cm^{-1})[a]	Reference
Mononitrosyl	$[Fe(H_2O)_5NO]^{2+}$	336/440, 451/265, 585/85 (water)	[57]
	$[Ru(NH_3)_5NO]^{3+}$	300/67.2 M^{-1} cm^{-1}, 460/ 14.4 M^{-1} cm^{-1}	[242]
	$[(NO)Fe(SC_9H_6N)_2]$	432/3,571, 512/2,967, 800/428 (THF)	[243]
	$[Fe(NO)(SPh)_3]^-$	500[b] (THF)	[244]
	$[Fe(NO)(SEt)_3]^-$	459[b] (THF)	[244]
	[Et4N][Fe(S*t*Bu)$_3$(NO)]$^-$	370, 475 (THF)	[245]
Dinitrosyl	$[Fe(CS)_2(NO)_2]^-$	392/3,580, 603/299, 772/312 (water)	[108]
	$[(PhS)_2Fe(NO)_2]^-$	479, 798 (THF)	[244]
	$[(EtS)_2Fe(NO)_2]^-$	436, 802 (THF)	[244]
	$[PPN][(S(CH_2)_3S)Fe(NO)_2]$	430, 578, 807 (THF)	[246]
Dinitrosyl, dimer	$[Fe_2(CS)_2(NO)_4]$	305, 362, 440 (sh), 755 (water)	[108]
Pseudo-heme	[Fe(L)(NO)]	303/17,000, 403/4,000, 613/600	[115]
Heme, 6C	[Fe(OEP)(SR)(NO)]	430, 536, 567	[49]
Heme, 5C	[Fe(T*o*-F$_2$PP)(NO)]	403, 475, 550	[50]

OEP octaethylporphyrin, *To-F$_2$PP* *tetrakis*-5,10,15,20-(*o*-difluorophenyl)porphyrin, *PPN* bis triphenylphosphineiminium cation, *CS* cysteamine, *L* Schiff-base tetradentate macrocyclic ligands
[a]Extinction coefficients given where available
[b]Spectra registered in the range 390–1,200 nm

ca. 770 nm is mainly of the *S*π → π* NO and *S*π → d. The transition at about 600 nm is composed of *S*π → π* NO and π* NO → d excitations, and the transitions at about 390 nm are predominantly of the π* NO → d character. Calculations indicate a transition with a large oscillator strength at 513.8 nm which is formed by Sπ → π* NO excitation and the two groups of transitions with oscillator strengths larger than 0.01, lying between 303 and 290 nm and the second one in between 280 and 270 nm. They are mostly of the π* NO → d and partly d → π* NO character. The transitions at wavelengths below 260 nm are of Rydberg and d → π* NO type [110]. The electronic transition spectra give an additional evidence that the $\{Fe(NO)_2\}^9$ complexes show a structure of trigonal bipyramid with a missing ligand. According to Costanzo [108] the molar extinction coefficient values of d–d transfer absorption bands for the DNICs are much higher than for the octahedral complexes and typical for the tetrahedral ones. Yellowish dimeric DNICs give two characteristic bands at ca. 310 and 360 nm in water in neutral pH [94]. For the dinuclear nitrosyl complexes of the type $[Fe_2(SR)_2(NO)_4]$, the transitions at the long wavelengths may be generally described as π_{NO}^* → d (LMCT), the most intense transitions at the short wavelengths (beyond 250 nm) as d → π_{NO}^* (MLCT). In the middle part of the spectra the both types of transitions are present, but the π_{NO}^* → d transitions still prevail. There are also some d → d transitions, but their intensities are relatively weak. A participation of sulfur in the transitions throughout the whole spectrum is evident [111]. The differences in the spectra of mono- and dinucleated form of $\{Fe(NO)_2\}^9$ DNIC are illustrated in Fig. 10.

UV–Vis spectroscopy is widely used in kinetic studies on the interaction of metalloporphyrins with NO. The electronic absorption spectra of metalloporphyrins are dominated by the strong π–π* intraligand transitions, which are, however, affected by the nature of the metal and the axial ligands coordinated to the metal center (Fig. 11) [112–114].

The new approach in the studies on the electronic properties of the 5-coordinate heme-type nitrosyl complexes of iron has been recently proposed by Sulok et al. [115]: These authors proposed 5C low-spin ferrous iron nitrosyls containing Schiff-base-type tetradentate macrocyclic ligands as the model complexes. Structural and vibrational data confirm the usefulness of these species as the spectroscopic models for 5C ferrous heme nitrosyls, and the advantage is that they give less obscure electronic spectra due to the absence of the very intense π → π* transitions of the heme macrocycle. These complexes were used to locate the d → π*(NO) CT transitions, which would then provide evidence where such transitions could be expected for the heme nitrosyls. Analysis of the electronic spectral features of the mentioned model systems indicated a somewhat weaker Fe–NO σ bond and a stronger Fe–NO π backbond, than in their heme-containing counterparts. Using the TD-DFT calculations employing the B3LYP/TZVP functional and basis set combination, the authors performed a detailed analysis, revealing four principle types of electronic transitions in the observed optical spectra: (1) ligand field (d → d), (2) ligand-to-metal (π → d) and (3) metal-to-ligand (d → π*) charge transfers (CT), and (4) inner-ligand (π → π*)-type transitions, as shown in the diagram B, in Fig. 12.

Of the special interest here are the transitions in the UV–Vis region that involve the NO ligand, which would be of the d → π* (NO) character. The TD-DFT calculations predicted that the absorption spectra would be dominated by two main bands in the Vis region between 20,000 and 30,000 cm^{-1} (500–330 nm, features I and II in Fig. 12c) and one band in the UV region above 30,000 cm^{-1} (i.e. below 330 nm, feature III) as evident from Fig. 12. The low-energy region (I) is dominated by the three types of transitions, viz., ligand field (d → d), π(lig) → d LMCT, and inner-ligand π(lig) → π*(lig) excitations. The minor contribution of the d → π*(NO) MLCT transitions is delocalized over several excited states and therefore of negligible absorption intensities. The second region (II) in the 25,000–30,000 cm^{-1} (400–330 nm) range is dominated by the d → d ligand field, π(lig) → d LMCT, d → π*(lig) MLCT, and inner-ligand π(lig) → π*(lig) transitions. Again, there are possible weak d → π*(NO) contributions to the excited states in this region, but as in the region I, there are no features that have a predominant d → π*(NO) character. The last region, III, is mostly dominated by the π(lig) → π*(lig) transitions. The fact that the regions I–III contain significant contributions of π(lig) → d LMCT, d → π*(lig) MLCT and in particular, inner-ligand π(lig) → π*(lig) transitions explains why the extinction coefficients in these

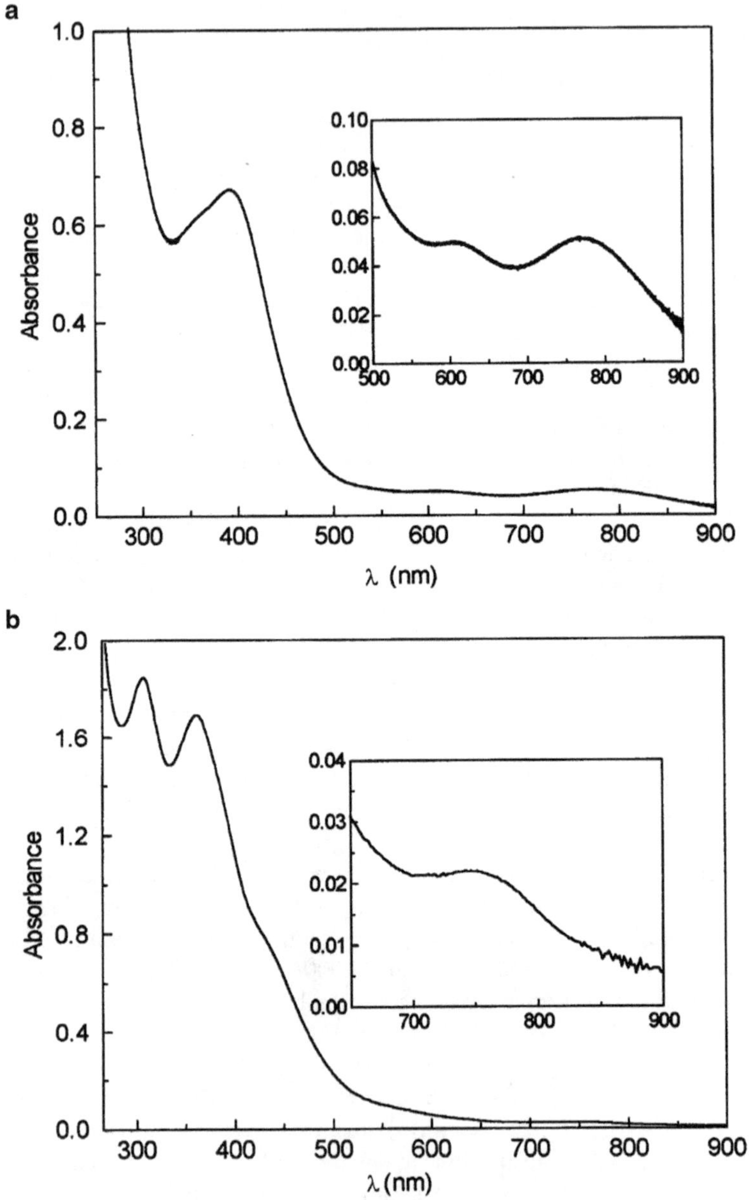

Fig. 10 Comparison between the UV–Vis spectra of LMW mononucleated complex ($[Fe(CS)_2(NO)_2]^-$) (**a**) and dinucleated ($[Fe_2(CS)_2(NO)_4]$) (**b**) form of thiolate DNIC of the $\{Fe(NO)_2\}^9$ type; CS = cysteamine hydrochloride; recorded in HEPES buffer (0.1 M, pH 7.8). Reprinted from the reference [108], with permission of Elsevier, copyright 2001

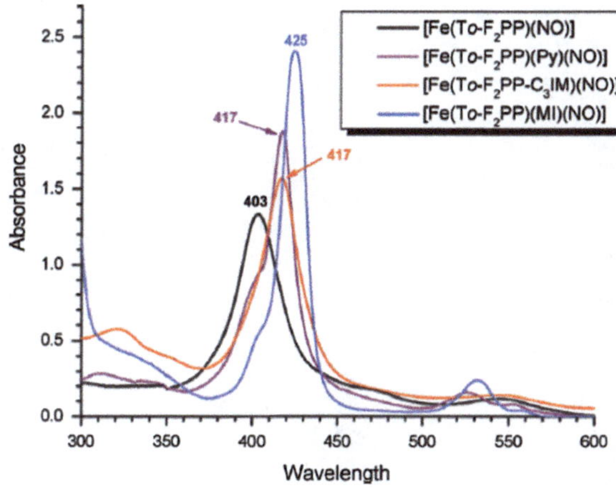

Fig. 11 Electronic absorption spectrum of [Fe(To-F_2PP-C_3Im)(NO)] (2, *red*) in comparison to 5C [Fe(To-F_2PP)(NO)] (*black*) and the 6C complexes [Fe(To-F_2PP)(MeIm)(NO)] (blue) and [Fe(To-F_2PP)(Py)(NO)] (*purple*, Py = free pyridine). Spectra were recorded in CH_2Cl_2 or toluene solution at room temperature. Soret B band visible at 400 nm and Q band at 550 nm (ligand $\pi \rightarrow \pi^*$ transitions). Reprinted from the reference [139], with permission of the American Chemical Society, copyright 2009

regions show such a dramatic increase when comparing the spectra of the above complexes containing ligands of increasing number of aromatic rings in the molecule: the presence of the aromatic π systems greatly enhances the orbital overlap between the ground and the corresponding excited states, which then increases the extinction coefficients of the corresponding transitions in complexes, along with the addition of aromatic moieties to the ligand structure.

5 Application of Mössbauer Spectroscopy in the Investigations of Some Transition Metal Nitrosyl Complexes

Mössbauer (MB) spectroscopy is an especially valuable experimental tool in bioinorganic chemistry because it has an extremely fine energy resolution and can detect even subtle changes in the nuclear environment of the relevant atoms. MB spectroscopy can provide information on the spin and oxidation state of the MB-active atom basing on chemical analysis of environmental influences on nuclear interactions. The method relies on the recoilless resonant absorption of γ-radiation by the MB-active nuclei from a source emitter isotope [116]. Mössbauer spectroscopy uses the fact that the excitation energy of the nucleus (caused by the absorption of gamma rays) is disturbed by its chemical environment. The disturbance is quantified by measuring the Doppler shift necessary to cause the resonant absorption between the sample and the reference. For some isotopes the nuclear transition energy can be determined with resolution of the order of 10^{12}. Generally,

Fig. 12 Schematic structure of the model 5-coordinate heme-mimicking nitrosyl complex (**a**), its orbital diagram showing the predicted electronic transitions (**b**), and its experimental and calculated electronic spectrum (**c**). R = $-C_2H_2-$ or *o*-phenolate; MO labels are derived from a B3LYP/TZVP calculation. Reprinted from the reference [115], with the permission of Elsevier Science Ltd., copyright 2012

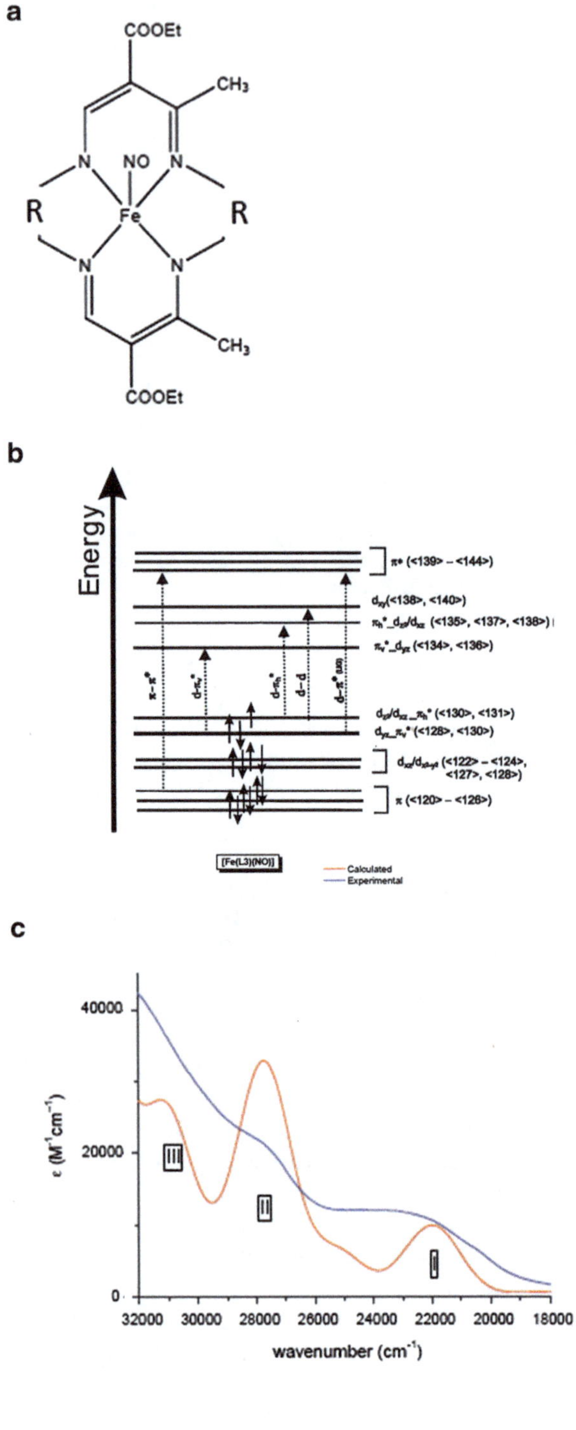

there are three examined effects providing a link to the electronic structure, i.e., isomer shift (or chemical shift), quadrupole splitting, and hyperfine splitting (or Zeeman splitting). Of the three, isomeric shift effect is a phenomenon encountered only in this type of spectroscopy [117]. The isomer shifts (δ) provide information on the metal–ligand bond covalency and on the spin and oxidation states of the MB-active atom. The magnitude of the isomer shift depends on the electron shielding covalency effects and the changes in bond lengths [118]. The Mössbauer isomer shifts are proportional to the *s*-electron density at the nucleus, which in turn is influenced by the d-electrons shielding, and, consequently, the isomer shift is a sensitive measure of the aforementioned properties of an MB-active atom.

As explicitly formulated by Krebs [119], the following simple rules are often useful for interpretation of the Mössbauer isomer shifts: (1) the isomer shifts of the low-spin complexes are lower than those of the high-spin complexes, (2) the isomer shift increases with the increasing coordination number (however, the associated lengthening of bonds must be also considered), (3) the isomer shift increases with the decreasing oxidation state, and (4) the isomer shifts of metals coordinated by soft (sulfur) ligands are less than those coordinated by hard (N, O) ligands. Broadly speaking, the higher ligand field strength, the greater δ observed. In this regard, the important considerations are the σ-donor and π-acceptor characteristics of ligands.

The quadrupole splitting (QS) reflects the interaction between the nuclear energy levels and the surrounding electric field gradient. Nuclei in states with nonspherical charge distributions, i.e., all those with the angular quantum number (I) greater than 1/2, produce an asymmetrical electric field which splits the nuclear energy levels. The QS parameters are very sensitive to the changes in the population and geometry of MB-active-atom derived molecular orbitals. The magnetic splitting (hyperfine splitting) is a result of the interaction between the nucleus and any surrounding magnetic field. A nucleus with the spin I splits into 2I + 1 sub-energy levels in the presence of magnetic field, giving a hyperfine structure of the MB spectrum. In atoms, a hyperfine structure occurs due to the energy of the nuclear magnetic dipole moment in the magnetic field generated by the electrons and the energy of the nuclear electric quadrupole moment in the electric field gradient due to the distribution of charge within the atom. Being dominated by the two described effects, the molecular hyperfine structure also includes the energy associated with the interaction between the magnetic moments of different magnetic nuclei in a molecule, as well as between the nuclear magnetic moments and the magnetic field generated by the rotation of the molecule. The most popular MB research relates to compounds of iron, ruthenium, tin, antimony, tellurium, xenon, tantalum, tungsten, iodine, iridium, gold, and europium. ^{57}Fe is by far the most common element studied using this technique, although ^{129}I, ^{119}Sn, and ^{121}Sb are also frequently studied; ^{61}Ni and ^{67}Zn are two other biologically relevant MB-active isotopes. MB studies on metal nitrosyls include those of iron [44, 120–131], ruthenium [132, 133], and some early data on iridium [134]. Of particular interest from the biological point of view are studies concerning cobalt [135, 136], due to the crucial role Co plays in many biological pathways, as a protein cofactor.

Fig. 13 Mössbauer spectrum of sodium nitroprusside, registered in the absence of magnetic field applied. Reprinted from the reference [235], with the permission of Elsevier, copyright 2009

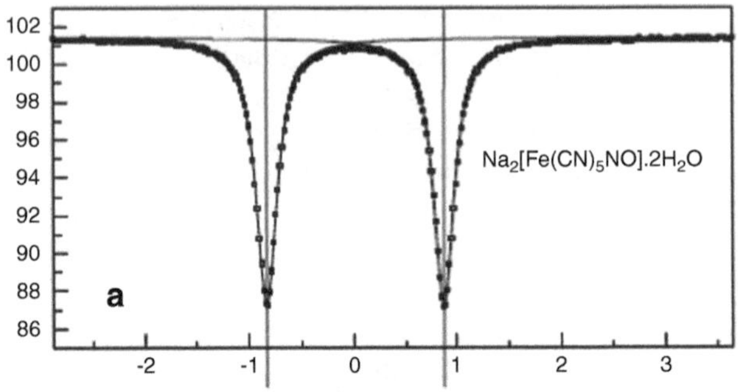

In case of ^{57}Fe, it is the resonant absorption of 14.4 keV γ-rays emitted from a radioactive ^{57}Co source that is measured. A typical sample for Mössbauer spectroscopy requires circa 10^{16}–10^{17} of MB-active atoms. Because the natural abundance of ^{57}Fe is only 2.2%, it is almost always necessary to selectively enrich the sample with ^{57}Fe [119]. The simplest type of the ^{57}Fe-Mössbauer spectrum is commonly referred to as a "quadrupole doublet" (Fig. 13 illustrates a simple spectrum for sodium nitroprusside, often used in MB measurements as a reference). It consists of two peaks associated with two spectroscopic transitions coming from the interaction of the quadrupole moment of the $I = 3/2$ excited state and the electric field gradient (EFG), which results in the splitting of the excited nuclear state into the $M_I = \pm 1/2$ and $M_I = \pm 3/2$ levels in the absence of a magnetic field at the ^{57}Fe nucleus. The transitions obey the selection rule of the Mössbauer spectroscopy, $\Delta M_I = 0, \pm 1$. The energy difference between the two doublets is called the quadrupole splitting parameter, ΔE_Q, and corresponds to the separation of the two transitions of the quadrupole doublet. The average position of the two lines is known as the isomer shift δ.

It must be noted, that in the MB spectroscopy all the MB-active nuclei in the sample are probed simultaneously. Challenges in using this technique include the requirement of ^{57}Fe (or other) enrichment and the need to deconvolute the complicated spectra when several active nuclei are present [121]. In conjunction with the rapid freeze-quench (FQ) techniques, MB spectroscopy is a particularly valuable tool in bioinorganic chemistry because the Fe environment can be monitored during the course of a biochemical reaction to give an insight into the nature of the short-lived intermediates and the reaction mechanisms they participate in. This permits the observation of reactions occurring on the millisecond-time scale [137]. Another advantage provided by the FQ method is the resolution of a problem with long-drawn character of the MB experiments.

The MB spectroscopy is applicable in investigating the structures of metal nitrosyls, as MNO electronic structure is extremely sensitive to its coordination environment. It is particularly effective when the spectroscopic measurements are combined with the results of DFT calculations. The application of the MB parameters in the isomer shift calibration allows one to precisely estimate the Fe and NO spin populations (see, e.g., [121]). Recently DFT has become an increasingly popular tool to calculate structures and Mössbauer properties [118, 138]. MB combined with DFT is presently applied to study metalloenzymes, mostly those containing iron (see, e.g., [119, 121, 139–141]). Combining these two methods to study enzymes offers an atomic-level understanding of their mechanisms of action. The interactive comparison of calculations with the experiment allows a rationalization of the existing data and calibration of the Mössbauer parameters for a variety of the commonly used DFT functionals. Such an approach has been successfully applied to a test set of nonheme iron nitrosyls including mono- and poly-nucleated low- ($S = 1/2$) and high-spin ($S = 3/2$) $\{FeNO\}^7$, $\{Fe(NO)_2\}^9$ Fe centers. The ligand field and geometry of the nitrosyl complexes strongly influence their MB spectra. Some MB data for the characteristic Fe and Ru nitrosyl compounds along with the oxidation states of the {FeNO} cores are given in Table 4.

The zero-field Mössbauer spectrum of the ^{57}Fe-enriched nitrosylated pentaaquairon $[Fe(H_2O)_5(NO)]^{2+}$ solution measured at 80 K shows (besides the characteristic peak for non-reacted hexaaquairon(II) at $\delta = 1.39$ mm s^{-1}, and $\Delta E_Q = 3.33$ mm s^{-1}) a peak at about 1.8 mm s^{-1}, $\delta = 0.76$ mm s^{-1}, $\Delta E_Q =$ 2.1 mm s^{-1}, according to [57] or 2.3 according to [142]. It was inferred that since the Mössbauer and EPR parameters of the $[Fe(H_2O)_5(NO)]^{2+}$ complex closely resembled those of the $\{FeNO\}^7$ units in any of the other well-characterized nitrosyl complexes, its electronic structure is best described as a high-spin ($S = 5/2$) Fe^{3+} antiferromagnetically coupled to the NO^- ($S = 1$) yielding the observed spin quartet ground state ($S = 3/2$). The calculated Mulliken Fe spin population in the $S = 3/2$ $[Fe(H_2O)_5(NO)]^{2+}$ complex was found between 3.4 and 3.8 [57]. Yet, in 2004 this view was revised by Cheng et al. [143]. On the basis of DFT calculations these authors postulated that the spin-quartet ground state of the $[Fe(H_2O)_5(NO)]^{2+}$ is best described rather as the Fe^{II} ($S = 2$) antiferromagnetically coupled to the NO ($S = {}^1/_2$), yielding the $[Fe^{II}(H_2O)_5(NO)]^{2+}$, and not the $[Fe^{III}(H_2O)_5(NO^-)]^{2+}$ or the conventional textbook $[Fe^{I}(H_2O)_5(NO^+)]^{2+}$ assignment. The Fe–N–O bond in the optimized structure of $[Fe(H_2O)_5(NO)]^{2+}$ was found linear.

In the case of the ruthenium(II) compounds, the isomer shifts ($\delta \approx -0.20$ mm s^{-1}) are nearly equal for the most known nitrosyls and consistent with a +2 charge of the central ion. However, the much more negative δ for $K_2[Ru(NO)Cl_5]$ (-0.43 mm s^{-1}) implies a much stronger ligand field in this compound than in the other ruthenium nitrosylates [132]. The two distinct MB lines for the $[Ru(NO)Cl(py)_4](PF_6)_2$, $[Ru(NO)\text{-}Br(py)_4](PF_6)_2$, and $[Ru(NO)(NH_3)_5]Cl_3$ ($\Delta E_Q \approx 0.40$ mm s^{-1}) illustrate the nonsymmetric environment of the central ion [144]. The back-donation to the π* orbitals of the NO ligand is very strong but approximately the same for each of the complexes studied, as can be seen by the relatively small variations in ν(NO). Thus the variations seen in the δ values are a result of the contribution to the ligand field from the remaining ligands. The ΔE_Q values imply an asymmetrical electronic environments around the Ru ion in the complexes corresponding to less electron density in the orbitals oriented toward the axial ligands (d_{z2}, d_{xz}, and d_{yz}) relative to those oriented in the *xy* plane. Such an orientation of the electric field tensor implies that the Ru–NO bonding interactions (σ and π) should weaken with the increasing values of ΔE_Q, and indeed, with the increasing quadrupole splitting, a trend toward higher ν(NO) values is observed [132].

Table 4 Isomeric shifts for $Fe(NO)^n$ motifs are highly dependent on the ligand field and geometry of the complex. Experimental isomer shifts obtained at higher temperatures are corrected to the value expected at 4.2 K by taking into account the second-order Doppler shift, unless otherwise indicated

Complex	Iron oxidation state	*T*-correlation δ ($mm\ s^{-1}$)	ΔE_Q ($mm\ s^{-1}$)	Reference
(a) *Nonheme complexes*				
$[Fe(H_2O)_5(NO)]^{2+}$	$\{FeNO\}^7$	0.790	2.10	[57]
$[Fe_4(NO)_4\ (\mu 3\text{-}S)_4]^-$	$2\times\{FeNO\}^{7,5}$	0.270	0.94	[247]
$[Fe_4(NO)_4\ (\mu 3\text{-}S)_4]$	$4\times\{FeNO\}^7$	0.180	1.47	[248, 249]
$[Fe(NO)_2\{Fe(NO)(N(CH_2CH_2S)_3)\}$-S, S′]	$\{FeNO\}^7$	0.210	1.04	[250]
$[Fe(NO)_2\{Fe(NO)(N(CH_2CH_2S)_3)\}$-S, S′]	$\{Fe(NO)_2\}^9$	0.400	1.04	[250]
$[Fe(SC_2H_3N_3)(SC_2H_2N_3)(NO)_2]$	$\{Fe(NO)_2\}^9$	0.310	1.12	[251, 252]
$[Fe(SPh)_2\ (NO)_2]^{1-}$	$\{Fe(NO)_2\}^9$	0.182	0.69	[253]
$[Fe(NO)(dtci\text{-}Pr_2)_2]$	$\{FeNO\}^7$	0.350	0.89	[254, 255]
$[Fe_2(NO)_2\ (EtHPTB)(O_2CPh)]^{2+}$	$2\times\{FeNO\}^7$	0.670	1.44	[256]
(b) *Heme-type complexes*				
$[Fe(TpivPP)(NO_2)(NO)]$-perpendicular form	$\{FeNO\}^7$	0.22 (200 K)	1.78	[146]
$[Fe(TpivPP)(NO_2)(NO)]$-parallel form	$\{FeNO\}^7$	0.35	1.2	[146]
Siroheme(NO)	$\{FeNO\}^7$	0.32	1.54	[257]
[Fe(TPP)-(NO)]	$\{FeNO\}^7$	0.35	1.24	[146]
Heme–NO (cytochrome cd_1)	$\{FeNO\}^7$	0.34	0.8	[149]
(c) *Ruthenium nitrosyls*				
$K_2[Ru(NO)(NO_2)_4\ (OH)]$	$\{Ru(NO)\}^6$	−0.22	<0.07	[132]
$K_2[Ru(NO)Cl_5]$	$\{Ru(NO)\}^6$	−0.43	0.11	[132]
$[Ru(NO)(NH_3)_5]Cl_3$	$\{Ru(NO)\}^6$	−0.2	0.41	[132]
$[Ru(NO)Cl(py)_4](PF_6)_2$	$\{Ru(NO)\}^6$	−0.2	0.37	[132]
$[Ru(NO)Br(py)_4](PF_6)_2$	$\{Ru(NO)\}^6$	−0.2	0.37	[132]

TpivPP α,α,α,α-*tetrakis*(*o*-pivalamidophenyl)-porphinato, *N-EtHPTB* *N,N,N'N'-tetrakis*(2-benzimidazolylmethyl)-2-hydroxy-1,3-diaminopropane anion, *TPP* tetraphenylporphyrin, *dtci-Pr*$_2$, di-isopropyl dithiocarbamate, *Ph* phenyl

The Mössbauer isomer shifts for 5C and 6C iron porphyrins usually range from 0.02 to 0.35 mm s^{-1} (4.2 K) [24, 145–147](Fig. 14, Table 4 section b). MB spectra can provide information on the electronic changes too subtle to be resolved with other structural methods alone [146]. For instance, Mössbauer investigations show that the iron centers in the two different crystalline forms of the complex $[Fe(TpivPP)(NO_2)(NO)]^-$ have different electronic properties (see Table 4). The two forms had similar isomer shifts but significantly different quadrupole splitting, implicating the differences in the symmetry of the charge distribution (see discussion in [146]). On the other hand, as demonstrated by Ellison et al. [148], the strong similarity between the MB parameters of the highly nonplanar porphyrin cores versus those of

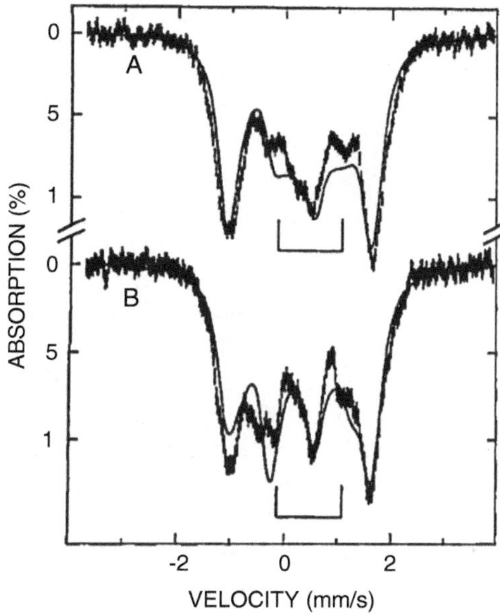

Fig. 14 Mössbauer spectra of the d_1 heme-NO complex in *T.* denitrificans cytochrome cd_1 in the presence of nitrite at pH 7.6. The data were recorded at 4.2 K with a magnetic field of 50 mT applied parallel (*A*) or perpendicular (*B*) to the γ-rays. The brackets indicate the positions of the quadrupole doublets of the reduced c heme which had been removed from the raw data. The *solid lines* are theoretical simulations. Reprinted from the reference [149], with the permission of John Wiley and Sons, copyright 1987

the $\{FeNO\}^6$ porphyrin complexes with planar structures and linear FeNO groups is an evidence that the heme core distortions do not lead to significant changes in the electronic structure at iron nor in the mode of the Fe–NO bending. Liu et al. [149] have analyzed the Mössbauer spectra of the ferrous d_1 heme-NO complex using a spin-Hamiltonian formalism, according to the early model for the electronic structure of hemoglobin-NO, proposed by Lang [150, 151]. The magnetic hyperfine coupling tensor was found to be consistent with the unpaired electron residing on a σ orbital. [149]. The detailed insight into the dependence of the isomer shift on the chemical nature of the iron site is presented in computational studies [118, 121, 152, 153].

6 Magnetic Circular Dichroism in the Study of Metalloproteins

A hybrid technique used more and more often as a probe of the electronic and geometric structures of the metal centers in metalloproteins is magnetic circular dichroism (MCD). This technique involves measurement of the CD spectra in a longitudinal magnetic field, the direction of which is parallel to the wave vector of circularly polarized light. MCD optical transitions in molecular species arise if (1) degenerate electronic states are split in the presence of a magnetic field (first-order-Zeeman effect) or (2) states are mixed together by the applied magnetic field (second-order-Zeeman effect). The electronic state splitting or mixing may occur in both the initial and the final states [115, 154–156]. The MCD signal ΔA is derived via the absorption of the LCP and RCP light [Eq. (1)].

Equation 1. The MCD signal ΔA is derived via the absorption of the LCP and RCP light.

$$\Delta A = \frac{A_{-} - A_{+}}{A_{-} + A_{+}} \tag{1}$$

MCD is observed in optically active materials at wavelengths with nonvanishing absorption. It occurs for diamagnetic, paramagnetic, and (anti)-ferromagnetic materials and has been observed from IR (infrared) to X-ray regions. MCD is a sensitive technique, of high resolution, characterized with site selectivity; it does not require isotopic enrichment and is not restricted to certain elements, neither to paramagnetic species. It can be used as an optical technique for the detection of the electronic structure of both the ground and excited states at the same time and it is applicable both in solution and in solid samples [157]. The term-related band shape puts severe constraints on possible band assignments. Compared to the classical optical absorption or circular dichroism spectra, MCD allows for many more electronic transitions to be identified.

The total MCD intensity can be written as (Eq. (2), [157]):

Equation 2. The intensity of the MCD signal is proportional to the three different contributions, designated as *A*, *B*, and *C* MCD terms.

$$I \sim \left[A_1\left(\frac{-\partial f(E)}{\partial E}\right) + \left(B_0 \frac{C_0}{kT}\right) f(E)\right] \tag{2}$$

From this equation the MCD intensity is proportional to the three different contributions, designated as A, B, and C MCD terms. The A and B terms are temperature independent and present in diamagnetic, paramagnetic, and (anti)-ferromagnetic materials. The function *f*(*E*) represents the shape of an absorption band. If degenerate excited state is available for a molecule, in the presence of a magnetic field, it is split due to the Zeeman effect. The small difference of the energies for the two resulting transitions generates bands shifted of a few wave numbers. The overlayed signals of the oppositely signed transitions for rcp and lcp light will almost cancel leading to a derivative band shape (A term; see Scheme 1A). The B-term MCD signal arises upon magnetic field induced mixing of an excited state or, in a second possible case, a ground state with an energetically close intermediate state. In the second case it is required that the energetic separation between the two states is large enough such that the intermediate state is not thermally populated. Otherwise, this leads to C-term intensity. As already stated, B-term signals have an absorption band shape. The C term, characteristic for paramagnetic compounds, is temperature dependent and originates from spin–orbit coupling of the degenerate ground state and target excited states with other intermediate excited states. The temperature dependency allows to extract the paramagnetic C-term contribution by subtracting MCD data taken at variable temperatures [158, 159]. Thus, the above-described features of MCD allow an excellent

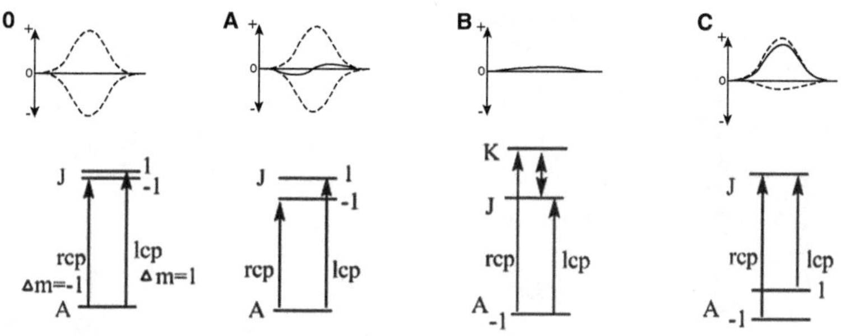

Scheme 1 MCD absorption mechanism. 0, no magnetic field applied; *A*, *B*, *C*, in the presence of a magnetic field, degenerate states are split due to the Zeeman effect giving rise to the *A*, *B*, and *C* terms, respectively

disambiguation between the paramagnetic and diamagnetic contributions to the electronic profile of the analyzed moiety [155, 159, 160].

Metalloproteins are the most likely candidates for the MCD measurements, as the presence of metals with degenerate energy levels leads to strong MCD signals. In the case of ferric heme proteins the MCD is capable of determining both oxidation and spin state to a remarkable degree [161]. For the integer spin metal ions the low-temperature MCD spectroscopy provides indispensable information about the details of the electronic structure of metal centers, as reviewed by Solomon and coauthors [162]. The MCD spectra offer an excellent "fingerprint" for defining the oxidation, spin and ligation states of hemoproteins [163–165]. MCD has provided significant insight into the nature of the axial donors at heme centers, and, more recently, sophisticated methods for the analysis of the MCD spectra have had a major impact on the study of the electronic structures of a range of the biological active sites containing Cu, nonheme iron, and Mo [70, 115, 156, 164, 166]. Application of the MCD spectroscopy greatly contributed to the understanding of the electronic configurations of ferrous nonheme systems allowing the direct observation of the d–d transitions. Due to the different selection rules, LMCT and d–d transitions are significantly more prominent in the MCD spectra compared to the UV–Vis [159, 167]. In particular, the large spin–orbit coupling constant associated with metal-based transitions can cause high-intensity d–d transitions in MCD, which are typically very weak in the UV–Visible absorption spectra [158].

In high-spin nonheme $\{Fe–NO\}^7$ iron nitrosyls (such as [Fe(EDTA)(NO)], $[Fe(Me_3TACN)(NO)(N_3)_2]$ (where Me_3TACN is *N,N',N''*-trimethyl-1,4,7-triazacyclononane), the NO adducts of protocatechuate 3,4-dioxygenase, 4-hydroxyphenylpyruvate dioxygenase, and superoxide dismutase), the NO^- (π^*) → Fe^{III} charge transfer (CT) and spin-forbidden d–d transitions give broad medium–intense ($\varepsilon < 2{,}000\ M^{-1}\ cm^{-1}$) features between 300 and 700 nm in the UV–Vis absorption spectra [56, 168–170]. Typically, the 6C species show at least three distinct transitions in the absorption spectrum: a single very low-intensity band near 600 nm and two more prominent peaks at ~350 and ~450 nm. In the 5C iron nitrosylates, such as iron nitrosyls with tris(*N*-R-carbamoylmethyl)amine-based ligands [171], these features

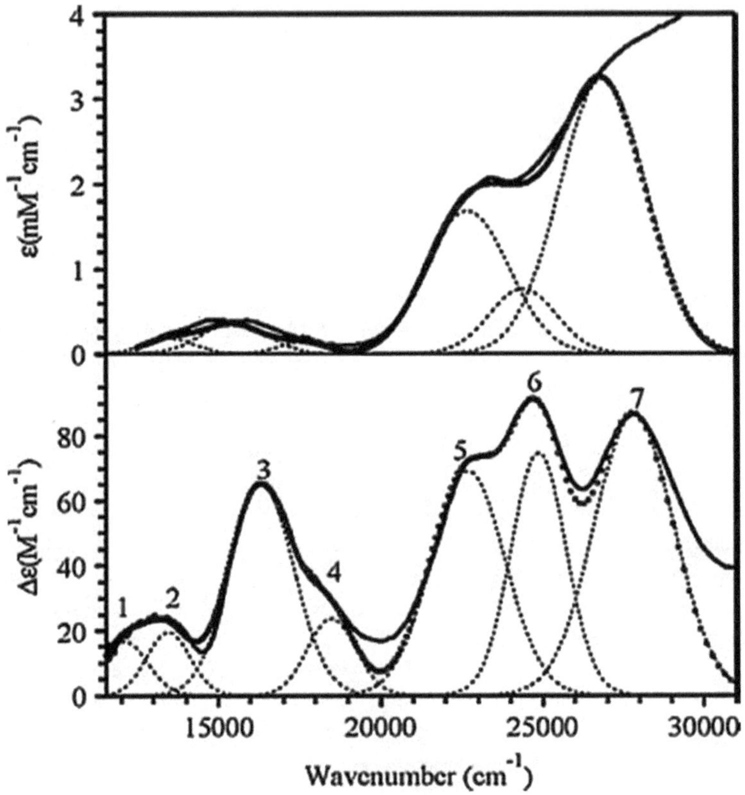

Fig. 15 Room temperature absorption spectrum (*top*) and 7 T 4.5 K MCD spectrum (*bottom*) of FeSOD–NO. *Dashed lines* represent individual Gaussian curves which make up the Gaussian fit (*dotted line*). Reprinted from reference [168] with permission of the American Chemical Society, copyright 2003

shift to lower energy and are observed at approximately 350, 550, and 850 nm, respectively. Similar absorption features are observed in both synthetic model complexes and the nitrosyl adducts of nonheme iron enzymes [168, 172–175]. The observation and distinction between the d–d type transitions becomes possible through MCD, as in the case of the 5C low-spin ferrous iron-nitrosyl model complexes with Schiff-base-type tetradentate macrocyclic ligands that are useful as models of 5C ferrous heme nitrosyls. While the UV–Vis spectra of these compounds show similar overall appearances, with two to three broad bands in the UV–Vis region [176], magnetic circular dichroism (MCD) spectroscopy reveals distinct differences in the optical spectra of the complexes [115]. Low-temperature MCD data for the biologically relevant complex of superoxide dismutase, NO–FeSOD, also allow to recognize several additional electronic transitions between 1,300 and 2,800 cm^{-1}. These bands resolved in MCD are temperature dependent and therefore may be assigned as *C* terms associated with a paramagnetic center (see Fig. 15) [168].

In five-coordinate [Fe(TPP)(NO)] and six-coordinate [Fe(TPP)(MeIm)(NO)] [50] the MCD spectra allowed to define and compare the spin density distributions. In the five-coordinate complex, a strong Fe–NO σ bond between π^*_h and d_{z2} leading to a large transfer of spin density from the NO ligand to Fe(II) was found. On coordination of the sixth ligand, the spin density withdrawal from the iron toward the NO ligand was reflected by the dominant diamagnetic contributions in the MCD spectrum. These results illustrate the change of the Fe–NO character from Fe(I)–NO^+ in the 5C to Fe(II)–NO· in the 6C porphyrinato complexes. In addition to strong porphyrin bands at 350–450 nm (Soret) and 500–600 nm (Q band), several weaker transitions at 600–1,300 nm were assigned to porphyrin–iron and iron–sulfur charge-transfer bands [154, 164].

7 Nuclear Resonance Vibrational Spectroscopy of Metal Porphyrinates

Nuclear resonance vibrational spectroscopy (NRVS, synonymous with nuclear inelastic X-ray scattering) is a synchrotron-based technique that probes vibrational energy levels [177]. The technique is specific for samples that contain nuclei that respond to Mössbauer spectroscopy [130]. NRVS is selective for vibrations involving displacement of Mössbauer-active nuclei. The prevailing number of the published NRVS results concerns isotope ^{57}Fe; this abundance of the data for iron results from the fact that iron is present in the heme and nonheme centers of proteins. The NRVS spectrum can be thought of as a Mössbauer signal with vibrational sidebands [130]. Distinct from Raman and infrared spectroscopy, NRVS is not subject to the optical selection rules, thus all of the iron–ligand modes can be observed. For heme, these include the in-plane Fe vibrations and the Fe–imidazole stretching for 6C porphyrinates, which was not reported in resonance Raman investigations. Other low-frequency vibrations, including the heme doming, can also be investigated by NRVS. Thus, the NRVS experiment can provide the complete set of bands corresponding to the modes that involve motion of the iron atom, also those not being subject to the optical selection rules of Raman or infrared spectroscopy. On the other hand, only iron vibrations are observed, which eliminates solvent interference [177]. NRVS is ideal for the identification of metal–ligand stretching vibrations because it samples the kinetic energy distribution of the vibrational modes, thus the band intensities are proportional to the amount and direction of the iron motion in a normal mode [178]. This makes the metal–ligand stretching vibrations very intense in NRVS [130, 179, 180].

An early approach to interpret the resonance Raman (RR) spectra of heme proteins was made by Kitagawa et al. [181–184] for Ni-octaethylporphyrin (NiOEP), which yields RR spectra similar to those of heme proteins, at the same time being soluble and highly symmetric. On the basis of the polarization properties, the observed bands were classified into the symmetry species (A_{1g}, A_{2g}, B_{1g}, B_{2g}, and E_u) of D_{4h} group [184]. The in-plane molecular vibrations were calculated with

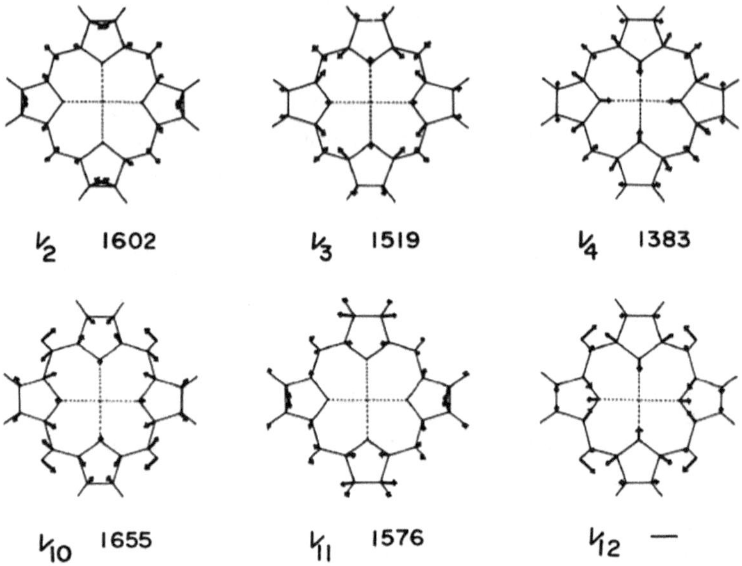

Fig. 16 Vibrational modes of marker bands used in resonance Raman spectra of heme proteins. $\nu_{10}(A_{1g})$ serves as the core size marker, $\nu_3(A_{1g})$ and $\nu_{10}(B_{1g})$ do as the coordination state marker, and ν_4 does as the oxidation state marker. Adopted from the reference [184] with the permission of American Institute of Physics, copyright 1978

the Urey–Bradley force field under the assumption of ethyl groups as a single mass unit, in which the force constants were adjusted to reproduce the observed frequencies. The observed bands were correlated with the calculated values, and thus, the vibrational modes of porphyrin skeleton have been established, although the low-frequency modes containing appreciable contributions from deformation vibrations of pyrrole rings involved some extent of uncertainty. The mode assignment proposed by Kitagawa is widely referred to (e.g., [185–189]). The vibrational modes of bands used as marker bands in resonance Raman spectra of heme proteins are depicted in Fig. 16.

The first mode assignments for the entire iron vibrational density of states (VDOS) of iron porphyrins utilized an empirically derived force field to carry out the normal mode analysis. Rai et al. [179] started the normal coordinate analysis for [Fe(TPP)(NO)] using the above-mentioned force constants developed by Kitagawa for nickel porphyrins (see Fig. 16) [181–184]. The experimentally observed modes were characterized by the classification scheme developed for porphyrin derivatives with D_{4h} symmetry [183, 190, 191]. Four categories of modes can be theoretically distinguished: (1) the in-plane modes, (2) the ligand modes, (3) the modes involving vibrations of the peripheral phenyl groups, and (4) the low-frequency out-of-plane modes. NRVS data for [^{57}Fe(TPP)(NO)] [130, 179], both in a powder sample and as an oriented single-crystal array, were obtained with all porphyrin planes at an angle of 6° to the incident X-ray beam. This allowed to enhance the in-plane modes intensity in the crystal relative to the powder sample while the out-of-plane mode intensities were reduced relative to the powder sample. The modes less intense than their equivalents in the powder spectrum represented out-of-plane vibration modes character. According to the mentioned rule, the remaining modes clearly had a strong in-plane character. The in-plane modes in D_{4h} symmetry are Raman inactive, and were previously unobserved. Although in the case of the [Fe(TPP)(NO)] the D_{4h} symmetry breaking caused by the axial NO ligand might allow Raman observation of these in-plane vibrations, they have not been reported. On the other hand, NRVS will always allow observation of these in-plane modes. The in-plane modes were found in the region 200–500 cm^{-1} for the series of 5C iron porphyrin nitrosylates [192]. Vibrations associated with hindered rotation/tilting of the NO and heme doming are predicted at low frequencies, where Fe motion perpendicular to the heme (out-of-plane vibrations) is identified experimentally for [Fe(TPP)(NO)] at 73 and 128 cm^{-1}. Identification of these two modes was predicted to be crucial for the reactive energetics of Fe porphyrins and heme proteins during binding and dissociation of the ligands to the active sites [178, 192–194].

Although the assignments for almost all modes could be made, it is also clear that the observed modes are delocalized to a great extent with significant mixing of the classification scheme modes. DFT calculations suggest that no independent Fe–N–O stretching mode can be assigned; the Fe–NO stretch modes have a significant Fe–N–O bend character as well as a Fe–NO stretch character [179, 195, 196]. The Fe–N–O bending mode was also calculated to contribute in the in-plane Fe vibrations. Measured ^{57}Fe excitation probabilities for a series of iron porphyrins [192] show that iron-nitrosyl porphyrinato complexes have Fe–NO stretch/bend modes in the 520–540 cm^{-1} region. The bands in the 300–400 cm^{-1} range were associated with stretching of the four in-plane $Fe-N_{pyr}$ bonds [178, 179]. It is noteworthy that in the case of the Fe(III) complex Fe(OEP)(Cl), IR measurements led to assignment of the stretching of the $Fe–N_{pyr}$ bonds to a mode observed at 275 cm^{-1} [197], coincident with a cluster of modes in the NRVS signal (Fig. 17 b). These results illustrate the considerable sensitivity of the in-plane $Fe–N_{pyr}$ bonds strength to the oxidation state of the Fe. Comparison among the nitrosyl complexes (see Fig. 17) reveals that peripheral groups strongly influence the vibrational frequencies and amplitudes of the central ion. The influence of peripheral groups resulting from both their electronic properties and symmetry features is described in detail in the chapter by Lehnert.

The recent results by Lehnert et al. [33] complement the above picture, presenting for the first time the detailed single-crystal NRVS-based theoretical analysis of the 6C heme-type [^{57}Fe(TPP)(MeIm)(NO)] complex vibrations. The results obtained by these authors show a very strong mixing between the vibrations of the axial (Im)N–Fe–NO unit and the porphyrin-based vibrations [33]. The study points to a number of important biological conclusions that may be drawn from the analysis and comparison of the data for high- and low-spin 6C and 5C ferrous heme nitrosyls. In particular, the questions can be addressed of (1) how strongly the vibrational dynamics of the ferrous heme active sites in proteins change upon NO binding as a result of changes in the Fe–NIm bond, the position of iron relative to the porphyrin ring, and the spin state and (2) how strongly the presence of the distal histidine in globins influences the properties of

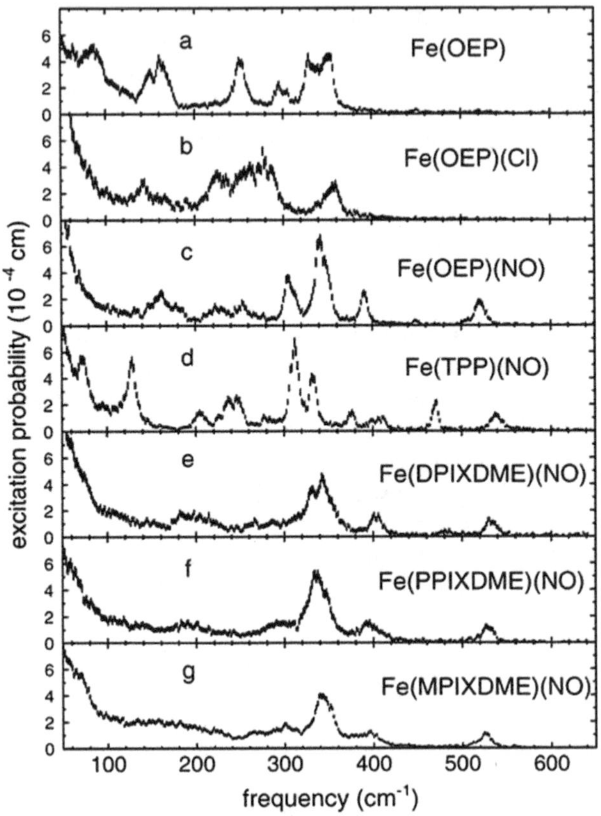

Fig. 17 Measured ^{57}Fe excitation probabilities for a series of iron porphyrins. All nitrosyl complexes have an Fe–NO stretch/bend mode in the 520–540 cm^{-1} region. Comparison among the nitrosyl complexes (*c*–*g*) reveals that peripheral groups strongly influence the vibrational frequencies and amplitudes of the central Fe. Sample temperatures were 34 K for Fe(OEP), 30 K for Fe(OEP)(NO), 80 K for Fe(TPP)(NO), 35 K for Fe(DPIXDME)(NO), 34 K for Fe(PPIXDME)(NO), and 64 K for Fe(MPIXDME)(NO). The Fe(OEP)(Cl) spectrum is an average over multiple scans with an estimated average temperature of 87 K. Reprinted from reference [192], with the permission of the American Chemical Society, copyright 2004

the bound NO via hydrogen bonding. The most obvious effect of NO binding to a 5C deoxy-heme site is caused by the change of the spin state of iron(II) center from high-spin to low-spin upon axial ligand binding. This strengthens the Fe–NPyr (Pyr=pyrrole) bonds, as manifested by the appearance of the intense features around 300–350 cm^{-1} in contrast to the 230 cm^{-1} for the high-spin [Fe(TPP)(2-MeIm)] [180]. In the early studies of Ogoshi et al. [197], this distinct difference in heme dynamics had also been related to a well-known change in the position of iron(II), which is pulled into the porphyrin plane upon coordination of a sixth ligand. Nevertheless, the fact that the NRVS spectrum of the 5C low-spin complex [Fe(TPP)(NO)] shows similar features in the 300–350 cm^{-1} region [180, 192] to the low-spin [Fe(TPP)(MeIm)(NO)] indicates that the spin state change is the more important contribution. This is further pronounced by the strong energy splitting of the intense NRVS features in the 300–350 cm^{-1} region (originating from selective mode mixing of the higher energy component of this mode with the Fe–N(O) torsion and ON–Fe–NPyr octahedral bends) observed for both the low-spin 5C and 6C nitrosylates. A similar splitting observed for six-coordinate [Fe(TPP)(MeIm)(CO)] [192], which cannot result from mixing with vibrations of the linear Fe–C–O unit, implies that the observed effect is caused by the imidazole ligand in the low-spin orbital configuration. The considerable shift of the Fe–NIm stretching mode upon NO complexation underlines the strength of the σ trans interaction of imidazole with the bound NO ligand. Reversibly, comparison of the NRVS spectra for 5C and 6C heme nitrosylates implies the pronounced effect of Im coordination on the properties of the Fe–N–O bonding. Upon the coordination of the Im *trans*-ligand the lowering of the force constants throughout the Fe–N–O moiety can be observed, and the electron density is pushed back from iron toward NO. This increase in radical character on NO, combined with the weakening of the Fe–NO bond, could be crucial for the activation of bacterial NO reductase ferrous heme/nonheme active site (for the detailed discussion see [33] and references therein).

8 X-ray Absorption Spectroscopy

The X-ray absorption spectroscopy (XAS) is a technique allowing determination of the local geometric and/or electronic structure of matter by tuning the photon energy to a range where the core electrons can be excited (0.1–100 keV photon energy). For theoretical considerations on the X-ray spectroscopic methods see [198]. XAS methods are useful in the studies on the structure and redox properties of metal coordination compounds. In particular, the extended X-ray absorption fine structure (EXAFS) spectroscopy can provide highly accurate metal–ligand bond distances, and the number and identity of coordinating atoms in metalloenzymes [199], as in the case of the thiolate-ligated heme of the cytochromes P450 [200]. In contrast with EXAFS, XANES spectroscopy focuses on the region ca. 100 eV around the Fe–K-edge absorption. By focusing on the edge region of the X-ray absorption envelope, the X-ray absorption near-edge spectroscopy (XANES) has been used to probe the efficient oxidation state of redox-active metals and the ligand coordination geometry/symmetry for selected metals. The XANES spectrum carries information about the geometry of heme iron, and the electronic structure of the iron coordination sphere [201]. In many cases X-ray absorption spectroscopy can provide the most accurate metal–ligand bond distances, and structural information can be obtained from samples in either frozen solution or solid state. In addition, this spectroscopy is very element selective with little interference from other centers. However, the method requires high-intensity X-rays from a synchrotron light source and often does not give a unique determination of ligand environment due to similarity of oxygen and nitrogen in scattering. Moreover, the conclusions often depend strongly on theoretical simulation and curve fitting.

Sulfur-K-edge X-ray absorption spectroscopy (XAS) was demonstrated to provide a superior level of insight on elucidating the electronic structure of the $\{Fe(NO)_2\}^9$ core of DNICs. XAS allows a direct probing of the chemical forms of sulfur, bonding mode, and covalency of the metal–sulfur bond which modulates the function and activity of the abundant metal–sulfur clusters [202–208]. The intensity of the pre-edge is known to serve as a probe of the ligand character of the unoccupied or partially occupied orbitals and a probe of the Fe–S bond covalency (see discussion in [209]). Furthermore, it was noticed that the Fe–S bond covalency perturbed by the hydrogen–bonding interaction within the protein regulates the redox potential of the [Fe–S] proteins [210]. The S–K-edge XAS helped to identify and distinguish the mononucleated and dinucleated DNICs. The electronic structures of the $\{Fe(NO)_2\}^9$ core of the two forms were assigned. In particular, the intensity of the S–K-edge pre-edge absorption peak of the dinuclear DNICs featuring the Fe–S bond covalency and rationalizing the binding preference of the $\{Fe(NO)_2\}$ motif toward thiolates was also delineated. For the series of thiolate DNICs, the higher than in the case of free RSH-type ligands, thiolate peak energies implicate that there is a significant charge contribution from terminal thiolates to the $\{Fe(NO)\}^7$ core of MNIC and the $\{Fe(NO)_2\}^9$ core of DNIC (as consistent with the Mulliken average charge population on S atoms). In the S–K-edge pre-edge region, the complexes of the type $[(NO)Fe(SR)_3]^-$ (where R = Ph, Et, Ph) and of the type $[(NO)_2Fe(SR)_2]^-$ (R = Et) exhibit the intense pre-edge features (2,470.2–2,471.3 eV) derived from the $S_{1s} \rightarrow Fe_{3d}$ transition. Since the thiolate peak energy ($S_{1s} \rightarrow S_{C\text{–}S\ \sigma^*}$ transition, C–S σ* orbitals are degenerated for the same thiolate ligand) reflects the relative energy of the S_{1s} orbital of [M–S] complexes (M=Fe and Ni) [211–214], the pre-edge energy $S_{1s} \rightarrow Fe_{3d}$ transition may be adopted to establish the relative energy of the Fe_{3d} manifold orbitals and to estimate the effective nuclear charge (Z_{eff}) of the Fe center of MNICs and DNICs [215–218]. On the basis of the linear relationship between the oxidation state and the Z_{eff} and following the method reported by Sun et al. [217], the oxidation state of Fe can be easily calculated. The relative d-manifold energy derived from the S–K-edge XAS as well as the Fe–K-edge pre-edge energy suggests that the electronic structure of the $\{Fe(NO)_2\}^9$ core of the mononuclear DNICs $[(NO)_2Fe(SR)_2]^-$ is best described as $\{Fe^{III}(NO^-)_2\}^9$ compared to $[\{Fe^{III}(NO^-)_2\}^9\text{-}\{Fe^{III}(NO^-)_2\}^9]$ for the dinuclear DNICs $[Fe_2(\mu\text{-}SR)(\mu\text{-}S)(NO)_4]^-$ and $[Fe_2(\mu\text{-}S)_2(NO)(4)]^{2-}$, but the existence of a resonance electronic structure of $\{Fe^{III}(NO^-)_2\}^9$ and $\{Fe^{II}(\bullet NO)(NO^-)\}^9$ could not be ruled out, since the replacement of two $[SPh]^-$ by two NO^- in DNICs may change the shielding effect [209].

9 Final Remarks

The impact of NO complexation by metal ions is of a crucial meaning to biological systems. NO reacts with all transition metals to give metal nitrosyls. Proteins containing transition metals are particularly prone to react with NO, since its unpaired electron can interact and bond with the d-orbitals of metal cofactors. These interactions are widely used by the nature and function in various cellular regulatory pathways. It is clear that the diversity of the NO-sensing proteins requires a combination of complementary spectroscopic methods to provide insights into the mechanisms of their action. As well, the detailed understanding of the mechanistic aspects responsible for NO biological functions allows to predict and design compounds with desired medical properties or applicable in biochemical/biophysical studies. The structural and electronic features, together with the modes of complexation, are decisive toward nitric oxide's biological action, the form in which it is liberated, influence on enzyme functioning and molecular mechanisms underlying its activity. Therefore structural aspects of NO binding shall be studied with a particular involvement. The available structural and spectroscopic methods give a wide range of aspects to be studied. Only the most popular and resultful of the modern methods have been discussed in this review, many other yet approaching in the fast developing issue of biophysics concerning interactions of small regulatory molecules with biological targets.

Acknowledgments The author wishes to thank Nicolai Lehnert from the Department of Chemistry at the University of Michigan for the invaluable meritorical discussion, which contributed to improve this review.

References

1. Ueno T, Suzuki Y et al (1999) In vivo distribution and behavior of paramagnetic dinitrosyl dithiolato iron complex in the abdomen of mouse. Free Radic Res 31:525–534
2. Ueno T, Yoshimura T (2000) The physiological activity and in vivo distribution of dinitrosyl dithiolato iron complex. Jpn J Pharmacol 82:95–101
3. Ueno T, Suzuki Y et al (2002) In vivo nitric oxide transfer of a physiological NO carrier, dinitrosyl dithiolato iron complex, to target complex. Biochem Pharmacol 63:485–493
4. Fichtlscherer B, Mülsch A (2000) MR imaging of nitrosyl-iron complexes: experimental study in Rats1. Radiology 216:225–231
5. Richter-Addo GB, Legzdins P (1992) Metal nitrosyls. Oxford University Press, Oxford
6. Siladke NA, Meihaus KR et al (2012) Synthesis, structure, and magnetism of an F element nitrosyl complex, $(C_5Me_4H)_3UNO$. J Am Chem Soc 134:1243–1249
7. McCleverty JA (2004) Chemistry of nitric oxide relevant to biology. Chem Rev 104:403–418
8. Mingos DMP, Sherman DJ (1989) Transition metal nitrosyl complexes. In: Sykes AG (ed) Advances in inorganic chemistry, vol 34. Academic Press: New York, NY, USA
9. Hayton TW, Legzdins P, Sharp WB (2002) Coordination and organometallic chemistry of metal-NO complexes. Chem Rev 102:935–992
10. Gans P (1965) The bonding of nitric oxide in transition-metal nitrosyl complexes. Chem Commun (London) 144–145
11. Gans P, Sabatini A, Sacconi L (1966) Infrared spectra and bonding in transition metal nitrosyl complexes. Inorg Chem 5:1877–1881
12. Weaver DL, Snyder DA (1970) Crystal and molecular structure of *trans*-chloronitrosylbis (ethylenediamine)cobalt (III) perchlorate. Inorg Chem 9:2760–2767
13. Haymore BL, Ibers JA (1975) Comparison of linear nitrosyl and singly bent aryldiazo complexes of ruthenium. structures of trichloronitrosylbis (triphenylphosphine)ruthenium, $RuCl_3$ $(NO)(P(C_6H_5)_3)_2$, and trichloro (P-Tolyl)Diazobis (triphenylphosphine)ruthenium-dichloromethane, $RuCl_3$ $(P\text{-}NNC_6H_4CH_3)(P(C_6H_5)_3)_2{\cdot}CH_2Cl_2$. Inorg Chem 14:3060–3070
14. De La Cruz C, Sheppard N (2011) A structure-based analysis of the vibrational spectra of nitrosyl ligands in transition-metal coordination complexes and clusters. Spectrochim Acta A Mol Biomol Spectrosc 78:7–28

15. Wang X, Zhou M, Andrews L (2000) Manganese carbonyl nitrosyl complexes in solid argon: infrared spectra and density functional calculations. J Phys Chem 104:7964–7973
16. Andrews L, Zhou M, Wang X (2000) Matrix infrared spectra and density functional calculations for GaNO, InNO, and TlNO. J Phys Chem A 104:8475–8479
17. Wang X, Zhou M, Andrews L (2000) Reactions of iron atoms with nitric oxide and carbon monoxide in excess argon: infrared spectra and density functional calculations of iron carbonyl nitrosyl complexes. J Phys Chem A 104:10104–10111
18. Citra A, Wang X et al (2001) Reactions of laser-ablated platinum with nitrogen: matrix infrared spectra of platinum nitride, complexes, and anions. J Phys Chem 105:7799–7811
19. Wang X, Andrews L (2001) Cobalt carbonyl nitrosyl complexes: matrix infrared spectra and density functional calculations. J Phys Chem A 105:4403–4409
20. Citra A, Wang X, Andrews L (2001) Reactions of laser-ablated gold with nitric oxide: infrared spectra and DFT calculations of AuNO and $Au(NO)_2$ in solid Argon and Neon. J Phys Chem A 106:3287–3293
21. Beck W, Lottes K (1965) Vergleichende IR-Spektroskopische Untersuchungen an Nitrosyl-Komplexen Der Übergangsmetalle: I. Diskussion Der NO-Valenzschwingungen. Chem Ber 98:2657–2673
22. Feltham RD, Enemark JH (1981) Structures of metal nitrosyls. In: Topics in stereochemistry. Vol. 12; Geoffroy, G. L., Ed. p 155 Wiley &Sons, Inc., Hoboken, NJ, USA
23. Hoffmann R, Chen MML et al (1974) Pentacoordinate nitrosyls. Inorg Chem 13:2666–2675
24. Scheidt WR, Lee YJ, Hatano K (1984) Preparation and structural characterization of nitrosyl complexes of ferric porphyrinates. Molecular structure of aquonitrosyl (meso-tetraphenylporphinato)iron (III) perchlorate and nitrosyl (octaethylporphinato)iron (III) perchlorate. J Am Chem Soc 106:3191–3198
25. Beck W, Melnikoff A, Stahl R (1966) Spektroskopische Untersuchungen An Komplexverbindungen, IX. IR-Intensitäten Von CO- Und NO-Valenzschwingungen Und Bindungswinkel in Metallcarbonylen Und –Nitrosylen. Chem Ber 99:3721–3727
26. Poletti A, Foffani A, Cataliotti B (1970) Infrared intensity measurements on nitrosyltriearbonylcobalt and monosubstituted derivatives. Spectrochim Acta A Mol Spectrosc 26:1063–1069
27. Atkinson FL, Blackwell HE et al (1996) Synthesis of the 17-electron cations $[FeI(L')(NO)_2]+$ (L, L' = PPh_3, $OPPh_3$): structure and bonding in four-co-ordinate metal dinitrosyls, and implications for the identity of paramagnetic iron dinitrosyl complex catalysts. J Chem Soc Dalton Trans 3491–3502
28. Gwost D, Caulton KG (1973) Reductive nitrosylation of group VIIIb compounds. Inorg Chem 12:2095–2099
29. Kaduk JA, Ibers JA (1975) Crystal and molecular structure of dinitrosylbis (triphenylphosphine)rhodium perchlorate, $[Rh (NO)_2(P(C_6H_5)_3)_2][ClO_4]$. Inorg Chem 14:3070–3073
30. Malatesta L, Angoletta M, Caglio G (1963) Nitrosyl (triphenylphosphine)iridium compounds. Angew Chem Int Ed Engl 2:739
31. Durig JR, Wertz DW (1968) Spectroscopic review of the normal vibrations of metal complexes with nitrogen-containing ligands. Appl Spectrosc 22:627–633
32. Dai RJ, Ke SC (2007) Detection and determination of the $Fe(NO)_2$ core vibrational features in dinitrosyl-iron complexes from experiment, normal coordinate analysis, and density functional theory: an avenue for probing the nitric oxide oxidation state. J Phys Chem B 111:2335–2346
33. Lehnert N, Sage JT et al (2010) Oriented single-crystal nuclear resonance vibrational spectroscopy of [Fe(TPP)(MI)(NO)]: quantitative assessment of the trans effect of NO. Inorg Chem 49:7197–7215
34. Obayashi E, Tsukamoto K et al (1997) Unique binding of nitric oxide to ferric nitric oxide reductase from *Fusarium oxysporum* elucidated with infrared, resonance Raman, and X-ray absorption spectroscopies. J Am Chem Soc 119:7807–7816

35. Linder DP, Rodgers KR (2005) Fe–N–O structure and bonding in six-coordinate {FeNO}6 porphyrinates containing imidazole: implications for reactivity of coordinated NO. Inorg Chem 44:1367–1380
36. Linder DP, Rodgers KR et al (2004) Five-coordinate $Fe^{III}NO$ and $Fe^{II}CO$ porphyrinates: where are the electrons and why does it matter? J Am Chem Soc 126:14136–14148
37. Paulat F, Lehnert N (2007) Electronic structure of ferric heme nitrosyl complexes with thiolate coordination. Inorg Chem 46:1547–1549
38. Brudvig GW, Stevens TH, Chan SI (1980) Reactions of nitric oxide with cytochrome C oxidase. Biochemistry 19:5275–5285
39. Henry Y, Lepoivre M et al (1993) EPR characterization of molecular targets for NO in mammalian cells and organelles. FASEB J 7:1124–1134
40. Van Faassen E, Vanin AF (2007) Radicals for life: the various forms of nitric oxide. Elsevier Science Ltd., Amsterdam
41. Cammack R, Joannou CL et al (1999) Nitrite and nitrosyl compounds in food preservation. Biochim Biophys Acta 1411:475–488
42. Mcdonald CC, Phillips WD, Mower HF (1965) An electron spin resonance study of some complexes of iron, nitric oxide, and anionic ligands. J Am Chem Soc 87:3319–3326
43. Vanin AF, Serezhenkov VA et al (1998) The 2.03 signal as an indicator of dinitrosyl-iron complexes with thiol-containing ligands. Nitric Oxide 2:224–234
44. Serres RG, Grapperhaus CA et al (2004) Structural, spectroscopic, and computational study of an octahedral, non-heme [Fe–NO] (6–8) series: $[Fe(NO)(Cyclam\text{-}Ac)]^{2+/+/0}$. J Am Chem Soc 126:5138–5153
45. Neyman KM, Ganyushin DI et al (2003) Electronic g values of Na^{+}–NO and Cu^{+}–NO complexes in zeolites: analysis using a relativistic density functional method. Phys Chem Chem Phys 5:2429–2434
46. Frantz S, Sarkar B et al (2004) EPR insensitivity of the metal-nitrosyl spin-bearing moiety in complexes $[L_nRu^{II}\text{-}NO^{\cdot}]^{k}$. Eur J Inorg Chem 2004:2902–2907
47. Wanner M, Scheiring T et al (2001) EPR characteristics of the $[(NC)_5M(NO)]^{3-}$ ions (M=Fe, Ru, Os). Experimental and DFT study establishing NO as a ligand. Inorg Chem 40:5704–5707
48. Pipes DW, Meyer TJ (1984) Comparisons between polypyridyl nitrosyl complexes of osmium (II) and ruthenium (II). Inorg Chem 23:2466–2472
49. Lehnert N, Galinato MG et al (2010) Nuclear resonance vibrational spectroscopy applied to [Fe(OEP)(NO)]: the vibrational assignments of five-coordinate ferrous heme-nitrosyls and implications for electronic structure. Inorg Chem 49:4133–4148
50. Praneeth VKK, Neese F, Lehnert N (2005) Spin density distribution in five- and six-coordinate iron (II)-porphyrin NO complexes evidenced by magnetic circular dichroism spectroscopy. Inorg Chem 44:2570–2572
51. D'Autreaux B, Tucker N et al (2008) Characterization of the nitric oxide-reactive transcriptional activator NORr. Methods Enzymol 437:235–251
52. D'Autreaux B, Tucker NP et al (2005) A non-haem iron centre in the transcription factor NORr senses nitric oxide. Nature 437:769–772
53. Li M, Bonnet D et al (2002) Tuning the electronic structure of octahedral iron complexes $[Fe^{I}$ (X)] (L=1-alkyl-4,7-bis (4-tert-butyl-2-mercaptobenzyl)-1,4,7-triazacyclononane, X=Cl, CH_3O, CN, NO). The S = 1/2, S = 3/2 spin equilibrium of $[Fe^{I}(Pr)(NO)]$. Inorg Chem 41:3444–3456
54. Pohl K, Wieghardt K et al (1987) Preparation and magnetism of the binuclear iron (II) complexes $[\{Fe(C_9H_{21}N_3)X_2\}_2]$ (X=NCS, NCO, or N_3) and their reaction with NO. Crystal structures of $[\{Fe(C_9H_{21}N_3)(NCS)_2\}_2]$ and $[Fe(C_9H_{21}N_3)(NO)(N_3)_2]$. J Chem Soc Dalton Trans 187–192
55. Zhang Y, Pavlosky MA et al (1992) Spectroscopic and theoretical description of the electronic structure of the S = 3/2 nitrosyl complex of non-heme iron enzymes. J Am Chem Soc 114:9189–9191

56. Brown CA, Pavlosky MA et al (1995) Spectroscopic and theoretical description of the electronic structure of S = 3/2 iron–nitrosyl complexes and their relation to O_2 activation by non-heme iron enzyme active sites. J Am Chem Soc 117:715–732
57. Wanat A, Schneppensieper T et al (2002) Kinetics, mechanism, and spectroscopy of the reversible binding of nitric oxide to aquated iron (II). An undergraduate text book reaction revisited. Inorg Chem 41:4–10
58. Joannou CL, Cui XY et al (1998) Characterization of the bactericidal effects of sodium nitroprusside and other pentacyanonitrosyl complexes on the food spoilage bacterium clostridium sporogenes. Appl Environ Microbiol 64:3195–3201
59. Clarke MJ, Gaul JB (1993) Chemistry relevant to the biological effects of nitric oxide and metallonitrosyls. In: Structures and biological effects, 81th edn. Springer, Berlin Heidelberg
60. Henry Y, Ducrocq C et al (1991) Nitric oxide, a biological effector. Electron paramagnetic resonance detection of nitrosyl-iron-protein complexes in whole cells. Eur Biophys J 20:1–15
61. Wayland BB, Olson LW (1974) Spectroscopic studies and bonding model for nitric oxide complexes of iron porphyrins. J Am Chem Soc 96:6037–6041
62. Kappl R, Hüttermann J (1989) An ENDOR study of nitrosyl myoglobin single crystals. Isr J Chem 29:73–84
63. Hüttermann J (1993) ENDOR of randomly oriented mononuclear metalloproteins. In: EMR of paramagnetic molecules 13th edn. Springer, USA
64. Hoff AJ (1989) Advanced EPR: applications in biology and biochemistry. Elsevier Science, Amsterdam
65. Tyryshkin AM, Dikanov SA et al (1999) Characterization of bimodal coordination structure in nitrosyl heme complexes through hyperfine couplings with pyrrole and protein nitrogens. J Am Chem Soc 121:3396–3406
66. Morse RH, Chan SI (1980) Electron paramagnetic resonance studies of nitrosyl ferrous heme complexes. Determination of an equilibrium between two conformations. J Biol Chem 255:7876–7882
67. Hori H, Ikeda-Saito M, Yonetani T (1981) Single crystal EPR of myoglobin nitroxide. Freezing-induced reversible changes in the molecular orientation of the ligand. J Biol Chem 256:7849–7855
68. Hüttermann J, Kappl R (2000) EPR and ENDOR of metalloproteins. Electron Paramagnetic Reson 17:246–304
69. Westcott BL, Enemark JH (1999) Transition metal nitrosyls. Wiley, New York
70. Goodrich LE, Paulat F et al (2010) Electronic structure of heme-nitrosyls and its significance for nitric oxide reactivity, sensing, transport, and toxicity in biological systems. Inorg Chem 49:6293–6316
71. Lehnert N (2008) Chapter 6 – EPR and low-temperature MCD spectroscopy of ferrous heme nitrosyls. In: Abhik G (ed) The smallest biomolecules: diatomics and their interactions with heme proteins. Elsevier, Amsterdam
72. Hayes RG, Ellison MK, Scheidt WR (2000) Definitive assignment of the G tensor of [Fe(OEP)(NO)] by single-crystal EPR. Inorg Chem 39:3665–3668
73. Henry Y (1996) Utilization of nitric oxide as a paramagnetic probe of the molecular oxygen binding site of metalloenzymes. In: Nitric oxide research from chemistry to biology. Springer, USA
74. Utterback SG, Doetschman DC et al (1983) EPR study of the structure and spin distribution at the binding site in human nitrosylhemoglobin single crystals. J Chem Phys 78:5874–5880
75. Ghosh A, Wondimagegn T (2000) A theoretical study of axial tilting and equatorial asymmetry in metalloporphyrin-nitrosyl complexes. J Am Chem Soc 122:8101–8102
76. Patchkovskii S, Ziegler T (2000) Structural origin of two paramagnetic species in six-coordinated nitrosoiron (II) porphyrins revealed by density functional theory analysis of the G tensors. Inorg Chem 39:5354–5364
77. Praneeth VK, Nather C et al (2006) Spectroscopic properties and electronic structure of five- and six-coordinate iron (II) porphyrin NO complexes: effect of the axial N-donor ligand. Inorg Chem 45:2795–2811

78. Singel DJ, Lancaster JR, Jr (1996) Electron paramagnetic resonance spectroscopy and nitric oxide biology. In: Feelisch M, Stamler JS (ed) Methods in nitric oxide research. Wiley, Chichester
79. Boese M, Mordvintcev PI et al (1995) S-nitrosation of serum albumin by dinitrosyl-iron complex. J Biol Chem 270:29244–29249
80. Henry Y, Giussani A, Ducastel B (1997) Nitric oxide research from chemistry to biology: EPR spectroscopy of nitrosylated compounds. Springer-Verlag, Berlin
81. Bouton C, Drapier JC (2003) Iron regulatory proteins as NO signal transducers. Sci STKE 2003:E17
82. Vanin AF (1998) Dinitrosyl iron complexes and S-nitrosothiols are two possible forms for stabilization and transport of nitric oxide in biological systems. Biochemistry (Mosc) 63:782–793
83. Mulsch A, Mordvintcev P et al (1991) The potent vasodilating and guanylyl cyclase activating dinitrosyl-iron (II) complex is stored in a protein-bound form in vascular tissue and is released by thiols. FEBS Lett 294:252–256
84. Vanin AF (1991) Endothelium-derived relaxing factor is a nitrosyl iron complex with thiol ligands. FEBS Lett 289:1–3
85. Keese MA, Bose M et al (1997) Dinitrosyl–dithiol–iron complexes, nitric oxide (NO) carriers in vivo, as potent inhibitors of human glutathione reductase and glutathione-S-transferase. Biochem Pharmacol 54:1307–1313
86. Stamler JS, Singel DJ, Loscalzo J (1992) Biochemistry of nitric oxide and its redox-activated forms. Science 258:1898–1902
87. Alencar JL, Chalupsky K et al (2003) Inhibition of arterial contraction by dinitrosyl-iron complexes: critical role of the thiol ligand in determining rate of nitric oxide (NO) release and formation of releasable NO stores by S-nitrosation. Biochem Pharmacol 66:2365–2374
88. Woolum JC, Tiezzi E, Commoner B (1968) Electron spin resonane of iron–nitric oxide complexes with amino acids, peptides and proteins. Biochim Biophys Acta 160:311–320
89. Woolum JC, Commoner B (1970) Isolation and identification of a paramagnetic complex from the livers of carcinogen-treated rats. Biochim Biophys Acta 201:131–140
90. Timoshin AA, Vanin AF et al (2007) Protein-bound dinitrosyl–iron complexes appearing in blood of rabbit added with a low-molecular dinitrosyl-iron complex: EPR studies. Nitric Oxide 16:286–293
91. Kleschyov AL, Sedov KR et al (1994) Biotransformation of sodium nitroprusside into dinitrosyl iron complexes in tissue of ascites tumors of mice. Biochem Biophys Res Commun 202:168–173
92. Vanin AF (1998) Biological role of nitric oxide: history, modern state, and perspectives for research. Biochemistry (Mosc) 63:731–733
93. Vanin AF, Sanina NA et al (2007) Dinitrosyl–iron complexes with thiol-containing ligands: spatial and electronic structures. Nitric Oxide 16:82–93
94. Vanin AF, Stukan RA, Manukhina EB (1996) Physical properties of dinitrosyl iron complexes with thiol-containing ligands in relation with their vasodilator activity. Biochim Biophys Acta 1295:5–12
95. Lee M, Arosio P et al (1994) Identification of the EPR-active iron–nitrosyl complexes in mammalian ferritins. Biochemistry 33:3679–3687
96. Butler AR, Megson IL (2002) Non-heme iron nitrosyls in biology. Chem Rev 102:1155–1166
97. Tsai FT, Chiou SJ et al (2005) Dinitrosyl iron complexes (DNICs) $[L_2Fe(NO)_2]^-$ (L=Thiolate): interconversion among $\{Fe(NO)_2\}^9$ DNICs, $\{Fe(NO)_2\}^{10}$ DNICs, and [2Fe-2S] clusters, and the critical role of the thiolate ligands in regulating NO release of DNICs. Inorg Chem 44:5872–5881
98. Ilyakina EV, Poddel'sky AI et al (2012) Binding of NO by nontransition metal complexes. Mendeleev Commun 22:208–210
99. Bordini J, Hughes DL et al (2002) Nitric oxide photorelease from ruthenium salen complexes in aqueous and organic solutions. Inorg Chem 41:5410–5416

100. Ridley J, Zerner M (1973) An intermediate neglect of differential overlap technique for spectroscopy: pyrrole and the azines. Theoret Chim Acta 32:111–134
101. Pople JA, Beveridge DL, Dobosh PA (1967) Approximate self-consistent molecular-orbital theory. V. Intermediate neglect of differential overlap. J Chem Phys 47:2026–2033
102. Borges SD, Davanzo CU et al (1998) Ruthenium nitrosyl complexes with N-heterocyclic ligands. Inorg Chem 37:2670–2677
103. Khan MMT, Srinivas D et al (1990) Synthesis, characterization, and EPR studies of stable ruthenium (III) schiff base chloro and carbonyl complexes. Inorg Chem 29:2320–2326
104. Ookubo K, Morioka Y et al (1996) Vibrational spectroscopic study of light-induced metastable states of ethylenediaminenitrosylruthenium (II) complexes. J Mol Struct 379:241–247
105. Schreiner AF, Lin SW et al (1972) Chemistry and optical properties of 4d and 5d transition metals. III. Chemistry and electronic structures of ruthenium acidonitrosylammines, [Ru $(NH_3)_4$ (NO)(L)]q+1a. Inorg Chem 11:880–888
106. Gorelsky SI, Da Silva SC et al (2000) Electronic spectra of *trans*-[Ru $(NH_3)_4$ $(L)NO]^{3+/2+}$ complexes. Inorg Chim Acta 300–302:698–708
107. Szaciłowski K, Oszajca J et al (2001) Photochemistry of the $[Fe(CN)_5N\ (O)SR]^{3-}$ complex: a mechanistic study. J Photochem Photobiol A 143:93–262
108. Costanzo S, Ménage S et al (2001) Re-examination of the formation of dinitrosyl–iron complexes during reaction of S-nitrosothiols with Fe(II). Inorg Chim Acta 318:1–7
109. Conrado CL, Bourassa JL et al (2003) Photochemical investigation of Roussin's red salt esters: $Fe_2\ (\mu\text{-}SR)_2\ (NO)_4$. Inorg Chem 42:2288–2293
110. Jaworska M, Stasicka Z (2004) Structure and UV–Vis spectroscopy of nitrosylthiolatoferrate mononuclear complexes. J Organometallic Chem 689:1702–1713
111. Jaworska M, Stasicka Z (2005) Structure and UV–Vis spectroscopy of the iron–sulfur dinuclear nitrosyl complexes $[Fe_2S_2\ (NO)_4]^{2-}$ and $[Fe_2\ (SR)_2\ (NO)_4]$. New J Chem 29:604–612
112. Hoshino M, Laverman L, Ford PC (1999) Nitric oxide complexes of metalloporphyrins: an overview of some mechanistic studies. Coord Chem Rev 187:75–102
113. Scheidt WR, Ellison MK (1999) The synthetic and structural chemistry of heme derivatives with nitric oxide ligands. Acc Chem Res 32:350–359
114. Berto TC, Praneeth VK et al (2009) Iron–porphyrin NO complexes with covalently attached N-donor ligands: formation of a stable six-coordinate species in solution. J Am Chem Soc 131:17116–17126
115. Sulok CD, Bauer JL et al (2012) A detailed investigation into the electronic structures of macrocyclic iron (II)-nitrosyl compounds and their similarities to ferrous heme-nitrosyls. Inorg Chim Acta 380:148–160
116. Gütlich P, Bill E et al (2011) Mössbauer spectroscopy and transition metal chemistry: fundamentals and applications. Springer Verlag, Berlin Heidelberg
117. Schünemann V, Paulsen H (2007) Mössbauer spectroscopy. In: Scott RA (ed) Applications of physical methods to inorganic and bioinorganic chemistry, 2nd edn. Wiley &Sons, Hoboken, NJ
118. Neese F (2002) Prediction and interpretation of the ^{57}Fe isomer shift in Mössbauer spectra by density functional theory. Inorg Chim Acta 337:181–192
119. Krebs C, Martin Bollinger J Jr (2009) Freeze-quench ^{57}Fe-Mössbauer spectroscopy: trapping reactive intermediates. Photosynth Res 102:295–304
120. Hsieh CH, Erdem OF et al (2012) Structural and spectroscopic features of mixed valent Fe(II) Fe(I) complexes and factors related to the rotated configuration of diiron hydrogenase. J Am Chem Soc 134 (31): 13089–13102
121. Sandala GM, Hopmann KH et al (2011) Calibration of DFT functionals for the prediction of Fe Mossbauer spectral parameters in iron–nitrosyl and iron–sulfur complexes: accurate geometries prove essential. J Chem Theory Comput 7:3232–3247
122. Pluth MD, Lippard SJ (2012) Reversible binding of nitric oxide to an Fe(III) complex of a tetra-amido macrocycle. Chem Commun (Camb) 48:11981–11983

123. Brothers SM, Darensbourg MY, Hall MB (2011) Modeling structures and vibrational frequencies for dinitrosyl iron complexes (DNICs) with density functional theory. Inorg Chem 50:8532–8540
124. Hess JL, Hsieh CH et al (2011) Self-assembly of dinitrosyl iron units into imidazolate-edge-bridged molecular squares: characterization including Mossbauer spectroscopy. J Am Chem Soc 133:20426–20434
125. Aquino F, Rodriguez JH (2009) Accurate calculation of zero-field splittings of (bio)inorganic complexes: application to an $FeNO^7$ (S = 3/2) compound. J Phys Chem A 113:9150–9156
126. Hopmann KH, Ghosh A, Noodleman L (2009) Density functional theory calculations on Mossbauer parameters of nonheme iron nitrosyls. Inorg Chem 48:9155–9165
127. Sanina N, Roudneva T et al (2009) Structure and properties of binuclear nitrosyl iron complex with benzimidazole-2-thiolyl. Dalton Trans 1703–1706
128. Behan RK, Hoffart LM et al (2007) Reaction of cytochrome P450BM3 and peroxynitrite yields nitrosyl complex. J Am Chem Soc 129:5855–5859
129. Sanina NA, Rudneva TN et al (2006) Influence of CH_3 group of μ-N–C–S ligand on the properties of $[Fe_2 (C_4H_5N_2S)_2 (NO)_4]$ complex. Inorg Chim Acta 359:570–576
130. Scheidt WR, Durbin SM, Sage JT (2005) Nuclear resonance vibrational spectroscopy (NRVS). J Inorg Biochem 99:60–71
131. Petrouleas V, Diner BA (1990) Formation by NO of nitrosyl adducts of redox components of the photosystem II reaction center. I. NO binds to the acceptor-side non-heme iron. Biochimica Et Biophysica Acta (BBA) Bioenergetics 1015:131–140
132. Goodman MS, Demarco MJ et al (2002) 99Ru Mossbauer effect study of ruthenium nitrosyls. J Chem Soc Dalton Trans 117–120
133. Callahan RW, Meyer TJ (1977) Reversible electron transfer in ruthenium nitrosyl complexes. Inorg Chem 16:574–581
134. Holsboer F, Beck W, Bartunik HD (1973) X-Ray photoelectron and Mössbauer spectroscopy of triphenylphosphine–iridium complexes. J Chem Soc Dalton Trans 1828–1829
135. Kamnev AA, Antonyuk LP et al (2003) Application of emission Mössbauer spectroscopy to the study of cobalt coordination in the active centers of bacterial glutamine synthetase. Dokl Biochem Biophys 393:321–325
136. Nath A, Harpold M et al (1968) Emission Mössbauer spectroscopy for biologically important molecules. Vitamin B_{12}, its analogs, and cobalt phthalocyanine. Chem Phys Lett 2:471–476
137. Krebs C, Price JC et al (2005) Rapid freeze-quench 57Fe Mössbauer spectroscopy: monitoring changes of an iron-containing active site during a biochemical reaction. Inorg Chem 44:742–757
138. Bochevarov AD, Friesner RA, Lippard SJ (2010) The prediction of ^{57}Fe Mössbauer parameters by the density functional theory: a benchmark study. J Chem Theory Comput 6:3735
139. Kurian R, Filatov M (2008) DFT approach to the calculation of Mössbauer isomer shifts. J Chem Theory Comput 4:278–285
140. Horner O, Oddou JL et al (2006) Mössbauer Identification of a protonated ferryl species in catalase from *Proteus mirabilis*: density functional calculations on related models. J Inorg Biochem 100:477–479
141. Friedle S, Reisner E, Lippard SJ (2010) Current challenges of modeling diiron enzyme active sites for dioxygen activation by biomimetic synthetic complexes. Chem Soc Rev 39:2768–2779
142. Mosbjek H, Poulsen KG (1971) Mössbauer investigation of the electronic structure of the "Brown-Ring" complex. Acta Chem Scand 25:2421–2427
143. Cheng HY, Chang S, Tsai PY (2003) On the "Brown-Ring" reaction product via density-functional theory. J Phys Chem A 108:358–361
144. Greatrex R, Greenwood NN, Kaspi P (1971) Ruthenium-99 Mossbauer spectra of some nitrosylruthenium (II) compounds. J Chem Soc A Inorg Phys Theor 1971:1873–1877

145. Ellison MK, Scheidt WR (1999) Synthesis, molecular structures, and properties of six-coordinate $[Fe(OEP)(L)(NO)]^+$ derivatives: elusive nitrosyl ferric porphyrins. J Am Chem Soc 121:5210–5219
146. Nasri H, Ellison MK et al (1997) Sharing the π-bonding. An iron porphyrin derivative with trans, π-accepting axial ligands. Synthesis, EPR and Mössbauer spectra, and molecular structure of two forms of the complex nitronitrosyl (α, α, α, α-tetrakis (*O*-pivalamidophenyl)-porphinato)Ferrate (II). J Am Chem Soc 119:6274–6283
147. Nasri H, Ellison MK et al (2006) Electronic, magnetic, and structural characterization of the five-coordinate, high-spin iron (II) nitrato complex $[Fe(Tpivpp)(NO_3)]^-$. Inorg Chem 45:5284–5290
148. Ellison MK, Schulz CE, Scheidt WR (2002) Nitrosyliron (III) porphyrinates: porphyrin core conformation and FeNO geometry. Any correlation? J Am Chem Soc 124:13833–13841
149. Liu MC, Huynh BH et al (1987) Optical, EPR and Mössbauer spectroscopic studies on the NO derivatives of cytochrome Cd1 from *Thiobacillus denitrificans*. Eur J Biochem 169:253–258
150. Lang G (1970) Mössbauer spectroscopy of haem proteins. Q Rev Biophys 3:1–60
151. Kelly M, Lang G (1970) Evidence from mossbauer spectroscopy for the role of iron in nitrogen fixation. Biochim Biophys Acta (BBA) Bioenergetics 223:86–104
152. Zhang Y, Mao J, Oldfield E (2002) 57Fe Mössbauer isomer shifts of heme protein model systems: electronic structure calculations. J Am Chem Soc 124:7829–7839
153. Zhang Y, Mao J et al (2002) Mossbauer quadrupole splittings and electronic structure in heme proteins and model systems: a density functional theory investigation. J Am Chem Soc 124:13921–13930
154. Thomson AJ, Cheesman MR, George SJ (1993) Variable-temperature magnetic circular dichroism. Meth Enzymol 226:199–232
155. Oganesyan VS, Sharonov YA (1998) Determination of zero-field splitting and evidence for the presence of charge-transfer transitions in the Soret region of high-spin ferric hemoproteins obtained from an analysis of low-temperature magnetic circular dichroism. Biochim Biophys Acta (BBA) Protein Struct Mol Enzymol 1429:163–175
156. McMaster J, Oganesyan VS (2010) Magnetic circular dichroism spectroscopy as a probe of the structures of the metal sites in metalloproteins. Curr Opin Struct Biol 20:615–622
157. Stephens PJ (1976) Magnetic circular dichroism. Adv Chem Phys 35:197–264
158. Solomon EI, Pavel EG et al (1995) Magnetic circular dichroism spectroscopy as a probe of the geometric and electronic structure of non-heme ferrous enzymes. Coord Chem Rev 144:369–460
159. Lehnert N, George SD, Solomon EI (2001) Recent advances in bioinorganic spectroscopy. Curr Opin Chem Biol 5:176–187
160. Oganesyan VS, George SJ et al (1999) A novel, general method of analyzing magnetic circular dichroism spectra and magnetization curves of high-spin metal ions: application to the protein oxidized rubredoxin, *Desulfovibrio gigas*. J Chem Phys 110:762–777
161. Zoppellaro G, Bren KL et al (2009) Review: studies of ferric heme proteins with highly anisotropic/highly axial low spin ($S = 1/2$) electron paramagnetic resonance signals with bis-histidine and histidine–methionine axial iron coordination. Biopolymers 91:1064–1082
162. Solomon EI, Bell CB (2010) Inorganic and bioinorganic spectroscopy. In: Physical inorganic chemistry. Principles, Methods, and Models (ed A. Bakac), John Wiley &Sons, Inc., Hoboken, NJ, USA
163. Dawson JH, Sono M (1987) Cytochrome P-450 and chloroperoxidase: thiolate-ligated heme enzymes. Spectroscopic determination of their active-site structures and mechanistic implications of thiolate ligation. Chem Rev 87:1255–1276
164. Cheesman MR, Greenwood C, Thomson AJ (1991) Magnetic circular dichroism of hemoproteins. Adv Inorg Chem 36:201–255
165. Cheek J, Dawson JH (1999) Magnetic circular dichroism spectroscopy of heme proteins and model systems. Porphyr Handb 7:339–369

166. Borisov V, Arutyunyan AM et al (1998) Magnetic circular dichroism used to examine the interaction of *Escherichia coli* cytochrome Bd with ligands. Biochemistry 38:740–750
167. Neese F, Solomon EI (1999) MCD C-term signs, saturation behavior, and determination of band polarizations in randomly oriented systems with spin S = 1/2. Applications to S = 1/2 and S = 5/2. Inorg Chem 38:1847–1865
168. Jackson TA, Yikilmaz E et al (2003) Spectroscopic and computational study of a non-heme iron $[Fe\text{-}NO]^7$ system: exploring the geometric and electronic structures of the nitrosyl adduct of iron superoxide dismutase. J Am Chem Soc 125:8348–8363
169. Diebold AR, Brown-Marshall CD et al (2011) Activation of α-Keto acid-dependent dioxygenases: application of an $\{FeNO\}^7/\{FeO_2\}^8$ methodology for characterizing the initial steps of O_2 activation. J Am Chem Soc 133:18148–18160
170. Orville AM, Lipscomb JD (1993) Simultaneous binding of nitric oxide and isotopically labeled substrates or inhibitors by reduced protocatechuate 3,4-dioxygenase. J Biol Chem 268:8596–8607
171. Ray M, Golombek AP (1999) Structure and magnetic properties of trigonal bipyramidal iron nitrosyl complexes. Inorg Chem 38:3110–3115
172. Berto TC, Speelman AL et al (2013) Mono- and dinuclear non-heme iron–nitrosyl complexes: models for key intermediates in bacterial nitric oxide reductases. Coord Chem Rev 257:244–259
173. Berto TC, Hoffman MB et al (2011) Structural and electronic characterization of non-heme Fe(II)-nitrosyls as biomimetic models of the Feb center of bacterial nitric oxide reductase. J Am Chem Soc 133:16714–16717
174. Chiou YM, Que J (1995) Model studies of α-keto acid-dependent nonheme iron enzymes: nitric oxide adducts of $[Fe^{II}(L)(O_2CCOPh)](ClO_4)$ complexes. Inorg Chem 34:3270–3278
175. Schneppensieper T, Finkler S et al (2001) Tuning the reversible binding of NO to iron (II) aminocarboxylate and related complexes in aqueous solution. Eur J Inorg Chem 491–501
176. Weber B, Görls H et al (2002) Nitrosyliron complexes of macrocyclic $[N_4{}^{2-}]$ and open-chain $[N_2O_2{}^{2-}]$ chelate ligands: influence of the equatorial ligand on the NO binding mode. Inorg Chim Acta 337:247–265
177. Hu C, Barabanschikov A et al (2012) Nuclear resonance vibrational spectra of five-coordinate imidazole-ligated iron (II) porphyrinates. Inorg Chem 51:1359–1370
178. Sage and JT (2001) Nuclear resonance vibrational spectroscopy of a protein active-site mimic. J Phys Condensed Matter 13:7707
179. Rai BK, Durbin SM et al (2002) Iron normal mode dynamics in (nitrosyl)iron (II)tetraphenylporphyrin from X-ray nuclear resonance data. Biophys J 82:2951–2963
180. Rai BK, Durbin SM et al (2003) Direct determination of the complete set of iron normal modes in a Porphyrin–Imidazole model for carbonmonoxy–heme proteins: [Fe(TPP)(CO)(1-MeIm)]. J Am Chem Soc 125:6927–6936
181. Kitagawa T, Mizutani Y (1994) Resonance raman spectra of highly oxidized metalloporphyrins and heme proteins. Coord Chem Rev 135–136:685–735
182. Kitagawa T, Ozaki Y (1987) Infrared and Raman spectra of metalloporphyrins. Metal complexes with tetrapyrrole ligands I. Struct Bond 64:71–114
183. Abe M, Kitagawa T, Kyogoku Y (1978) Resonance Raman spectra of octaethylporphyrinato-Ni (II) and meso-deuterated and N substituted derivatives. II. A normal coordinate analysis. J Chem Phys 69:4526
184. Kitagawa T, Abe M, Ogoshi H (1978) Resonance Raman spectra of octaethylporphyrinato-Ni (II) and meso-deuterated and 15N substituted derivatives. I. Observation and assignments of nonfundamental Raman lines. J Chem Phys 69:4516
185. Nakamoto K (2009) Infrared and Raman spectra of inorganic and coordination compounds, applications in coordination, organometallic, and bioinorganic chemistry. Wiley, Hoboken, NY
186. Paulat F, Praneeth VK et al (2006) Quantum chemistry-based analysis of the vibrational spectra of five-coordinate metalloporphyrins [M (TPP)Cl]. Inorg Chem 45:2835–2856

187. Hu S, Kincaid JR (1991) Resonance Raman spectra of the nitric oxide adducts of ferrous cytochrome P450cam in the presence of various substrates. J Am Chem Soc 113:9760–9766
188. Sitter AJ, Reczek CM, Terner J (1985) Observation of the Fe^{IV}-O stretching vibration of ferryl myoglobin by resonance Raman spectroscopy. Biochimica Et Biophysica Acta (BBA) Protein Struct Mol Enzymol 828:229–235
189. Boldt NJ, Donohoe RJ et al (1987) Chlorophyll model compounds: effects of low symmetry on the resonance Raman spectra and normal mode descriptions of nickel (II) dihydroporphyrins. J Am Chem Soc 109:2284–2298
190. Spiro TG, Li XY (1988) Resonance Raman spectroscopy of metalloporphyrins, 3rd edn. Wiley, New York
191. Procyk AD, Bocian DF (1992) Vibrational characteristics of tetrapyrrolic macrocycles. Annu Rev Phys Chem 43:465–496
192. Leu BM, Zgierski MZ et al (2004) Quantitative vibrational dynamics of iron in nitrosyl porphyrins. J Am Chem Soc 126:4211–4227
193. Zhu L, Sage JT, Champion PM (1994) Observation of coherent reaction dynamics in heme proteins. Science (New York) 266:629
194. Klug DD, Zgierski MZ et al (2002) Doming modes and dynamics of model heme compounds. Proc Natl Acad Sci 99:12526–12530
195. Scheidt WR, Duval HF et al (2000) Intrinsic structural distortions in five-coordinate (nitrosyl) iron (II) porphyrinate derivatives. J Am Chem Soc 122:4651–4659
196. Wyllie GR, Schulz CE, Scheidt WR (2003) Five- to six-coordination in (nitrosyl)iron (II) porphyrinates: effects of binding the sixth ligand. Inorg Chem 42:5722–5734
197. Ogoshi H, Watanbe E et al (1973) Synthesis and far-infrared spectra of ferric octaethylporphine complexes. J Am Chem Soc 95:2845–2849
198. Rehr JJ, Albers RC (2000) Theoretical approaches to X-ray absorption fine structure. Rev Mod Phys 72:621–654
199. Andersson L, Dawson J (1991) EXAFS spectroscopy of heme-containing oxygenases and peroxidases. Metal complexes with tetrapyrrole ligands II. Struct Bond 74:1–40
200. Luthra A, Denisov IG, Sligar SG (2011) Spectroscopic features of cytochrome P450 reaction intermediates. Arch Biochem Biophys 507:26–35
201. Shiro Y, Makino R et al (1991) Structural and electronic characterization of heme moiety in oxygenated hemoproteins by using XANES spectroscopy. Biochimica Et Biophysica Acta (BBA) Gen Subj 1115:101–107
202. Glaser T, Hedman B et al (2000) Ligand K-edge X-ray absorption spectroscopy: a direct probe of ligand-metal covalency. Acc Chem Res 33:859–868
203. Solomon EI, Hedman B et al (2005) Ligand K-edge X-ray absorption spectroscopy: covalency of ligand-metal bonds. Coord Chem Rev 249:97–129
204. Rompel A, Cinco RM et al (1998) Sulfur K-edge X-ray absorption spectroscopy: a spectroscopic tool to examine the redox state of S-containing metabolites in vivo. Proc Natl Acad Sci 95:6122–6127
205. Pickering IJ, Prince RC et al (1998) Sulfur K-edge X-ray absorption spectroscopy for determining the chemical speciation of sulfur in biological systems. FEBS Lett 441:11–14
206. Szilagyi RK, Schwab DE (2005) Sulfur K-edge X-ray absorption spectroscopy as an experimental probe for S-nitroso proteins. Biochem Biophys Res Commun 330:60–64
207. Martin-Diaconescu V, Kennepohl P (2007) Sulfur K-edge XAS as a probe of sulfur-centered radical intermediates. J Am Chem Soc 129:3034–3035
208. Tenderholt AL, Wang JJ et al (2010) Sulfur K-edge X-ray absorption spectroscopy and density functional calculations on Mo (IV) and Mo (VI)-O bis-dithiolenes: insights into the mechanism of Oxo transfer in DMSO reductase and related functional analogues. J Am Chem Soc 132:8359–8371
209. Lu TT, Lai SH et al (2011) Discrimination of mononuclear and dinuclear dinitrosyl iron complexes (DNICs) by S K-edge X-ray absorption spectroscopy: insight into the electronic structure and reactivity of DNICs. Inorg Chem 50:5396–5406

210. Dey A, Jenney FE et al (2007) Solvent tuning of electrochemical potentials in the active sites of Hipip versus Ferredoxin. Science 318:1464–1468
211. Dey A, Chow M et al (2005) Sulfur K-edge XAS and DFT calculations on nitrile hydratase: geometric and electronic structure of the non-heme iron active site. J Am Chem Soc 128:533–541
212. Lugo-Mas P, Dey A et al (2006) How does single oxygen atom addition affect the properties of An Fe-nitrile hydratase analogue? the compensatory role of the unmodified thiolate. J Am Chem Soc 128:11211–11221
213. Szilagyi RK, Bryngelson PA et al (2004) S K-edge X-ray absorption spectroscopic investigation of the Ni-containing superoxide dismutase active site: new structural insight into the mechanism. J Am Chem Soc 126:3018–3019
214. Shearer J, Dehestani A, Abanda F (2008) Probing variable amine/amide ligation in $Ni^{II}N_2S_2$ complexes using sulfur K-edge and nickel L-edge X-ray absorption spectroscopies: implications for the active site of nickel superoxide dismutase. Inorg Chem 47:2649–2660
215. Glaser T, Rose K et al (2000) S K-edge X-ray absorption studies of tetranuclear iron–sulfur clusters: α-Sulfide bonding and its contribution to electron delocalization. J Am Chem Soc 123:442–454
216. Shadle SE, Hedman B et al (1994) Ligand K-edge X-ray absorption spectroscopy as a probe of Ligand–Metal bonding: charge donation and covalency in copper–chloride systems. Inorg Chem 33:4235–4244
217. Sun N, Liu LV et al (2010) S K-edge X-ray absorption spectroscopy and density functional theory studies of high and low spin $\{FeNO\}^7$ thiolate complexes: exchange stabilization of electron delocalization in $\{FeNO\}^7$ and $\{FeO_2\}^8$. Inorg Chem 50:427–436
218. Shadle SE, Hedman B et al (1995) Ligand K-edge X-ray absorption spectroscopic studies: Metal–Ligand covalency in a series of transition metal tetrachlorides. J Am Chem Soc 117:2259–2272
219. Chen LX, Bowman MK et al (1994) Structural studies of photoinduced intramolecular electron transfer in cyclopentadienylnitrosylnickel. J Phys Chem 98:9457–9464
220. Crichton O, Rest AJ (1978) Photochemistry of carbonyltrinitrosylmanganese in frozen gas matrices at 20 K. Infrared spectroscopic evidence for trinitrosylmanganese, (dinitrogen) trinitrosylmanganese, and a species formed by metal-to-nitrosyl photoelectron transfer. J Chem Soc Dalton Trans 202–207
221. Woike T, Zöllner H et al (1990) Raman-spectroscopic and differential scanning calorimetric studies of the light induced metastable states in $K_2[RuCl_5NO]$. Solid State Commun 73:149–152
222. Hedberg L, Hedberg K et al (1985) Structure and bonding in transition-metal carbonyls and nitrosyls. 1. Gas-phase electron-diffraction investigations of tetranitrosylchromium (Cr $(NO)_4$, carbonyltrinitrosylmanganese (MnCo $(NO)_3$), and dicarbonyldinitrosyliron ($Fe(CO)_2$ $(NO)_2$). Inorg Chem 24:2766–2771
223. Johnson BFG, McCleverty JA (1966) Nitric oxide compounds of transition metals. In: Progress in inorganic chemistry. (2007) Vol 7 (ed F. A. Cotton), John Wiley &Sons, Inc., Hoboken, NJ
224. Crichton O, Rest AJ (1978) Photochemistry of tetracarbonylnitrosylmanganese in frozen gas matrices at 20 K. Infrared spectroscopic evidence for tricarbonylnitrosylmanganese, tricarbonyl (dinitrogen)nitrosylmanganese, and a species formed by metal-to-nitrosyl photoelectron transfer. J Chem Soc Dalton Trans 208–215
225. Frenz BA, Enemark JH, Ibers JA (1969) Structure of nitrosyltetracarbonylmanganese (0), Mn $(NO)(CO)_4$. Inorg Chem 8:1288–1293
226. Treichel PM, Pitcher E et al (1961) Chemistry of the metal carbonyls. X. Tetracarbonylnitrosylmanganese (0)1,2. J Am Chem Soc 83:2593–2594
227. Crichton O, Rest AJ (1977) Photochemistry of dicarbonyldinitrosyliron in frozen gas matrices at 20 K. Infrared spectroscopic evidence for carbonyldinitrosyliron and carbonyl (dinitrogen)dinitrosyliron. J Chem Soc Dalton Trans 656–661

228. Crichton O, Rest AJ (1977) Photochemistry of tricarbonylnitrosylcobalt in frozen gas matrices at 20 K. Infrared spectroscopic evidence for dicarbonylnitrosylcobalt and dicarbonyl (dinitrogen)nitrosylcobalt. J Chem Soc Dalton Trans 536–541
229. Hedberg K, Hedberg L et al (1985) Structure and bonding in transition-metal carbonyls and nitrosyls. 2. Gas-phase electron diffraction reinvestigation of tricarbonylnitrosylcobalt. Inorg Chem 24:2771–2774
230. Mcdowell RS, Horrocks J, Yates JT (1961) Infrared spectrum of Co (CO)[Sub 3]NO. J Chem Phys 34:530–534
231. Fischer EO, Kuzel P (1962) Ein Kationsicher Cyclopentadienyl–Nitrosylcarbonyl–Komplex Des Chroms. Z Anorg Allg Chem 317:226–229
232. Praneeth VK, Haupt E, Lehnert N (2005) Thiolate coordination to Fe(II)-porphyrin NO centers. J Inorg Biochem 99:940–948
233. Armstrong RN (1997) Structure, catalytic mechanism, and evolution of the glutathione transferases. Chem Res Toxicol 10:2–18
234. Lewandowska H, Kalinowska M et al (2011) Nitrosyl iron complexes-synthesis, structure and biology. Dalton Trans 40:8273–8289
235. Rusanov V, Stankov S et al (2009) Determination of the Mössbauer parameters of rare-earth nitroprussides: evidence for new light-induced magnetic excited state (LIMES) in nitroprussides. Journal of Solid State Chemistry 182:1252–1259
236. Pierpont CG, Van Derveer DG et al (1970) Ruthenium complex having both linear and bent nitrosyl groups. J Am Chem Soc 92:4760–4762
237. Grundy KR, Laing KR, Roper WR (1970) Dinitrosyl complexes of ruthenium and osmium and their reaction with oxygen. J Chem Soc D 1500–1501
238. Waters JM, Whittle KR (1971) The crystal structure of a dinitrosyl complex of osmium (II). J Chem Soc D 518
239. Li Kam Wah H, Postel M, Pierrot M (1989) Structure-activity correlation in iron nitrosyl complexes: crystal structures of [$Fe(NO)_2$ (Cl)]$_2$ (α-Dppe) and $Fe(NO)_2$ (Dppe). Inorg Chim Acta 165:215–220
240. Brock CP, Collman JP et al (1973) Bent vs. linear nitrosyl paradox. Infrared and X-ray photoelectron spectra of dichloronitrosylbis (L)cobalt (II) and crystal structure with L=diphenylmethylphosphine. Inorg Chem 12:1304–1313
241. Alnaji O, Peres Y et al (1986) Cox2 (NO)(PMe_3)$_2$ complexes: an example of the influence of X (X=CI, Br, I, NO_2) on the stereochemistry and electronic structure of the five-coordinate $\{Co\text{-}NO\}^8$ complexes. Inorg Chim Acta 114:151–158
242. Armor JN, Scheidegger HA, Taube H (1968) A bimolecular mechanism for substitution. J Am Chem Soc 90:5928–5929
243. Lu TT, Yang LB, Liaw WF (2010) Trigonal bipyramidal $\{Fe(NO)\}^7$ complex [(NO)Fe(SC_9H_6N)$_2$] containing an equatorial nitrosyl ligand: the critical role of chelating ligands in regulating the geometry and transformation of mononitrosyl iron complex (MNIC). J Chin Chem Soc 57:909
244. Lu TT, Chiou SJ et al (2006) Mononitrosyl tris (thiolate) iron complex $[Fe(NO)(Sph)_3]^-$ and dinitrosyl iron complex $[(EtS)_2Fe(NO)_2]^-$: formation pathway of dinitrosyl iron complexes (DNICs) from nitrosylation of biomimetic rubredoxin $[Fe(SR)_4]^{2\text{-}/1\text{-}}$ (R=Ph, Et). Inorg Chem 45:8799–8806
245. Harrop TC, Song D, Lippard SJ (2006) Interaction of nitric oxide with tetrathiolato iron (II) complexes: relevance to the reaction pathways of iron nitrosyls in sulfur-rich biological coordination environments. J Am Chem Soc 128:3528–3529
246. Hung MC, Tsai MC et al (2006) Transformation and structural discrimination between the neutral $\{Fe(NO)2\}^{10}$ dinitrosyliron complexes (DNICs) and the anionic/cationic $\{Fe(NO)_2\}^9$ DNICs. Inorg Chem 45:6041–6047
247. Ting-Wah Chu C, Yip-Kwai Lo F, Dahl LF (1982) Synthesis and stereochemical analysis of the [$Fe_4(NO)_4$ (.Mu.3-S)$_4$]N Series (N=0, −1) which possesses a Cubanelike Fe_4S_4 core: direct evidence for the antibonding tetrametal character of the unpaired electron upon a one-electron reduction of a completely bonding tetrahedral metal cluster. J Am Chem Soc 104:3409–3422

248. Baird P, Bandy JA et al (1991) Charge-transfer salts formed from redox-active cubane cluster cations $[M_4(\eta\text{-}C_5H_4R)_4(\mu_3\text{-}E)_4]^{n+}$(M=Cr, Fe or Mo; E=S or Se) and various anions. J Chem Soc Dalton Trans 2377–2393
249. Sedney D, Rieff WM (1979) A Mössbauer spectroscopy investigation of iron chromophore equivalence in three tetrairon clusters. Inorg Chim Acta 34:231–236
250. Davies SC, Evans DJ et al (2002) Mononuclear, binuclear, trinuclear and tetranuclear iron complexes of the $N(CH_2CH_2S)_3^{3-}$ (NS_3) ligand with nitrosyl co-ligands. J Chem Soc Dalton Trans 2473–2482
251. Sanina NA, Rakova OA et al (2004) Structure of the neutral mononuclear dinitrosyl iron complex with 1,2,4-triazole-3-thione $[Fe(SC_2H_3N_3)(SC_2H_2N_3)(NO)_2]\cdot 0.5\ H_2O$. Mendeleev Commun 14:7–8
252. Sanina NA, Aldoshin SM (2004) Functional models of [Fe–S] nitrosyl proteins. Russian Chem Bull 53:2428–2448
253. Harrop TC, Tonzetich ZJ et al (2008) Reactions of synthetic [2Fe-2S] and [4Fe-4S] clusters with nitric oxide and nitrosothiols. J Am Chem Soc 130:15602–15610
254. Butcher RJ, Sinn E (1980) Proof of a linear Fe–NO linkage, perpendicular to the S_4 plane in $Fe(NO)(S_2CN\{CH(CH_3)$ 2–2) 2. A new (aerobic) synthesis of the complex and its crystal structure. Inorg Chem 19:3622–3626
255. Johnson CE, Rickards R, Hill HAO (1969) Mössbauer-effect study of nitrosyliron bis (*N*, *N*-Diethyldithiocarbamate). J Chem Phys 50:2594–2597
256. Feig AL, Bautista MT, Lippard SJ (1996) A carboxylate-bridged non-heme diiron dinitrosyl complex. Inorg Chem 35:6892–6898
257. Christner JA, Janick PA et al (1983) Mössbauer studies of *Escherichia coli* sulfite reductase complexes with carbon monoxide and cyanide. Exchange coupling and intrinsic properties of the [4Fe-4S] cluster. J Biol Chem 258:11157–11164

Struct Bond (2014) 153: 167–228
DOI: 10.1007/430_2013_97

Published online: 10 December 2013

Nitrosyl Complexes in Homogeneous Catalysis

Yanfeng Jiang and Heinz Berke

Abstract In this chapter the application of transition metal nitrosyl complexes in homogeneous catalysis has been reviewed. Particular attention was paid to the function of nitrosyl as: (1) a π-accepting ancillary ligand; (2) a non-innocent ligand capable of reversible linear/bent transformations triggering catalytic reaction courses; (3) a redox-active ligand functioning as a "nitrosyl/nitro" redox couple for oxygen atom transfer reactions. The catalytic performance and the reaction mechanisms are discussed in terms of the respective function of the NO ligand. Group 6 molybdenum and tungsten mononitrosyl hydride complexes are reviewed with respect to their activity in hydrogenations proceeding with heterolytic splitting of H_2. Group 7 rhenium mononitrosyl and dinitrosyl hydride complexes are reviewed with respect to their highly efficient performance in hydrogen-related catalyses, which is highlighted by alkene hydrogenations based on Re(I) mononitrosyl complexes and efficient transfer hydrogenations based on bifunctional Shvo-type Re–H/OH and Noyori-type Re–H/NH complexes. Group 8 iron mononitrosyl- and dinitrosyl-based catalysts are surveyed with respect to their function as Lewis acids inducing nucleophilic activity in Lewis acid type organic transformations. Then the "catalytic nitrosyl effect" is described boosting catalyses by reversible nitrosyl bending, which plays a crucial role in Re(−I) dinitrosyl and Re(I) mononitrosyl catalytic systems. The oxygen atom transfer reaction based on the "nitrosyl/nitro" redox couple involving the Co, Pd, Rh, and Ru metal centers was also reviewed.

Keywords Ancillary ligand · Bifunctional · Homogeneous catalysis · Nitrosyl · Non-innocent · Redox-active · *Trans*-effect · *Trans*-influence

---

Y. Jiang and H. Berke (✉)
Anorganisch-Chemisches Institut, Universität Zürich, Winterthurerstr. 190, 8037 Zurich, Switzerland
e-mail: jiang@aci.uzh.ch; hberke@aci.uzh.ch

Contents

Abbreviations

BAr^F_4	$B[3,5\text{-}(CF_3)_2C_6H_3]_4$
Cy	Cyclohexyl
Dcype	1,2-Bis(dicyclohexylphosphino)ethane
Dippe	1,2-Bis-(diisopropylphosphino)ethane
Dippf	1,1′-Bis(diisopropylphosphino)ferrocene
Dippp	1,3-Bis(diisopropylphosphino)-propane
Diprpfc	1,1′-Bisdiisopropylphosphinopherrocene
DMF	Dimethylformamide
Dmpe	1,2-Bis-(dimethylphosphino)ethane
Dpephos	Bis(2-(diphenylphosphino)phenyl) ether
Dppfc	1,1′-Bis(diphenylphosphino)ferrocene
DQCC	Deuterium quadrupole coupling constants
edta	Ethylenediaminetetraacetate
Homoxantphos	10,11-Dihydro-4,5,-bis(diphenylphosphino)dibenzo[b,f]oxepine
IMes	1,3-Bis-(2,4,6-trimethylphenyl)imidazol-2-ylidene
KOtAm	Potassium 2-methylbutan-2-olate
LA	Lewis acid
MS	Molecular sieves
MTBE	Methyl *tert*-butyl ether
Ph	Phenyl

Pr	Isopropyl
Py	Pyridine
Saloph	*N,N′*-bisalicylidene-*o*-phenylenediamino
SIMes	1,3-Bis(2,4,6-trimethylphenyl)imidazolin-2-ylidene
Sixantphos	4,6-Bis(diphenylphosphino)-10,10-dimethylphenoxasilin
TBABr	Tetrabutylammonium bromide
TBAF	Tetrabutylammonium fluoride
TOF	Turn over frequency
TON	Turn over number
TPP	Tetraphenylporphyrin

1 Introduction

Processes of homogeneous catalysis are playing a pivotal role in chemical industry and academic research, which are to the main part catalyzed by transition metal compounds. Because of numerous publications and text books surveying the field of transition metal-catalyzed homogeneous catalysis, this area appears to be at first glance completed and a fully explored research topic of chemistry. Despite such impressions there is still great interest in this field and attracts even increasing attention.

Homogenous catalysis possesses lower catalytic activities in comparison with heterogeneous catalysis. But homogeneous catalysis revealed advantageously high selectivities including chemo-, regio-, diastereo-, and enantioselectivity, and therefore it became particularly important in fine chemical production. Progress in the field is grounded on sophisticated transition metal organometallic or coordination chemistry, which can often readily be adjusted to specific needs by "tuning" efforts varying the physical reaction conditions or the chemical conditions by variation of the metal center in conjunction with the surrounding ligands of the catalysts. Prominent representative processes, which have found wide industrial applications, are the Ziegler–Natta titanium- and zirconium-based catalyses for olefin polymerizations [1, 2], Noyori ruthenium-based asymmetric hydrogenations and transfer hydrogenations [3–7], Grubbs/Schrock ruthenium/molybdenum catalyses for olefin metathesis reactions [8–10], and palladium-based C–C coupling reactions. All fields, for which Nobel Prizes had been given and which seem as yet not completed, are developed further in worldwide research efforts by various groups [11–13].

In this context homogeneous catalysis with transition metal nitrosyl compounds bearing the nitrosyl ligand as an ancillary or functional ligand has as yet not played a significant role. Homogeneous catalytic processes were mainly developed based on platinum group metals and these looked as if they would not possess the ideal catalysis match with the NO ligand. Within organometallic or coordination chemistry the NO ligand was more regarded as a candidate to beneficially fertilize structural and material's research. However the discovery of several fundamentally important biological functions of NO led to a change of the image of NO boosting respective research interests in biochemistry, bioinorganic coordination chemistry, and biomedical research, particularly in the fields of enzyme research, biological signaling, and immune defense, and also to develop drugs based on NO [14, 15]. Except for the biological area, the catalytic potential of transition metal nitrosyl complexes for homogeneous organic transformations is still relatively unexplored, despite the fact that gaseous NO itself has been known to be "catalytic" since ca. 1750 in the historically quite important "lead chamber process" oxidizing SO_2 to SO_3 and sulfuric acid in the presence of oxygen [16].

This quite general and highly selective introduction on nitric oxide chemistry had the aim to stress the importance of reviewing the organometallic and coordination chemistry of NO, whenever this molecule may get connected to a function in homogeneous catalysis. This chapter will describe well-established examples of the versatility of nitrosyl-based homogeneous catalyses and maybe it will help to identify "nitrosyl niches" worth for exploitation and development to accomplish new and better performing transition metal catalysts, in particular those with middle transition elements.

2 Nitrosyl Ligand Effects in Homogeneous Catalysis

Nitrosyl (NO) is one of the prototype examples of a non-innocent ligand able to bind to metal centers in two structurally different forms, the linear and the bent form [17, 18]. Simplified MO pictures of M–NO binding are depicted in Scheme 1. The linear nitrosyl is considered as a 2e donor NO^+ (nitrosonium) isoelectronic to CO. The linear M–NO bond consists mainly of σ-donation from a filled σ-orbital on the N_{NO} atom to an empty σ type d orbital of the metal center, and of π-back donation, which occurs by the interaction of two filled π type metal d orbitals with the two perpendicular π*-orbitals of the NO ligand providing on total a conical shape of the π bonds [19–21]. The bent nitrosyl (nitroxyl) is regarded as a 2e ligand NO^- isoelectronic to the diazenido ligand –N=NR, or in rare cases also as a 2e ligand NO possessing a single electron "lone pair" (nitric oxide). In both these later cases, the nitrosyl functions as a σ-donor and a single-faced π-donor (Scheme 1) [20]. The non-bonding electron pair or to a much lesser extent the single electron "lone pair" on the N_{NO} atom can endow a *cis*-labilizing effect to repel (or destabilize) neighboring M–L bonds. With an electron-deficient or low d-electron count metal center, the linear nitrosyl prevails functioning merely as a σ-donor. Vice versa, the bent nitrosyl is thermodynamically more favored than the linear nitrosyl when the metal center becomes electron-rich or has a high d-electron count. This structural change can be viewed as a formal redox process transferring a pair of electrons originally assigned to the metal center in the NO^+ state to the N_{NO} atom lone pair electrons in the NO^- state. Similarly a single electron can be transferred

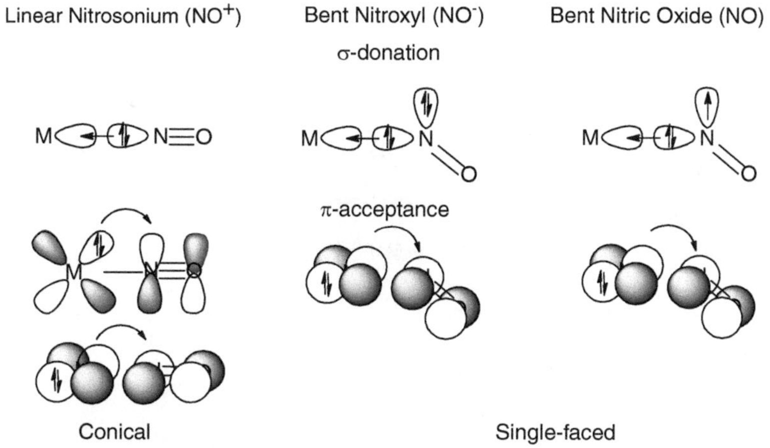

Scheme 1 Simplified MO picture of M–NO binding

from the metal center to the N_{NO} atom. This process describes the non-innocence property [22, 23]. It should be noted that the linear nitrosyl coordination is dominating in transition metal nitrosyl chemistry [19]. Therefore, when "nitrosyl" is stated in this chapter it refers to a linear NO^+, unless otherwise specified.

The NO ligand may possess three different functions in transition metal-based catalyses:

- Serving as a strongly bound ancillary ligand with certain electronic properties and effects to be imposed on the residual coordination sphere, functioning then mainly as 2-electron NO^+ ligand or as a 3-electron donor in the neutral form
- Providing temporary vacant sites by the reversible $NO^+ \leftrightarrow NO^-$ ligand transformation
- Acting as a reversible oxene $\langle O \rangle$ source via the $L_nM\text{–}NO + \langle O \rangle \leftrightarrow L_nM\text{–}NO_2$ equilibrium

2.1 NO as an Ancillary Ligand

A linear nitrosyl ligand is slim, monodentate, and binds strongly to the metal center. The resonance forms contributing to linear metal nitrosyl binding are shown in Scheme 2 [14]. Due to electron delocalization into the M–N bond, the M–N bonds are in general thermodynamically strong when double bonded or triply bonded forms prevail [19]. Therefore the NO ligand can often not easily be released from the metal center and can then offer function as an ancillary ligand and ligand placeholder also in transition metal nitrosyl catalysts.

As an ancillary ligand, the linear nitrosyl is a strong π-acceptor, even stronger in this respect than CO. However, the ancillary NO has also specific influence on

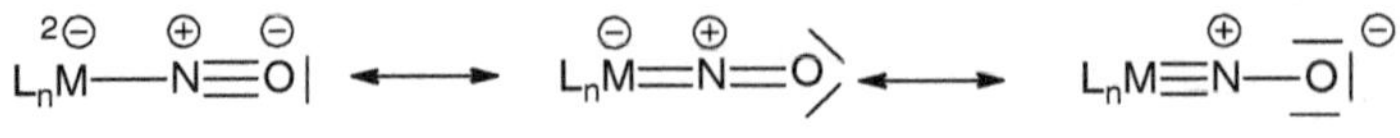

Scheme 2 Resonance forms of a linear M–NO unit

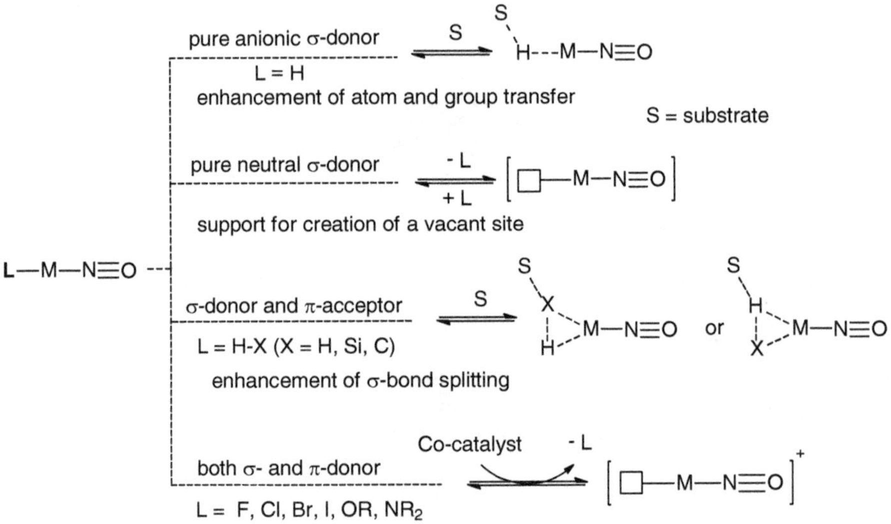

Scheme 3 Activation of σ-donor, σ-donor/π-acceptor and σ-donor/π-donor ligands by linear *trans* nitrosyls

σ-bonded ligands *trans* to nitrosyl. They can be activated by a catalytically relevant strong *trans*-effect and the *trans*-influence of the NO ligand. Depending on the type of the *trans*-ligand L, three scenarios can get involved concerning the activation function of the ancillary NO ligand, as depicted in Scheme 3:

(a) A pure anionic σ-donor, for instance a hydride ligand, possessing no π-donating property. In this case, the nitrosyl ligand could, via its *trans*-influence, weaken the M–H bond and enhance concomitantly the ionicity of the *trans* M–H bond. It would become more reactive toward external electrophiles facilitating M–H transfer reactions. This NO function is particularly prominent in secondary coordination sphere activation processes, wherein the reaction occurs on the hydride ligand instead of at the metal site. A pure neutral σ-donor L *trans* to the nitrosyl ligand is subjected to facile exchange or removal along the *trans*-effect [24]. The effect is based on σ bond weakening by the *trans*-influence and stabilization of the reactive intermediate obtained after elimination of L.

(b) σ-donor and π-acceptor η^2-bonded ligands, such as H–X (X = H, Si, C). In this case, the ancillary NO ligand is prone to assist heterolytic H–X bond activation [25, 26]. For instance, the H–H bond *trans* to nitrosyl possesses a classical dihydrogen character due to the reduced σ- and π-binding strength between M and H–H. This increases the acidity of the H–H ligand, thus promoting the heterolytic splitting of the H–H bond via deprotonation by an external basic substrate or an internal basic ligand [27, 28]. This circumstance can get pivotal to the so-called catalytic ionic hydrogenation reactions occurring with heterolysis of the H–H bond [29]. In the same manner, the heterolytic cleavage of a Si–H bond could be facilitated by a *trans* disposed linear nitrosyl. Even the highly stable sp^3-C–H bond can get susceptible to splitting fostered by the π-accepting NO ligand.

(c) Both σ-donor and π-donor, such as halide (F, Cl, Br, I), alkoxide (−OR), and amide ($-NR_2$) ligands. In this case, the *trans*-effect and *trans*-influence of nitrosyl ligands can be at least partly compensated by a push–pull π effect between the *trans* aligned π-donor X and the π-acceptor NO [30]. The *trans*-ligand L reveals enhanced bonding to the metal center, and the nitrosyl shows features of the chemical properties of isoelectronic metal carbonyls. Special chemical efforts have to be undertaken in order to render catalytic activation. For instance, the employment of appropriate co-catalysts capable of trapping the *trans*-L ligand would facilitate the generation of a highly electron-deficient metal centers bearing one vacant site. This function is especially suited to initiate catalysis in primary coordination sphere mechanisms, wherein a vacant site on an electron-poor metal center is generally a prerequisite.

2.2 The (Reversible) $NO^+ \leftrightarrow NO^-$ Ligand Transformation

As mentioned before the NO ligand is considered as a "non-innocent" ligand capable of binding to a metal center either as NO^+ (nitrosonium cation), a ·NO radical (nitric oxide, neutral), or NO^- (nitroxide anion) [19]. Based on this diversity of ligand types the presence of nitrosyl ligands always creates ambiguity with respect to the oxidation state assignment of the metal center. The different binding modes cause different oxidation states and changes in the covalent nature of the M–N–O interaction. Enemark and Feltham proposed a formalism to describe metal nitrosyl complexes, which treated the metal nitrosyl as a single entity [17]. This was represented as $\{M(NO)_x\}^n$, in which n is the total number of electrons associated with the metal d and π* (NO) orbitals. The number of d-electron is determined by the formal oxidation state of the metal atom assuming no charge on the nitrosyl group. A clear differentiation in the M–N–O bond angle is observed in the series $\{MNO\}^6 \rightarrow \{MNO\}^7 \rightarrow \{MNO\}^8$ (Scheme 4). The M–N–O moiety is essentially linear for the $\{MNO\}^6$ configuration describing nitrosyl bound as nitrosonium (NO^+). In contrast, the bond angles for $\{MNO\}^7$ and $\{MNO\}^8$ have been estimated to be $145 \pm 10°$ and $125 \pm 10°$ for bound nitric oxide (·NO) and nitroxyl (NO^-), respectively. A difference in N–O bond length has also been observed for these three different configurations.

The non-innocent character of the NO ligand is a basic feature of transition metal nitrosyl chemistry. Indeed, this feature may account also for the fact that nitrosyl complexes are rarely found in applications of homogeneous catalysis. Namely, the

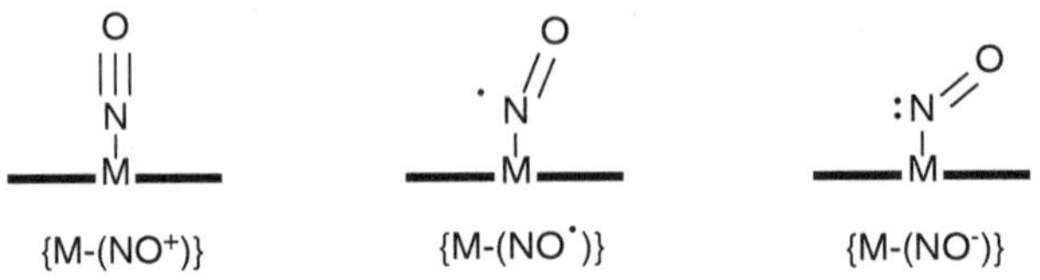

Scheme 4 The M–N–O bond angles in mononitrosyl complexes

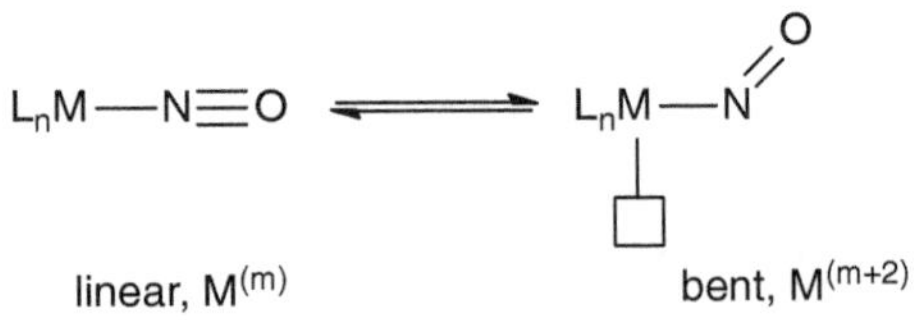

Scheme 5 Transformation of linear/bent nitrosyls with reversible availability of a vacant site

nitrosyl ligand is in principal flexible in stabilizing various types of organometallic species involved in catalytic processes by functioning as either an electron-reservoir (to supply certain electron density to the metal center) or an electron scavenger (to suppress and "hideaway" excess of electron density on metal centers). Therefore both the thermodynamics and kinetics of catalytic processes can be counteracted by the non-innocence of the NO ligand. This capability of flexible adjustment hinders catalysis in many cases. However, if these effects can be exploited in a rational design approach, they may serve and support homogeneous catalysis.

Indeed, when the equilibrium between the linear and bent form of a nitrosyl ligand could be "perturbed" by external influences so that only one coordination mode prevails, catalytic activity of the nitrosyl complex might be triggered and enhanced establishing the so-called "catalytic nitrosyl effect" (Scheme 5) [31]. The bending of a nitrosyl ligand would create a vacant site accompanied by a *formal* 2e oxidation of the metal center without the requirement of ligand dissociation. Such a feature is particularly useful for middle transition metal-based catalytic processes, which generally stick strongly to the 18e rule showing strong reluctance for ligand dissociation. Furthermore, nitrosyl possesses an oxygen atom that shows certain Lewis basicity [32, 33]. Interaction at the O_{NO} atom with Lewis acids would also increase the π-back donation from the metal center to the nitrosyl ligand, which could in principle perturb the linear/bent equilibrium by retaining only the bent nitrosyl form for a specific activation course. However, in spite of such a sounded-promising "catalytic nitrosyl effect," the establishment of a transition metal nitrosyl system showing appropriate electronic properties at the metal center remains still challenging.

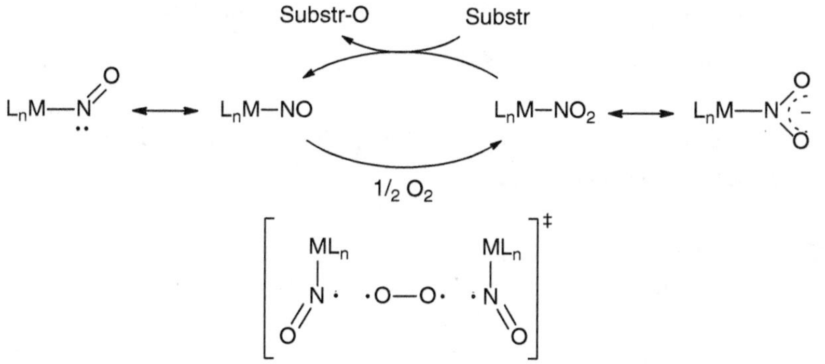

Scheme 6 The nitrosyl–nitro redox couple in homogeneous oxidation catalysis

2.3 The L_nM–NO + ⟨O⟩ ↔ L_nM–NO_2 Equilibrium

Transition metal nitrosyl complexes can also be involved as nitrosyl/nitro redox couples to promote secondary coordination sphere oxo- or oxene-transfer reactions, which has great perspective in dioxygen activation and industrial oxene-transfer catalyses [34]. Such ligand-oriented redox transformations appear to be rare, but are crucial to drive special catalyses, like the synthesis of epoxides, ketones, and other related "O-containing" oxidized species [35, 36]. As depicted in Scheme 6, certain NO ligands (mainly of the nitric oxide type) can react with dioxygen to generate nitro complexes mostly with concomitant oxidation of the metal center. In case a bent nitrosyl is initially involved, the oxidation state of the metal would remain unchanged, and the change in oxidation state occurs only on the nitrogen atom of the NO group. Subsequently the formed nitro complex undergoes the secondary coordination sphere oxo-transfer from the nitro ligand to an organic substrate affording oxidized products with completion of the catalytic cycle.

Considering the pathway of oxidation from the nitrosyl to the nitro ligand in the reaction with dioxygen, it can be rationalized that the nitrosyl should initially exist in the moderately bent form exhibiting free-radical nitric oxide type character. In this way the dioxygen, which in the ground state is a triplet diradical, reacts with the L_nM–NO radical through a transition state with "SOMO–SOMO" interaction to afford the oxidized nitro ligand via a peroxydinitro species. It has also been proposed that the reaction of the nitrosyl ligand with dioxygen could initially produce a mononuclear peroxynitro intermediate, which would then transfer one oxygen atom to the substrate forming the oxidized product along with the generation of a nitro complex.

In the following sections, we will review the known homogeneous transition metal nitrosyl catalyses with regard to the described three major NO functions. Not only the catalytic performance, but also the reaction mechanisms with reference to the various roles of the nitrosyl group in catalytic reaction courses will be evaluated.

3 The Ancillary Ligand Function of Nitrosyl Ligands in Homogeneous Catalysis

3.1 *Group 6 Molybdenum and Tungsten Nitrosyl Hydrides for Ionic Hydrogenations*

In hydrogen-related homogeneous catalyses the activation of L_nM–H bonds may play the decisive role. Insertion of unsaturated molecules such as alkene, imines, and ketones into transition metal hydride bonds are strongly dependent on the electronic properties of the metal center, which are determined by stereo electronic factors of the complexes, including the d-electron count and the oxidation state of the metal center and the respective properties of the ancillary ligand sphere [37]. The presence of strong σ-donating alkylphosphines is expected to enhance the electron density at the metal center and concomitantly at the hydride ligand. The resulting increased hydride transfer activity can be promoted further via the activation effect by *trans*-positioned strong π-acceptors including the nitrosyl ligand, which are attributed to the strong *trans*-effect and *trans*-influence of this group weakening the M–H bond [38]. In line with this thought, the catalytic reactivity of low valent group 6 molybdenum and tungsten complexes possessing *trans* aligned hydride and NO ligands with trialkylphosphine ligands is discussed in this section.

3.1.1 Activation of M–H and M–(H_2) Bonds in ON–M–H or ON–M–(H_2) (M = Mo, W) Complexes

For years the Berke group has systematically investigated the ligand sphere tuning changing the character of group 6 L_nM–H bonds [39]. A series of octahedral molybdenum and tungsten hydrides $[M(CO)_{4-n}(H)(NO)(PR_3)_n]$ (M = W, Mo, $n = 3$–4) with nitrosyl groups as ancillary, but also *trans*-ligands and strongly σ-donating alkylphosphines as *cis*-ligands have been studied [40–44]. Representative examples are listed in Scheme 7. The L_nM–H bonds in these compounds are characterized by bond ionicity determinations and the insertion behavior as a chemical measure for their hydride transfer capabilities.

Dating back to 2000, the organometallic chemistry of *mer*-Mo(CO)(H)(NO)$(PMe_3)_3$ (**1**) was reported with respect to the insertion capabilities into the Mo–H bond [40]. This hydride shows an unusual high propensity to react with carbonyl groups of aldehydes, ketones, $Fe(CO)_5$, $Re_2(CO)_{10}$, and CO_2 resembling closely that of metallocene hydride compounds with early transition metal centers (Scheme 8). This was to a great extent attributed to the linear nitrosyl ligand *trans* to the hydride and the three σ-donating trimethylphosphine groups. However, the insertion of relatively inert C=N bond of imines into the Mo–H bond failed, which suggested further tuning of the coordination sphere to increase the hydridicity of the M–H bonds.

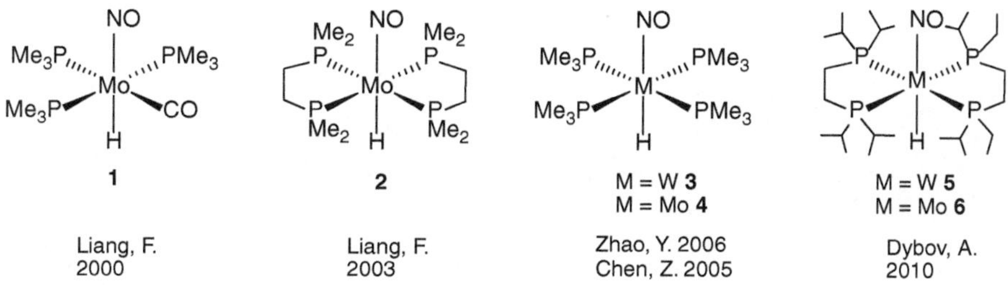

Scheme 7 Group 6 ON–M–H (M = W, Mo) complexes

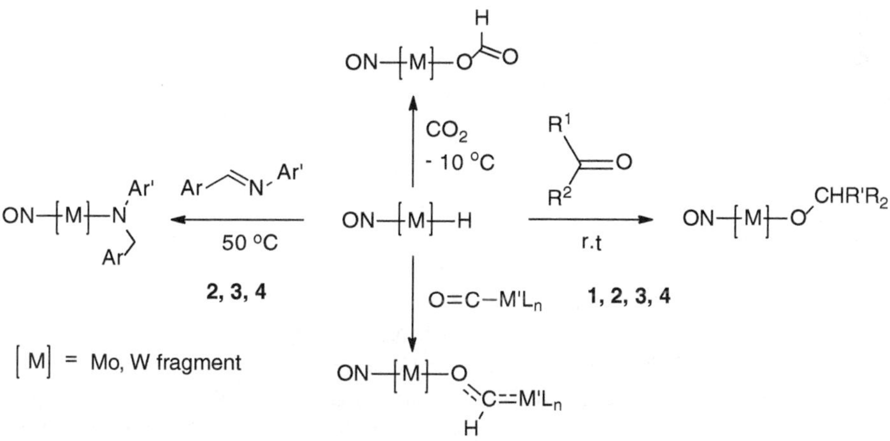

Scheme 8 Reactivities of Group 6 ON–M–H (M = W, Mo) complexes

As demonstrated in the nitrosyl-free $Re(CO)_nH(PR_3)_{5-n}$ series, an increasing number of σ-donating phosphine ligands enhances the ionicity or hydridicity at the M–H bond, leading eventually to studies of the tetraphosphine-substituted Mo–H complex *trans*-$Mo(dmpe)_2(H)(NO)$ (**2**) bearing 1,2-bis-(dimethylphosphino)ethane (dmpe) ligands [41]. The enhanced polarity of the Mo–H bond of **2** compared to that of **1** was verified by determinations of the deuterium quadrupole coupling constants (DQCC) in solution from $T_{1\ min}$ measurements of the ^2H nucleus and in solid state from static ^2H-NMR spectra of the Mo–D substituted analogues. This is also in agreement with the higher propensity of **2** for hydride transfer reactions. Disubstituted aromatic imines RCH=NR′ (R, R′ = Ar) undergo insertion into the Mo–H bond under mild conditions to afford the corresponding amido complexes (Scheme 8).

Further coordination sphere tuning focused on the tetraphosphine tungsten and molybdenum nitrosyl hydrides $M(PMe_3)_4(H)(NO)$ (M = W **3**, Mo **4**) [42, 43]. Similar to **2**, both **3** and **4** showed a strong propensity toward insertion of carbonyls and imines. Due to the fact that the radius and the electronegativity of tungsten are bigger than those of molybdenum, the W–H bond of **3** has a higher covalent

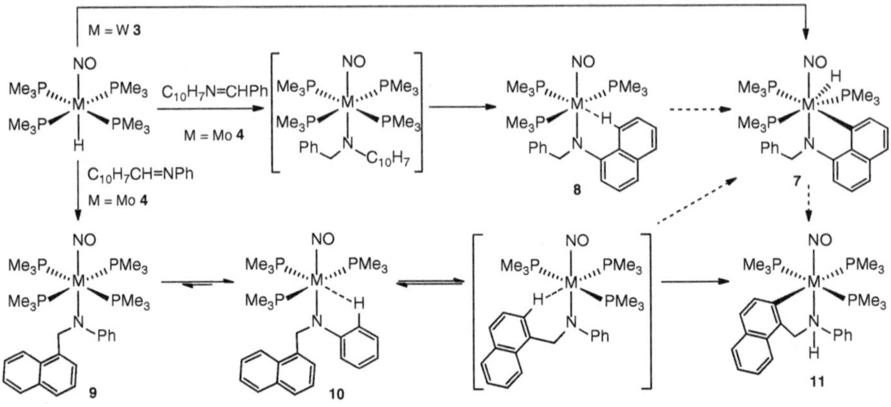

Scheme 9 Reaction pathway of sp^2-C–H bond activation in the course of imine insertion into the M–H (M = W, Mo) bond

contribution and is stronger than the Mo–H bond of **4**, and hence shows less tendency for hydride reactivity. Noteworthy is the fact that sp^2-C–H bond could be activated, as revealed by the reaction of **3** with *N*-benzylidene-1-naphthylamine ($C_{10}H_7N{=}CHPh$) affording the C–H bond splitting in the case of the tungsten hydride [W(NO)(H)$(PMe_3)_3(C_{10}H_6NCH_2Ph)$] (**7**), which is in line with the idea put forward in Scheme 4. A plausible reaction pathway is proposed in Scheme 9. Insertion of the C=N bond into the W–H bond leads to the loss of one PMe_3 ligand and at the same time a strong agostic interaction (sp^2-C–H···W), which is followed by oxidative addition of the C–H bond to the tungsten center. The agostic interaction was structurally confirmed in the reaction of the *N*-benzylidene-1-naphthylamine with the molybdenum hydride analogue **4**, wherein the 16e C–H···Mo agostic complex [Mo(NO)$(PMe_3)_3(C_{10}H_7NCH_2Ph)$] (**8**) could be fully characterized by NMR spectroscopy. Moreover, the reaction of **4** with *N*-1-naphthylideneaniline afforded the same type of agostic complex [Mo(NO)$(PMe_3)_3(C_{10}H_7CH_2NPh)$] (**10**), which was generated from the initially formed Mo–H insertion product **9** via isomerization. At elevated temperatures, **10** was isomerized into an alternative agostic species, which underwent further oxidative addition of the agostic C–H bond followed by the hydride transfer to the amide ligand eventually forming the amino complex [Mo(NO)$(PMe_3)_3(C_{10}H_6CH_2NHPh)$] (**11**). In comparison to the tungsten case, the seven-coordinated Mo–H species of type **7** might be spectroscopically observable. The possibility to change substitution of the Mo–H complex **4** in imine activation actually implied potential catalysis applications, as discussed in the next section.

Alternative tuning of **2** by replacing the dmpe ligand with the more electron-donating and steric demanding bidentate ligand dippe ($iPr_2PCH_2CH_2PiPr_2$) led to tungsten and molybdenum mononitrosyl hydride complexes $M(dippe)_2(H)(NO)$ (M = W **5**, Mo **6**) [44]. Protonation and hydrogen activation based on **5** and **6** were studied (Scheme 10). The tungsten hydride **5** reacted with $[H(Et_2O)_2][BAr^F{}_4]$ to form initially the unstable dihydrogen intermediate, which underwent oxidative

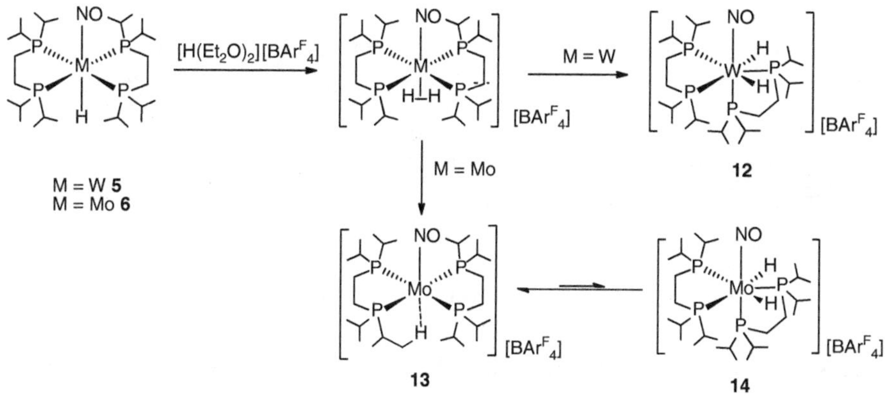

Scheme 10 Protonation and hydrogen activation based on **5** and **6**

addition to afford eventually the seven-coordinated pentagonal bipyramidal dihydride [W(dippe)$_2$(NO)H$_2$][BAr$^F{}_4$] (**12**). In contrast, the reaction of the molybdenum hydride **6** with [H(Et$_2$O)$_2$][BAr$^F{}_4$] furnished the formally 16e five-coordinated complex [Mo(dippe)$_2$(NO)][BAr$^F{}_4$] (**13**) presumably due to the too short-lived dihydrogen intermediate. X-ray diffraction revealed that **13** has a weak agostic Mo···H–C interaction *trans* to the NO ligand that can be readily displaced by 2e donors, such as acetone and THF. Remarkably, **13** was found to react rapidly with dihydrogen under ambient conditions to form the dihydride complex [Mo (dippe)$_2$(NO)H$_2$][BAr$^F{}_4$] (**14**), which is unstable in the absence of H_2. As established by Kubas in the molybdenum carbonyl series [25], the H_2/dihydride equilibrium of a weakly coordinated H_2 ligand can be triggered by the "*trans* ligand." Further labilization of the H_2 ligand can be induced by the nitrosyl group, which ought to bring about enhanced catalytic reactivity. However, for catalysis the sterically congested coordination sphere with the two bulky dippe ligands had to be reduced in steric demand.

3.1.2 Catalytic Applications in Ionic Hydrogenations

Wilkinson- and Osborn-type homogenous hydrogenations are mainly based on platinum group transition metal catalysts involving formal homolytic splitting of a H_2 ligand [45, 46]. In comparison, non-precious metal-based ionic hydrogenations involving heterolytic cleavage of H_2 is followed by the transfer of a hydride (H^-) and a proton (H^+) to the substrate as the H_2 equivalent [29]. A proper balance of proton or hydride transfer capabilities is essential in such ionic hydrogenations. Bullock and co-workers have reported various molybdenum and tungsten carbonyl complexes for ionic hydrogenations of alkenes, alkynes, and ketones [47–49]. In this context, the analogous nitrosyl derivatives are expected to exhibit also potential for ionic hydrogenations. In this section, the catalytic properties of the 16e agostic

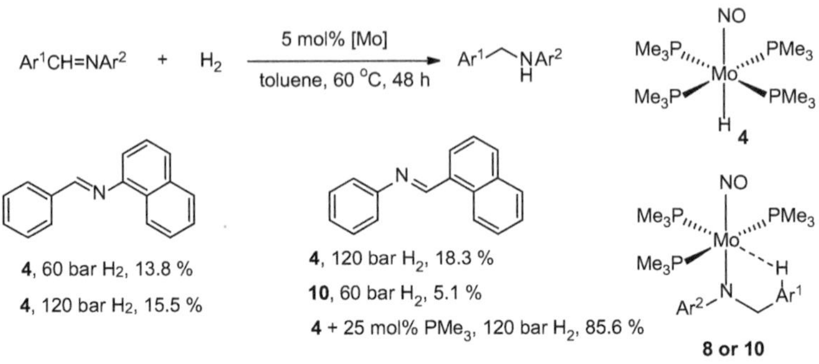

Scheme 11 Hydrogenations of imines catalyzed by ON–Mo–H complexes

complex (**10**) [43] and the diphosphine molybdenum nitrosyl complex (**15**) [50] in ionic hydrogenations of imines with a proton-before-hydride transfer characteristics will be discussed. It should be emphasized that hydrogenation of imines catalyzed by transition metal complexes is more challenging to be achieved than hydrogenations of alkenes due to several reasons [51]: (1) there is a smaller thermodynamic gain from the reduction of C=N bonds ca. −60 kJ/mol) relative to C=C bonds ca. −130 kJ/mol); (2) the typical end-on η^1 binding mode of the C=N groups contrasts with the side-on η^2 binding of C=C moieties, which results in less effective orbital overlaps with the metal center; (3) the catalytically active species is suspected to be poisoned by the hydrogenated products (amines) via competitive coordination to the metal center.

The 16e agostic complex **10**, which is derived from the insertion of a C=N bond into the Mo–H bond of **4**, is expected to show catalytic activity in the ionic hydrogenation of imines, not only because of its unsaturated nature of the metal center, but also because it shows oxidative addition capability of the sp^2-C–H bond [43]. Under a H_2 pressure of 120 bar and at 60°C, the hydrogenation of *N*-benzylidene-1-naphthylamine with 5 mol% of the catalyst **4** afforded a conversion of 15.5% within 48 h (Scheme 11).

Despite the rather low TON value, the reaction disclosed useful aspects, among others it showed ways for further tuning. The Mo–H bond becomes activated by the *trans*-nitrosyl group specifically for ionic hydrogenation, as proposed in Scheme 12. The insertion of the imine C=N double bond into the Mo–H bond led to the PMe_3 dissociated 16e agostic species of type **8** or **10**. Under conditions of high H_2 pressures, the agostic C–H ligand was substituted by a H_2 ligand. Due to the π-accepting property of the nitrosyl ligand the H_2 moiety is acidified so that in the presence of a basic imine substrate proton transfer can occur, which is equivalent to heterolytic H–H cleavage producing a Mo–H intermediate along with the amine cation. This process was followed by a hydride transfer from the molybdenum center to the amine cation giving the coordinated amine species, which eventually led to elimination of the amine and regeneration of the type **8** or **10** species, thus closing the catalytic cycle. Quite unexpected, the presence of catalytic amounts of PMe_3 boosted the

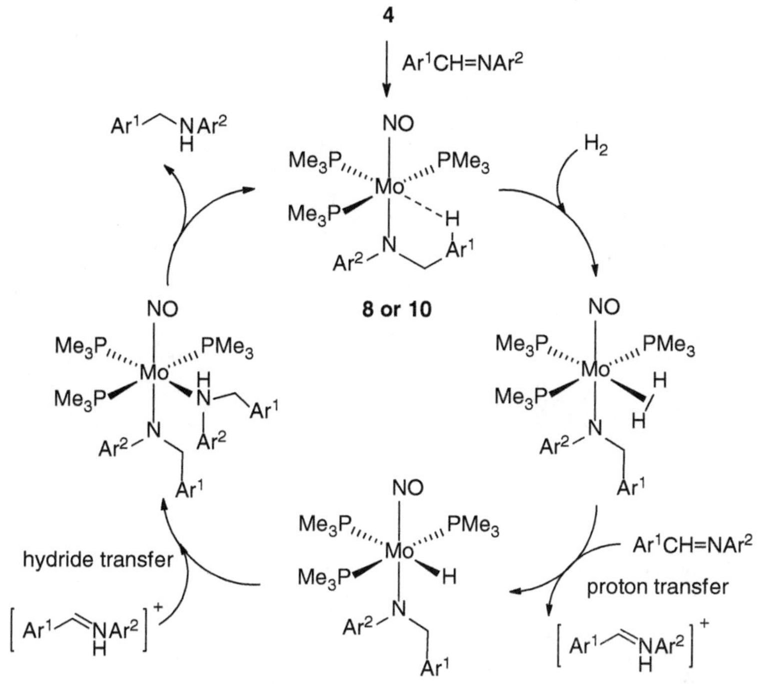

Scheme 12 Proposed mechanism for Mo–H hydrides (**4**, **8**, **10**) catalyzed hydrogenation of imines

hydrogenation performance. Assisted by 25 mol% of PMe_3 a conversion of 85.6% was accomplished in the hydrogenation of *N*-1-naphthylideneaniline under the same reaction conditions. This can be interpreted in terms of the fact that excess of free phosphine could mediate the proton transfer process, or alternatively, the free phosphine could stabilize various unsaturated intermediates as resting states of the catalytic cycle.

Enhanced catalytic performance of ionic hydrogenations was also accomplished with the molybdenum hydride system **15** [50]. Based on the reactivity of **6** in dihydrogen activation, ligand tuning was carried out to circumvent the presence of two sterically hindered bidentate phosphine ligands. The (diphosphane)nitrosyl molybdenum hydrides of the type $Mo(NO)(P{\cap}P)(CO)_2H$ (**15a–d**) were prepared by the reaction of $[Mo(CO)_4(NO)ClAlCl_3]$ with four diphosphine ligands including dippp (**a**), dippe (**b**), dippf (**c**), and dcype (**d**) in THF followed by treatment with excess $LiBH_4$ in Et_3N. In combination with 1 equiv. of $[H(Et_2O)][B(C_6F_5)]$ as a co-catalyst the hydrides **15a–d** showed valuable catalytic activity in the hydrogenation of imines. For instance, the hydrogenation of PhCH=N(α-naphtyl) catalyzed by **15a** and $[H(Et_2O)][B(C_6F_5)]$ afforded under 30 bar of H_2 at 23°C almost full conversion within 13 h showing an initial TOF of 106 h^{-1}. Testing **15b–d** as catalysts in the hydrogenation of PhCH=N(α-naphtyl) under the same conditions did not reveal a clear dependence of the activity on the bite angle of the

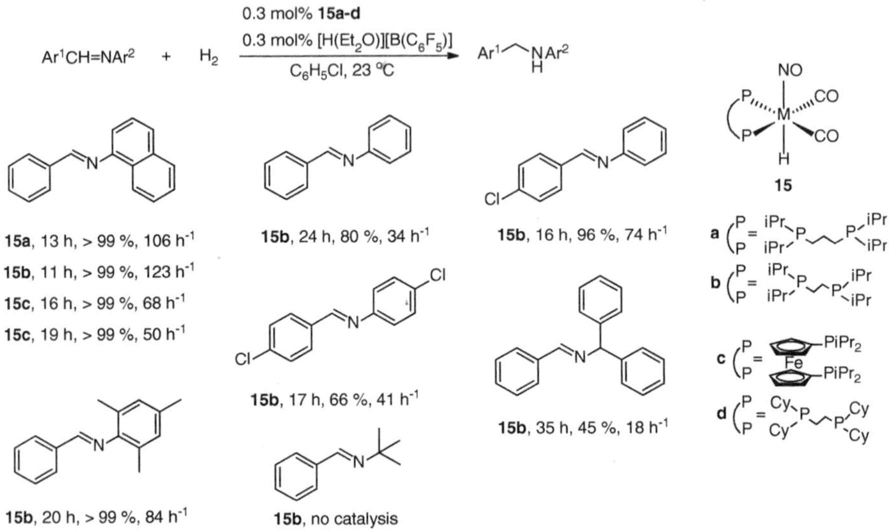

Scheme 13 Hydrogenation of imines catalyzed by Mo–H system **15**

phosphine ligands. The lowest hydrogenation rate was found for **15d**, presumably due to the steric bulk of the cyclohexyl substituent. The combination of **15b** as the catalyst with $[H(Et_2O)][B(C_6F_5)]$ as the co-catalyst displayed catalytic activity with an initial TOF of 123 h^{-1}, and based on this promising result the system was additionally tested in ionic hydrogenations of various other imines. The results are shown in Scheme 13. The rate of such ionic hydrogenation depends mainly on the size of the substituent at the N_{imine} atom. Highest rates were observed with more bulky substrates.

A mechanism for the ionic hydrogenation of imines catalyzed by the "Mo–H/[H $(Et_2O)][B(C_6F_5)]$" system is proposed in Scheme 14. The catalysis is initiated by protonation of the hydride ligand of the type **15** compounds forming a short-lived dihydrogen complex, which then undergoes proton transfer via heterolytic cleavage of the H–H bond by the imine substrate affording an iminium cation along with the regeneration of the Mo–H species **15**. Due to the effect of the *trans*-nitrosyl, subsequent hydride transfer to the iminium cation occurs in the secondary coordination sphere affording an amine-coordinated molybdenum intermediate. The amine ligand is eventually released from the Mo center by substitution with H_2 to reestablish the dihydrogen complex, and by this the catalytic cycle gets closed. The amine ligand is then expected to be replaced by the imine in equilibrium. Apparently with bulky substrates the amine dissociation is more feasible enabling reentry of H_2 into the cycle.

Crucial to the course of the hydrogenation reaction is the formation of the iminium cation, which competes with the formation of ammonium salts. The hydrogenation of aniline derivatives could be readily accomplished, which copes most probably with the fact that aryl-substituted imines are more basic than aryl-substituted amines. The iminium cations are therefore always accessible in kinetically relevant

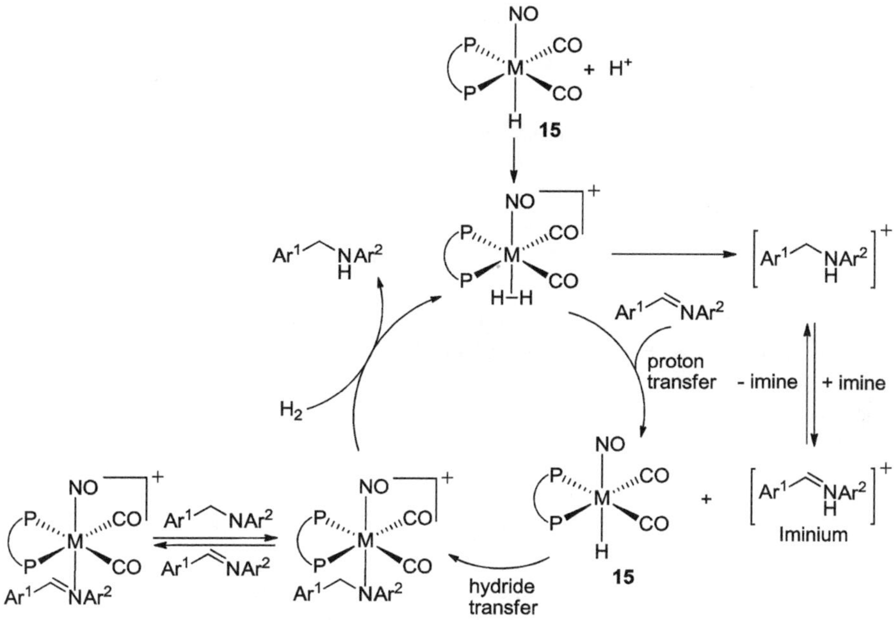

Scheme 14 Proposed mechanism for **15** catalyzed hydrogenation of imines

concentrations. In contrast, satisfactory conversions cannot be achieved for alkylimine derivatives presumably due to the fact that alkyl-substituted amines are the stronger bases than the corresponding imines. In the case of $PhCH{=}NCHPh_2$ the basicities of the amine and the imine are supposed to be comparable and the reaction stops when a certain amount of the amine is produced acting as a too efficient proton scavenger.

3.2 Group 7 Rhenium Nitrosyl Complexes in Homogeneous Catalysis

Within the notion of the development of non-platinum metal hydrogenation catalyses, the Berke group also focused on studies of group 7 transition metal rhenium nitrosyl chemistry, particularly on the organometallic chemistry of mononitrosyl and dinitrosyl complexes and their applications in hydrogen-related homogeneous catalysis, such as dehydrogenation, transfer hydrogenation, dehydrogenative silylation reactions, hydrosilylation, and hydrogenation. Two main aspects were considered: (1) the respective catalytic behavior of low valent d^6 Re(I) mononitrosyl complexes bearing either *trans*-phosphine or *cis*-bidentate phosphine ligands; (2) bifunctional catalyses with "Re–H/OH" and "Re–H/NH" complexes, which are isoelectronic to group 8 Ru(II) carbonyls. They are anticipated to be active catalysts in transfer hydrogenations applying a hydrogen donor as hydrogen carrier.

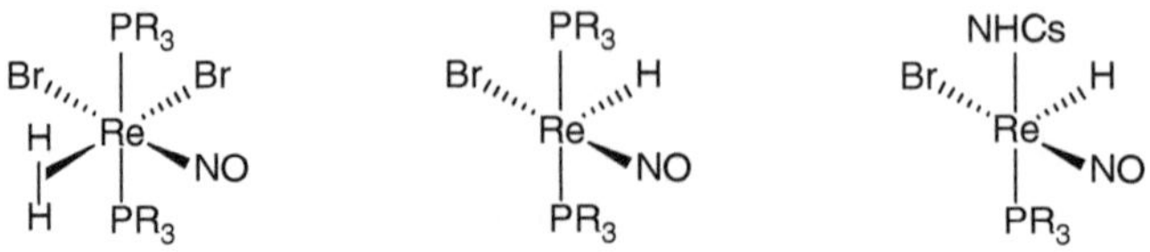

16 **17** **18** (NHCs = IMes)

19 (NHCs = SIMes)

R = *i*Pr (**a**), Cy (**b**)

Scheme 15 Representatives of Re(I) mononitrosyl complexes

In these rhenium nitrosyl complexes the nitrosyl ligand takes different functions: (1) as an ancillary NO^+ ligand mimicking isoelectronic metal carbonyls; (2) as a *trans*-influence ligand to activate σ-donor ligands for atom or group transfer reaction.

3.2.1 Re(I) Mononitrosyl Complexes for Homogeneous Catalysis

As a general entry synthetic access to all this chemistry has been achieved starting from a paramagnetic mononitrosyl Re(II) precursor to obtain a series of diamagnetic Re(I) mononitrosyl species bearing either *trans*-phosphines or *cis*-phosphines ligand patterns [33].

Treatment of the Re(II) complex $[NEt_4]_2[Re(Br)_5(NO)]$ with the bulky alkyl phosphine ($PiPr_3$ and PCy_3) in ethanol, which acts as a solvent and as a reductant, yields for instance the dihydrogen mononitrosyl Re(I) complex $[Re(Br)_2(NO)(PR_3)_2(\eta^2\text{-}H_2)]$ (**16**, R = *i*Pr **a**, Cy **b**). DFT calculation demonstrated propensity of Re(I) centers to bind H_2. The interaction with the rhenium center is relatively strong showing an elongated dihydrogen ligand, which lies in plane with the P–Re–P axis (Scheme 15) [33].

The bromide *trans* to the dihydrogen ligand is labile. Treatment of complexes **16** with excess of Et_3SiH afforded the five-coordinate 16e Re(I) hydride complexes $[Re(Br)(H)(NO)(PR_3)_2]$ (**17**, R = *i*Pr **a**, Cy **b**) in high yields. Noteworthy is the fact that in the ^1H-NMR spectrum the hydride resonance of **17** appears at ca. −17 ppm speaking for a hydridic character of the Re–H bond [52]. X-ray diffraction studies revealed a square-pyramidal geometry with the hydride ligand located in apical position, *trans* to the coordinative vacancy. The *trans* disposed π-donating bromide and the π-accepting nitrosyl occupy an axis of the basal sites with additional stabilization by a "push–pull" effect [30]. The catalytically important stabilization of the vacant site is at least partly due to the strong *trans*-influence of the hydride ligand, which leads to weakening any σ-type interaction at this position in such open-shell 16e Re(I) hydride complex.

In the presence of less bulky 2e donors, such as H_2, O_2, ethylene, CO, and CH_3CN, the vacant site of **17** could be occupied affording saturated 18e species [53]. However, ligand tuning with replacement of one phosphine of **17** by sterically hindered NHC ligands, such as IMes and SIMes, leads to the formation of a

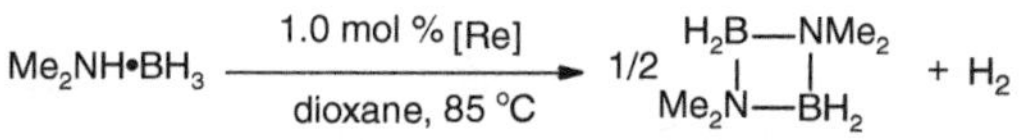

16a	**16b**	**17a**	**17b**
4 h, 96 %, 24 h^{-1}	4 h, 92 %, 23 h^{-1}	1.5 h, 82 %, 55 h^{-1}	1.3 h, 100 %, 77 h^{-1}
18a	**18b**	**19a**	**19b**
1 h, 92 %, 92 h^{-1}	1 h, 100 %, 100 h^{-1}	2 h, 88 %, 44 h^{-1}	3 h, 99 %, 33 h^{-1}

Scheme 16 Dehydrogenation of amine boranes catalyzed by Re(I) complexes **16–19**

NHC-substituted monophosphine rhenium hydrides [Re(Br)(H)(NO)$(NHCs)_2$] (**18** IMes, **19** SIMes, R = *i*Pr **a**, Cy **b**). As both IMes and SIMes ligands exhibit better electron-donating capabilities than alkylphosphines, complexes **18** and **19** are expected to be more active catalysts than the diphosphine derivatives **17** due to a labilized phosphine ligand *trans* to the NHC.

Dehydrogenation of Amine Borane and Transfer Hydrogenation Using Amine Borane as a Hydrogen Source

One remarkable application of the Re(I) complexes of type **16–19** is the catalytic dehydrogenation of amine boranes [54]. In recent years, the research on amine borane $R^1R^2NH{\cdot}BH_3$ (R^1, R^2 = alkyl, H) became an important topic based on their capability of chemical hydrogen storage [55–57]. In this respect transition metal-catalyzed dehydrocoupling reactions of amine borane offer great potential for control over both extent and rate of hydrogen release [58, 59].

When a solution of $Me_2NH{\cdot}BH_3$ in dioxane was treated with 1.0 mol% of **16a** and stirred at 85°C, evolution of hydrogen gas was observed. After 4 h the ^{11}B-NMR spectrum indicated that the conversion to the cyclic aminoborane dimer $[Me_2N\text{–}BH_2]_2$ was almost complete revealing a yield of 92%. Under the same condition, a 96% conversion was achieved by **16b** corresponding to a TOF of 24 h^{-1} (Scheme 16) [54]. Stoichiometric reactions of **16a(b)** with 10 equiv. of $Me_2NH{\cdot}BH_3$ afforded a complex mixture containing a small amount (less than 33%) of the **17a(b)** complexes, which were generated via heterolytic H–H splitting by the B–H bond of $Me_2NH{\cdot}BH_3$ or the dissociated base Me_2NH. When **17a(b)** were tested as catalysts in the dehydrogenation of $Me_2NH{\cdot}BH_3$ under the same condition, TOFs of 55 h^{-1} (**17a**) and 77 h^{-1} (**17b**) were obtained corresponding to a threefold enhancement in reactivity compared with the performance of **16a(b)**. This proves that complexes of type **17** are indeed catalytically active species in dehydrogenations with type **16** compounds taking the role of a catalyst precursors [53]. The most active hydride species for dehydrogenation of $Me_2NH{\cdot}BH_3$ proved to be the IMes Re(I) hydrides **18a** and **18b**, which exhibited TOFs of 92 h^{-1} and 100 h^{-1}, respectively. This is nearly four times higher than the catalytic activity accomplished with the dibromide complexes **16a** or **16b**. Surprisingly, the Re(I)

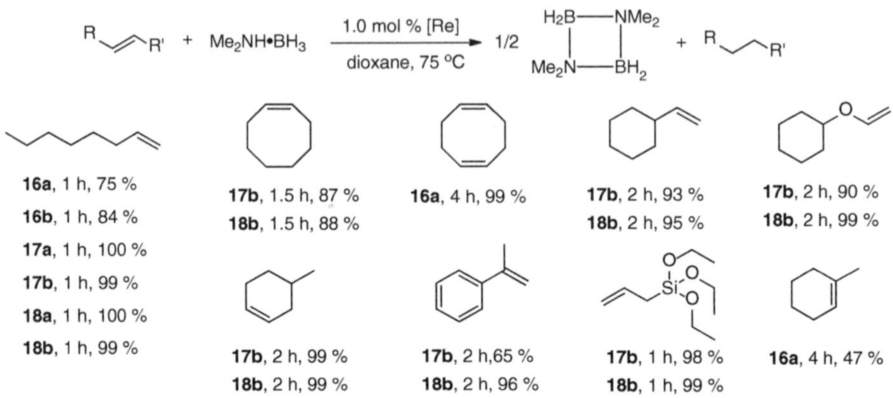

Scheme 17 Transfer hydrogenation of alkenes catalyzed by **16–18** using amine borane as a hydrogen source

hydrides **19a** and **19b** bearing the stronger electron-donating SIMes carbene showed an increase in rates of only a factor of 2 with respect to **16a** and **16b**.

Based on the fact that 1 equiv. of H_2 could be released from $Me_2NH{\cdot}BH_3$ using the Re(I) complexes as catalysts, we proceeded to investigate $Me_2NH{\cdot}BH_3$ as a hydrogen source in olefin transfer hydrogenations [53, 54, 60]. The mixture of equal amounts of $Me_2NH{\cdot}BH_3$ and 1-octene with 1 mol% of **16a** afforded at 85°C within 1 h the hydrogenated octane product in 93% yield. $Me_2NH{\cdot}BH_3$ was completely consumed with the formation of $[Me_2N–BH_2]_2$ as the main dehydrogenation product. The analogue reaction carried out in an open system afforded the same results suggesting the absence of free H_2. This observation speaks more for a transfer hydrogenation mechanism than for a tandem-reaction path. The other Re(I) complexes of Scheme 15 were all effective catalysts in such reactions (Scheme 17). Similar to the activity for the dehydrocoupling of $Me_2NH{\cdot}BH_3$, the bis(phosphine) hydride complexes **17a, b** and the IMes rhenium(I) hydrides **18a**, **b** showed the highest activities with a TOF of over 99 h^{-1}. Other olefins such as cyclooctene, 1,5-cyclooctadiene, vinylcyclohexane, and ally(triethoxyl)silane have also been tested as substrates in such transfer hydrogenations. Both catalysts performed in most cases quite similar. In the case of the sterically hindered olefin α-methylstyrene, the NHC compound **18a** displayed a distinctly higher activity than the bis(phosphine) hydride **17a**.

A mechanism for the dehydrogenation of $Me_2NH{\cdot}BH_3$ and the transfer hydrogenation of alkenes using $Me_2NH{\cdot}BH_3$ as a hydrogen donor is proposed in Scheme 18. Complexes **17** serve as the catalytically active species. $Me_2NH{\cdot}BH_3$ coordinates initially to the vacant site of **17** leading to the σ-amine borane complex, which undergoes oxidative addition of the B–H bond with phosphine dissociation to form a Re(III) dihydride species. Then reductive elimination of hydrogen occurs followed by a β-H shift affording initially the monomeric $[Me_2N{=}BH_2]$ species with closing the dehydrogenation cycle. Similarly, the olefin adds to the vacant site of **17** giving an 18e complex, which undergoes intramolecular insertion into the

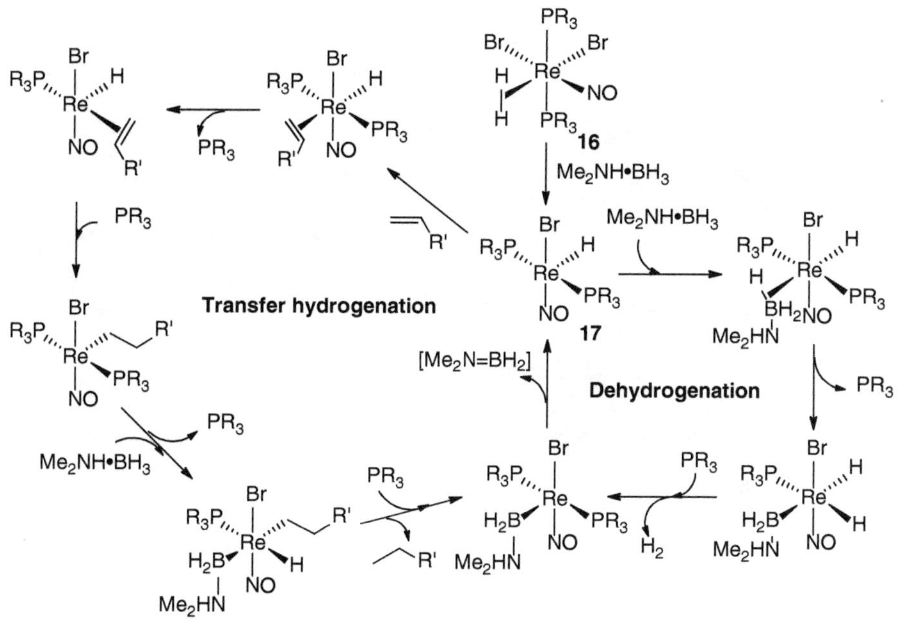

Scheme 18 Proposed mechanism for Re(I) complexes catalyzed dehydrogenation and transfer hydrogenation reactions

Re–H bond affording a Re–alkyl species. Then oxidative addition of a B–H bond occurs leading to the Re(III) species, which undergoes reductive elimination to form the hydrogenated products. Finally the monomeric [$Me_2N{=}BH_2$] species was released via a β-H shift, which closes up the transfer hydrogenation cycle.

Dehydrogenative Silylation Reactions of Alkenes

Dehydrogenative silylation are defined by the reaction of silanes with 2 equiv. of an alkene affording vinylsilanes and alkanes [61]. This reaction is competitive with the classical hydrosilylation reaction. Since it allows the direct formation of unsaturated silyl compounds from silanes and alkenes, dehydrogenative silylation has drawn a great deal of attention and becomes a useful method to prepare important synthetic precursors in organosilicon chemistry, although the main drawback is the formation of a mixture of compounds [62–65]. DFT calculations have shown that dehydrogenative silylations and hydrosilylations are thermodynamically sound reactions with dehydrogenative silylations somewhat favored energetically over hydrosilylations [52]. However, the thermodynamic difference of both reaction channels is too small to be the decisive factor for chemoselectivity in catalytic processes.

Quite unexpected was the discovery that the Re(I) complexes **16a, b** catalyzed dehydrogenative silylation reactions of alkenes in a highly selective manner [52].

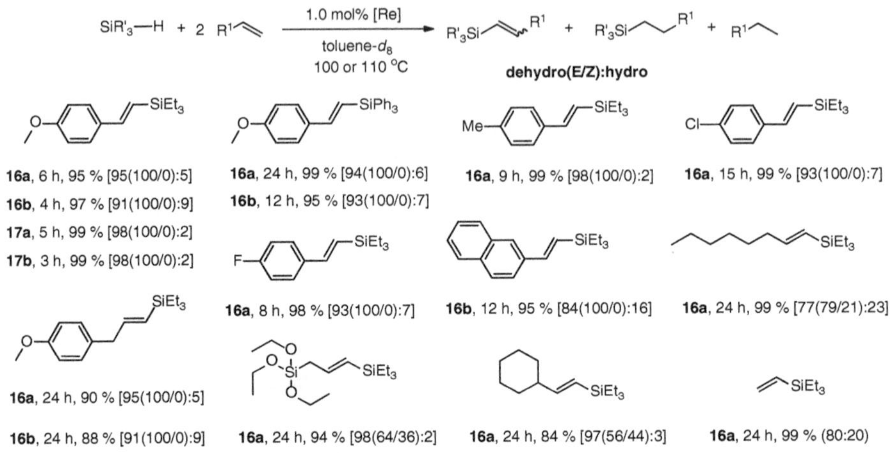

Scheme 19 Dehydrogenative silylation of alkenes catalyzed by Re(I) complex **16**

For instance, the reaction of Et_3SiH and 2 equiv. of *p*-methoxystyrene in toluene with 1.0 mol% of **16a** afforded at 100°C within 6 h the dehydrogenative silylation product (*E*)-1-(*p*-methoxystyryl)-2-(triethyl-silyl)ethylene in 95% yield. The reaction is of high selectivity that neither (*Z*)-isomers, nor branched dehydrogenative silylation products were seen. Less hydridic silanes, such as triphenylsilane, were less efficient than for instance Et_3SiH. Other substituted styrenes such as *p*-methyl, *p*-chloro-, and *p*-fluorostyrene also afforded the corresponding *trans*-vinylsilanes in high yields and selectivities (up to 98%). In the case of aliphatic alkenes, such as *n*-octene, allyltriethoxysilane, vinylcyclohexane, and ethylene, dehydrogenative silylations were still preferred, but showed less *E*/*Z* selectivity. Cyclic olefins, such as cyclooctene, furnished low conversions under the same reaction conditions. The results are summarized in Scheme 19.

The stoichiometric reaction of **16** with the substrate Et_3SiH and alkenes revealed the initial formation of the hydride complex **17** followed by alkene coordination to the Re center at the vacant site. The 18e species $[Re(Br)(H)(NO)(PR_3)_2(\eta^2\text{-}CH_2{=}CHR^1)]$ is formed. Sophisticated mechanistic studies were carried out, which supported the mechanism proposed in Scheme 20. The 18e species generated from **16** via **17** serves as a catalyst precursor. Thermally induced phosphine dissociation occurs affording the 16e monophosphine complex, which undergoes alkene insertion into the Re–H bond producing the Re–alkyl species in two isomeric forms. This is consistent with the observation that deuterium is incorporated at both the 1- and the 2-position of the vinyl group. Subsequent oxidative addition of the silane to the rhenium center leads to the Re(III) intermediate, in which the Re–alkyl ligand is positioned *cis* to the less-crowded H end of the R'_3Si–H bond rather than the R'_3Si end. At this stage the selectivity for dehydrogenative silylation is determined by preferential reductive elimination of the $R^1CH_2CH_3$ alkane yielding a Re–silyl species. Subsequent elementary steps involve substitution of one phosphine ligand by an alkene, alkene insertion into the Re–silyl bond, and eventually β-H elimination

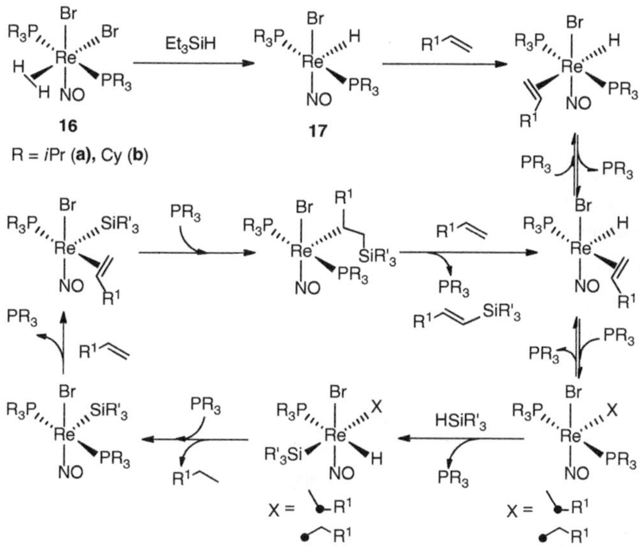

Scheme 20 Proposed mechanism for **16** catalyzed dehydrogenative silylation of alkenes

yielding the dehydrogenative silylation product. Deuterium isotope kinetics for the dehydrogenative silylation of *p*-methylstyrene with Et_3SiH and Et_3SiD was also studied, which revealed an inverse KIE value of 0.84. This suggested that a reductive elimination step of the catalytic cycle contributes to the overall reaction rate.

"Re(I) Hydride/Lewis Acid" Co-catalyzed Hydrogenations of Alkenes

In extension to the dehydrogenation studies of amine boranes and the transfer hydrogenations of alkenes catalyzed by the Re(I) complexes **16** and **17**, highly efficient hydrogenations of alkenes were established based on the co-catalytic systems of "**16**/$Me_2NH{\cdot}BH_3$" or "**17**/Lewis acid," which exhibited catalytic activities comparable to those of Wilkinson- or Schrock–Osborn-type hydrogenations accomplished with platinum group metal catalysts [24].

For instance, **16** alone showed only moderate activity in the hydrogenation of alkenes. Under 10 bar of H_2 at 90°C, a 60% conversion of 1-hexene was achieved by adding catalytic amounts of **16a**, and within 3 h a TOF value of 803 h^{-1} was obtained. In comparison, when 5 equiv. of $Me_2NH{\cdot}BH_3$ (ratio to **16**) was added as a co-catalyst, an 82% conversion was achieved under the same condition otherwise. Within only 5 min a TON of 2,987 and a TOF of 3.5×10^4 h^{-1} was accomplished, which is equivalent to a 43 times acceleration compared to the reaction of **16a** alone (Scheme 21). Using other alkenes, such as 1-octene, cyclooctene, styrene, and also dienes, such as 1,5-cyclooctadiene and 1,7-octadiene, the "**16**/$Me_2NH{\cdot}BH_3$" (1:5) system showed excellent performance in the hydrogenations leading to satisfactory TONs and TOFs with full conversions of the starting components. Only in the

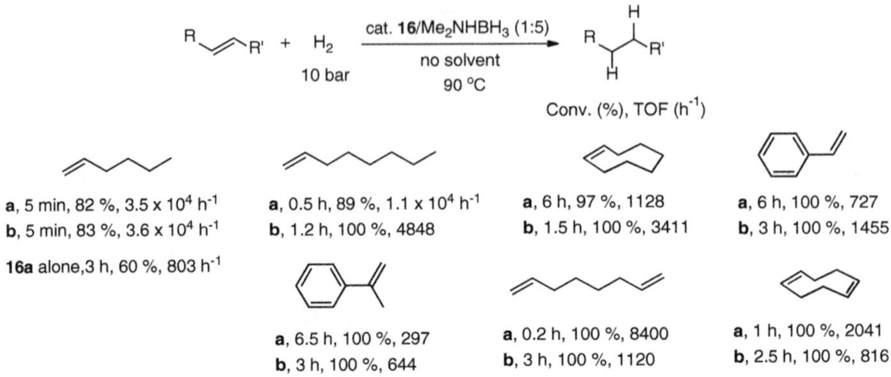

Scheme 21 Hydrogenation of alkenes catalyzed by the "**16**/Me_2NHBH_3" system

case of 1,1-disubstituted alkenes, such as α-methylstyrene, the hydrogenations proceeded slower. It's also important to note that not only full conversions were accomplished in all cases, but also the catalyses could be carried out under solvent-free conditions. Thus, the hydrogenated products could be easily isolated by distillation.

The extraordinary catalytic performance of the "**16**/$Me_2NH{\cdot}BH_3$" systems was attributed to formation of the five-coordinate Re(I) hydrides **17** and the solvent coordinated Lewis acid BH_3 generated by dissociation from $Me_2NH{\cdot}BH_3$. Under the same condition the related "**17**/$BH_3{\cdot}THF$" catalytic system also exhibited high activity in hydrogenations of various alkenes. For instance, a 92% conversion of 1-hexene (5 mL) was achieved by 0.03 mol% "**17a**/$BH_3{\cdot}THF$" at 90°C within 5 min corresponding to a TON of 3,083 and a TOF of $3.7 \times 10^4\ h^{-1}$. The active species of the **17**/$BH_3{\cdot}THF$ system showed longer lifetimes than those of the **16**/$Me_2NH{\cdot}BH_3$ systems, since re-addition of 5 mL of 1-hexene results in a 98% conversion within 10 min corresponding to a TON of 3,350 and TOF of $2.0 \times 10^4\ h^{-1}$. This procedure was repeated twice leading to some decrease in activity, but in absolute terms still high TOFs were seen (Scheme 22). Due to the living character of the **17**/$BH_3{\cdot}THF$ system, the loading of the rhenium catalyst could be decreased down to the very low value of 0.006 mol% (60 ppm) affording a maximum TON of 1.2×10^4 and a TOF of $4.0 \times 10^4\ h^{-1}$ within 18 min. Using other substrates, such as 1-octene, cyclooctene, α-methylstyrene, 1,5-cyclooctadiene, and 3,3-dimethylbutene, the **17**/$BH_3{\cdot}THF$ system performed also extremely well resulting in both high TONs and TOFs.

Further tuning of the hydrogenation catalysis with **17** made it necessary to study the effect of the boron Lewis acid in greater detail applying in the hydrogenation of 1-hexene different Lewis acids of various strengths: $BCl_3 > BH_3 > BEt_3 \approx BF_3 > B(C_6F_5)_3 > BPh_3 \gg B(OMe)_3$ [66, 67]. The order in catalytic activity was found to be $B(C_6F_5)_3 > BEt_3 \approx BH_3{\cdot}THF > BPh_3 \gg BF_3{\cdot}OEt_2 > B(OMe)_3 \gg BCl_3$, as depicted in Fig. 1. The stability of the catalytic systems was checked by changes of the TOFs vs time, which revealed that the effect of the boron Lewis acids is in an

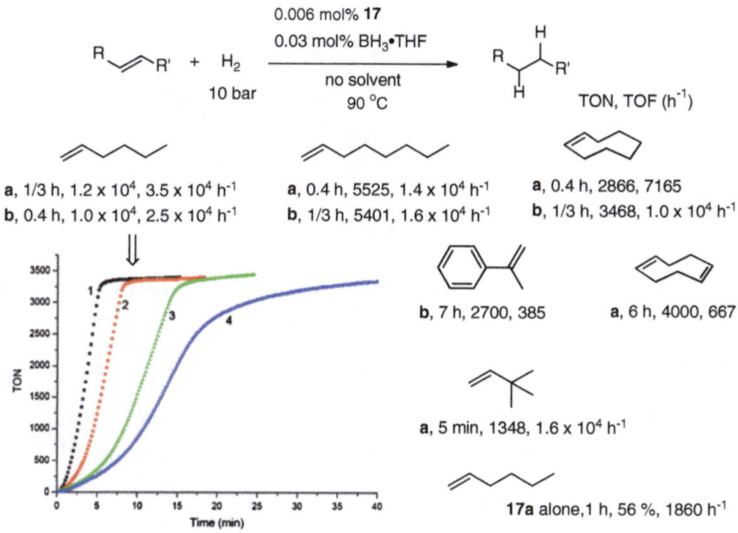

Scheme 22 Hydrogenation of alkenes catalyzed by the "**17**/BH_3·THF" system

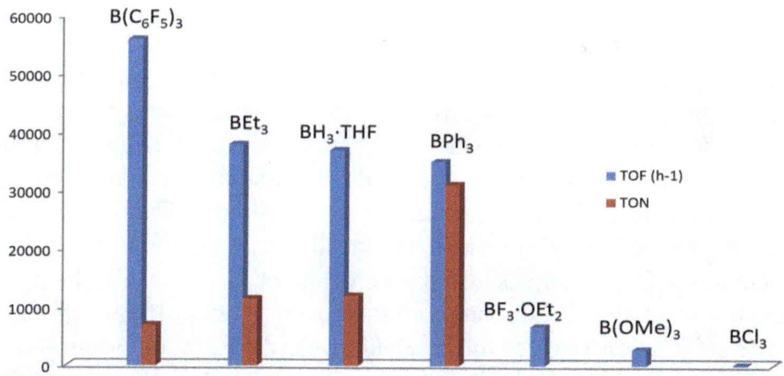

Fig. 1 Boron Lewis acid co-catalyst effect in hydrogenations of 1-hexene catalyzed by 17a at 23°C under 10 bar H_2

approximate inverse order: $BPh_3 > BEt_3 \approx BH_3$·THF $> B(C_6F_5)_3$. The weak Lewis acid BPh_3 turned out to be the most efficient co-catalyst demonstrating also the longest lifetime of the active species. The relatively bulky Lewis acid $B(C_6F_5)_3$ of medium strength appeared to be the most active co-catalyst demonstrating the highest activity of the catalytic species. A maximum TON of 3.1×10^4 was obtained for the **17a**/BPh_3 system and a maximum TOF of 5.6×10^4 h^{-1} was achieved for the **17**/$B(C_6F_5)_3$ system in the prototypic hydrogenation of 1-hexene.

On the basis of studies of the kinetic isotope effect, H_2/D_2 scrambling and halide exchange experiments, filtration experiments, Lewis acid variations, and isomerization studies of terminal alkenes, a mechanism including reversible bromide abstraction by the Lewis acid is proposed for the "**16**/Me_2NH·BH_3" or the

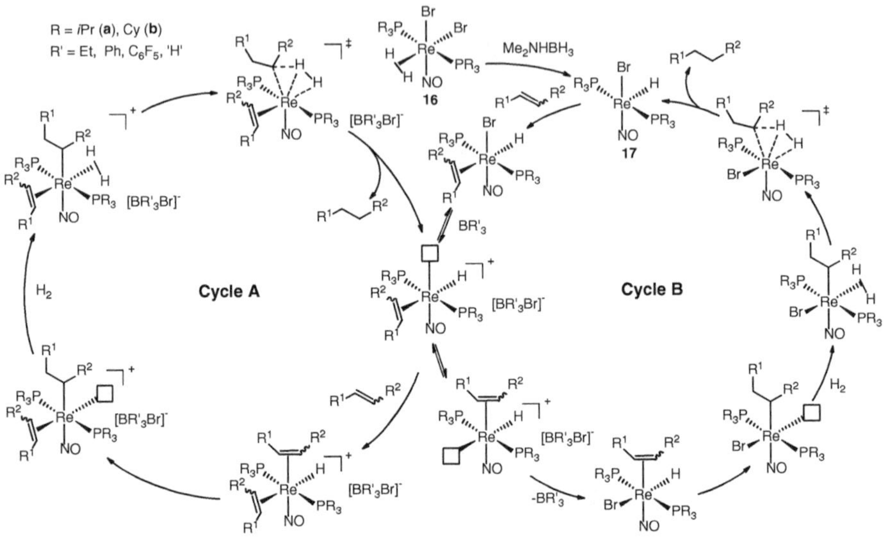

Scheme 23 Proposed mechanism for rhenium hydride and Lewis acid co-catalyzed hydrogenation of alkenes

"**17**/Lewis acid" catalyzed hydrogenations of alkenes (Scheme 23). Two catalytic cycles have to be discussed depending on whether the Lewis acid promoted reversible bromide abstraction is involved as an initiation step (**A**) and the cycle is turning-over "bromide-free" or whether the bromide is coming back in a "on–off" manner during the course of the cycle (**B**). Invariably the alkene coordinated 18-electron Re(I) hydride is first generated either directly from **17** or indirectly from **16** with the $Me_2NH{\cdot}BH_3$ reagent and the alkene. Then the Lewis-acidic BR'_3 component reversibly abstracts the bromide ligand affording the Re(I) species bearing one vacant site and the bromo borate counter anion $[BBrR'_3]^-$. This step serves as the initiation step for the catalytic cycle **A**. Then an Osborn-type cycle gets into operation with alkene before H_2 addition reaching the transition state, through which the alkane is eventually released by reductive elimination closing the catalytic cycle. The alternative catalytic cycle **B** is different from the cycle **A** as it includes the Lewis acid promoted reversible bromide abstraction and re-addition at the certain other stages of the catalytic cycle. Otherwise the sequence of an Osborn-type hydrogenation cycle is again followed with alkene before H_2 addition. The cycle **B** is presumably slower than the cycle **A**, because the bromide "on–off" process is assumed to retard the overall reaction rate.

3.2.2 Re(I) Mononitrosyl Complexes with *Cis*-Phosphine Arrangements for Hydrogen-Related Catalyses

The *trans*-phosphine arrangement is presumed to support strong binding of olefin ligands and thus impede respective homogeneous catalyses. In order to systematically investigate the phosphine effect on the catalytic performance of Re(I)

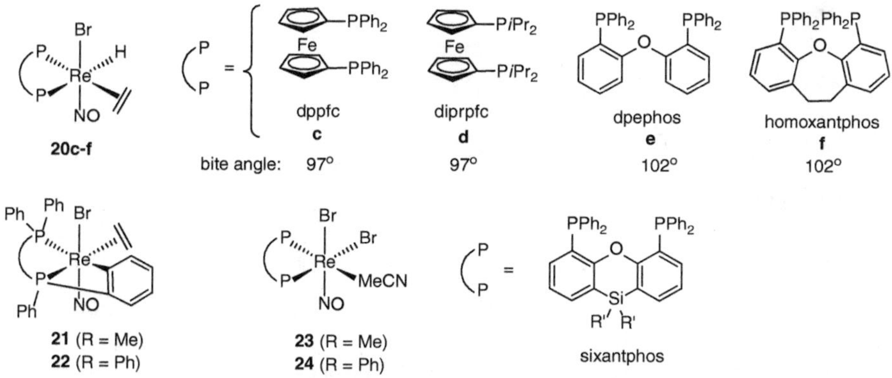

Scheme 24 Representatives of Re(I) mononitrosyl complexes bearing *cis*-phosphine ligand

mononitrosyl systems, the change of the coordination pattern from a *trans*- to a *cis*-phosphine arrangement was pursued (Scheme 24). Chelating diphosphine ligands are characterized in electronic terms by their donicity, in size by their bulk, and by their bite angle. The state of the art bidentate diphosphine ligands possessing large-bite-angles revealed in many catalytic processes to be superior in performance over those of large-cone angle monophosphine ligands [68, 69].

The preparation of the Re(I) precursors [$ReBr_2$(MeCN)(NO)(P∩P)] bearing large-bite-angle diphosphine ligands P∩P including dppfc, diprpfc, dpephos, and homoxantphos (see Scheme 24) was achieved via ligand substitution reactions starting from [NEt_4]$_2$[$ReBr_5$(NO)] [70]. Subsequent involvement of Et_3SiH and ethylene afforded the Re(I) hydride complexes [ReBrH(η^2-C_2H_4)(NO)(P∩P)] (**20c-f**). Interestingly, in the case of the chelating ligands sixantphos or sixantphos-ph, the *ortho* metalated complexes [ReBr(η^2-C_2H_4)(NO)(η^3-C_6H_4-*o*-Sixantphos)] (**21**) or [ReBr(η^2-C_2H_4)(NO)(η^3-C_6H_4-*o*-Sixantphos-ph)] (**22**) were obtained from the reaction of the dibromo precursor [$ReBr_2$(η^2-C_2H_4)(NO)(P∩P)(CH_3CN)] (**23** or **24**) with the Et_3SiH/ethylene reagents, most probably via *o*-C–H bond oxidative addition followed by H_2/ethylene exchange reactions. Both these products **21** and **22** were obtained as a mixture of two diastereomers differing in the axial arrangement of Br and NO.

Hydrogenation of Alkenes Catalyzed by *Cis*-Diphosphine Re(I) Mononitrosyl Complexes

In spite of the extremely efficient catalytic performance exhibited by the *trans*-phosphine Re(I) mononitrosyl complexes in the presence of a Lewis acid co-catalyst, the system was seen to suffer versatility in hydrogenation catalyses, exactly due to the employment of Lewis acids that prohibited the hydrogenation of alkenes bearing functional groups, such as carbonyl functions. The well-defined diphosphine complexes **20c–f** and **21** turned out to overcome this disadvantage.

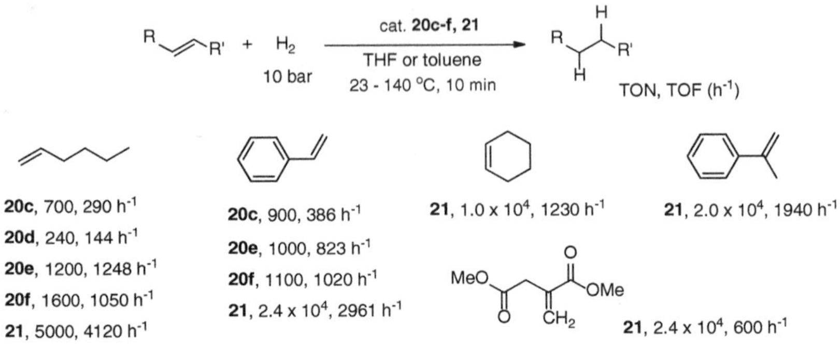

Scheme 25 Hydrogenation of alkenes catalyzed by Re(I) complexes **20** and **21**

They were found to be efficient catalysts for the hydrogenation of alkenes comparable in activity to Wilkinson- or Osborn-type Rh catalysts (Scheme 25) [71]. Variation of the bidentate ligand demonstrated a crucial influence of the large-bite-angle on the catalytic performance. In the case of hydrogenations of 1-hexene, very slow time-dependent decreases in activity were observed under the given conditions indicating the catalytic systems to be fairly stable. The catalytic performance of the rhenium catalyst was sensitive to the choice of the reaction medium. THF and toluene were screened out to be the best solvents. Complex **21** turned out to be the best catalyst showing the best longevity in the hydrogenation of alkenes. TONs of more than 24,000 could be reached for styrene hydrogenations without decrease in activity. Other sterically more demanding disubstituted substrates, such as cyclohexene, α-methylstyrene, and dimethylitaconate, were also tested in such hydrogenations. **20c–f** showed virtually no activity for these substrates, while **21** was found to be moderately active. Remarkably, a functional group tolerance was revealed, for instance, by the hydrogenation of dimethylitaconate.

The following sophisticated mechanistic studies were carried out: (1) evaluation of the chemoselectivity in the catalytic hydrogenation of substituted olefins; (2) quantitative kinetic studies of the catalytic cycle; (3) stoichiometric studies of parts of the catalytic cycle and the interception of intermediates; (4) dynamic NMR analysis of the β-hydride shift step; (5) DFT calculations. In the hydrogenation of alkenes an Osborn-type catalytic cycle with olefin before H_2 addition was thus proposed for **20c–f** and **21** (Scheme 26). The crucial steps are the formation of the alkyl complexes and the subsequent hydrogenolysis step of the alkyl rhenium bonds, which is expected to occur via formation of dihydrogen complexes and subsequent formation of seven-coordinate Re(III) dihydride intermediates. Octahedral Re(I) d^6 complexes normally possess high kinetic barriers for ligand exchange steps. In the presented systems this problem could be circumvented by appropriately tuned ligand sets enabling a facile β-hydride shift, which subsequently opens a vacant coordination site for the oxidative addition of H_2. The employed large-bite-angle diphosphines also support Re(I)/Re(III) redox changes favoring the generation of seven-coordinate Re(III) intermediates occurring after oxidative additions. From kinetic measurements

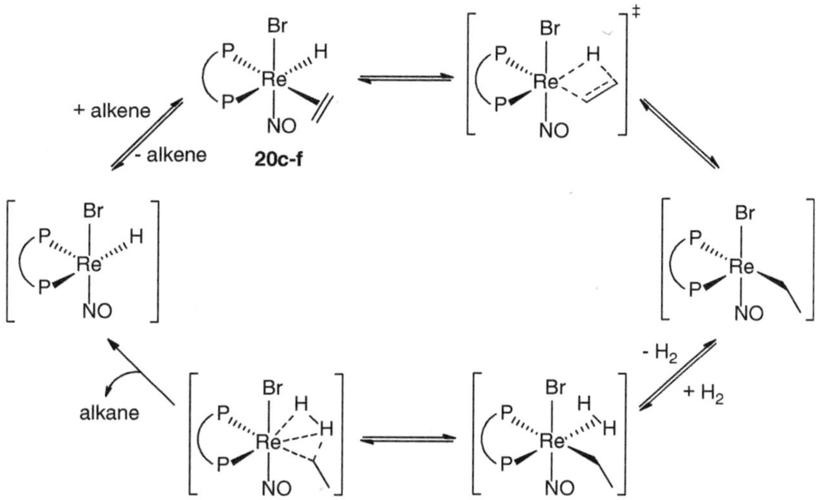

Scheme 26 Proposed Osborn catalytic cycle for **20** and **21** catalyzed hydrogenation of alkenes

we concluded that the overall rates of the hydrogenation is first order in the Re complex and the H_2 pressure, but zeroth order in the alkene. No significant kinetic isotope effect could be determined.

Hydrogenation of Nitriles Catalyzed by *Cis*-Diphosphine Re(I) Mononitrosyl Complexes

Hydrogenation of nitriles has great potential in synthetic organic chemistry and in the production of pharmaceuticals, agrochemicals, textile, and rubber chemicals. Generally the C≡N triple bonds of nitriles are more difficult to hydrogenate than the C=C double bonds of alkenes [72]. Amines as the hydrogenation products may quench the catalyses by stabilizing the active species. Moreover, the hydrogenation of nitriles often suffers from the problem of poor selectivities. A mixture of primary, secondary, and tertiary amines, and even intermediate imines might be generated concomitantly with products of full reduction. Among various products, the secondary amines are of particular significance in view of their role as biologically active molecules and versatile ligands. Up to date, the homogeneous hydrogenation reactions of nitriles are mainly achieved by late transition metal catalysts [73, 74].

So it became very surprising to see that the middle transition metal complexes **21–24** with Re(I) centers bearing chelate *cis*-diphosphine ligands were efficient catalysts in homogeneous hydrogenations of nitriles [75]. For instance, the hydrogenation of benzonitrile furnished the products dibenzylimine (**P1**), dibenzylamine (**P2**), and tribenzylamine (**P3**) with **P2** as the major product (Scheme 27). Formation of benzylimine or benzylamine was not observed. Dibenzylimine was formed by the hydrogenation of dibenzylimine, while the

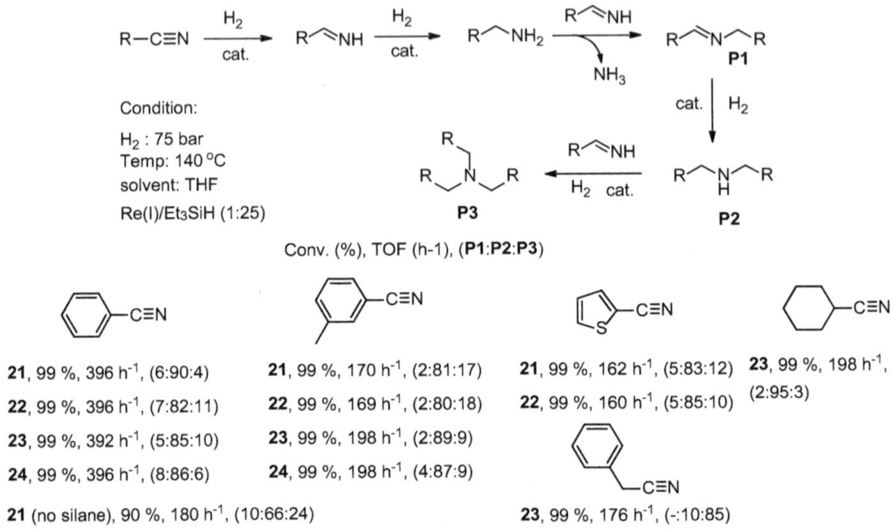

Scheme 27 Hydrogenation of nitriles catalyzed by Re(I) complexes **21–24**

formation of tribenzylamine was explained by hydrogenolysis of the *gem*-diamine followed then by hydrogenation. Aliphatic nitriles, such as phenylacetonitrile, led to the formation of tertiary amines via hydrogenation of the respective enamine followed by hydrogenolysis.

The hydrogenation of benzonitrile catalyzed by **21** afforded under 75 bar of H_2 and at 140°C in THF a 90% conversion revealing a TOF of 180 h^{-1} forming 66% dibenzylamine, 24% tribenzylamine, and 10% dibenzylimine within 1 h. The presence of Et_3SiH as a co-catalyst improves the catalytic performance with respect to both activity and selectivity. When 25 equiv. of Et_3SiH with respect to the rhenium catalyst was applied, the same reaction showed a conversion of 99% with a TOF of 396 h^{-1} within 0.5 h forming 90% dibenzylamine, 4% tribenzylamine, and 6% dibenzylimine. The generality of the reaction was probed by applying the Re(I) catalysts **22–24** under the same conditions in the hydrogenation of 3-methyl-benzonitrile, thiophen-2-carbonitrile, cyclohexanenitrile, and benzyl nitrile (Scheme 27).

3.2.3 Phosphine-Free Re(I) Mononitrosyl Complexes for Hydrosilylation of Organic Carbonyl Derivatives

In recent years hydrosilylation of various organic carbonyl compounds has made considerable progress and became a major tool of synthetic organic and organosilicon chemistry [76–78], in particular with respect to the functionalization of polymers [79]. It could be demonstrated in the previous section that mononitrosyl Re(I) complexes are active in dehydrogenative silylation reactions of alkenes. In comparison, the hydrosilylation of the polar double bonds of ketones revealed great activities using mononitrosyl Re(I) complexes [80].

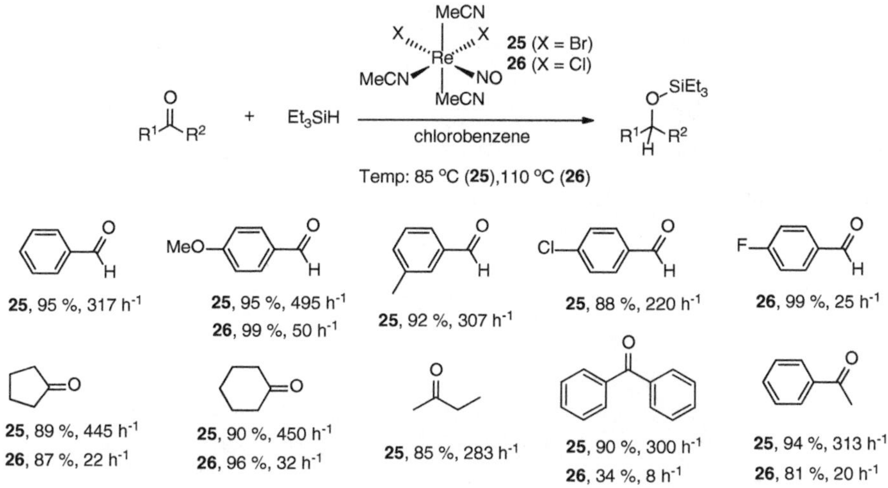

Scheme 28 Hydrogenation of carbonyls catalyzed by phosphine-free Re(I) complexes

For instance, the phosphine-free tris(acetonitrile) Re(I) complexes [$ReX_2(NO)(CH_3CN)_3$] (X = Br **25**, Cl **26**), which could be also used as synthetic precursors for the preparation of phosphine substituted Re(I) derivatives, were found to be efficient catalysts in the hydrosilylation of various ketones and aryl aldehydes under mild conditions (Scheme 28) [81, 82]. This is of great significance by taking into account the easy access to **25** and **26**, and their relative air- and water-stability in solid state in comparison with other more complicated phosphine- or carbene-substituted Re(I) compounds. The bromo compound **25** showed much better catalytic performance than the chloro derivative **26**. It is most likely that in these hydrosilylation reactions the rhenium catalyst functions as a Lewis acid coordinating initially the ketones or aryl aldehydes. Subsequent addition of the silane to the activated carbonyl function affords the hydrosilylation product. The rate-determining step is presumed to be the dissociation of one acetonitrile ligand.

3.2.4 Bifunctional Mononitrosyl Re(I) Catalysts for Transfer Hydrogenations

Re–H/OH Systems

Ligand–metal bifunctional catalysis provides an efficient method for the hydrogenation of various unsaturated organic compounds. Shvo-type [83–85] Ru–H/OH and Noyori-type [3–7] Ru–H/NH catalysts have demonstrated "bifunctionality" with excellent chemo- and enantioselectivities in transfer hydrogenations and hydrogenations of alkenes, aldehydes, ketones, and imines. Based on the isoelectronic analogy of H–Ru–CO and H–Re–NO units, it was anticipated that rhenium nitrosyl-based bifunctional complexes could exhibit catalytic activities comparable to the ruthenium carbonyl ones (Scheme 29) [86].

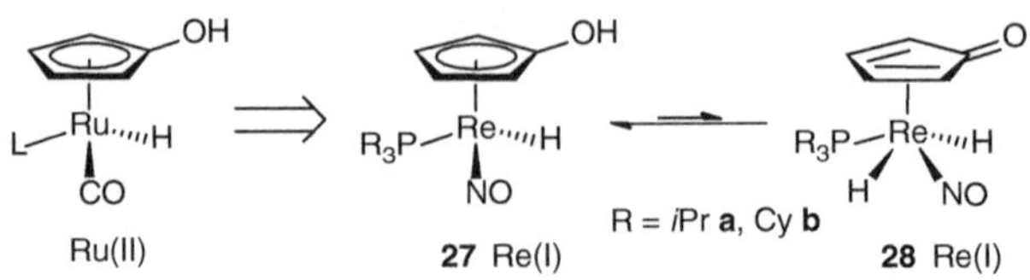

Scheme 29 Bifunctional mononitrosyl Re–H/OH systems

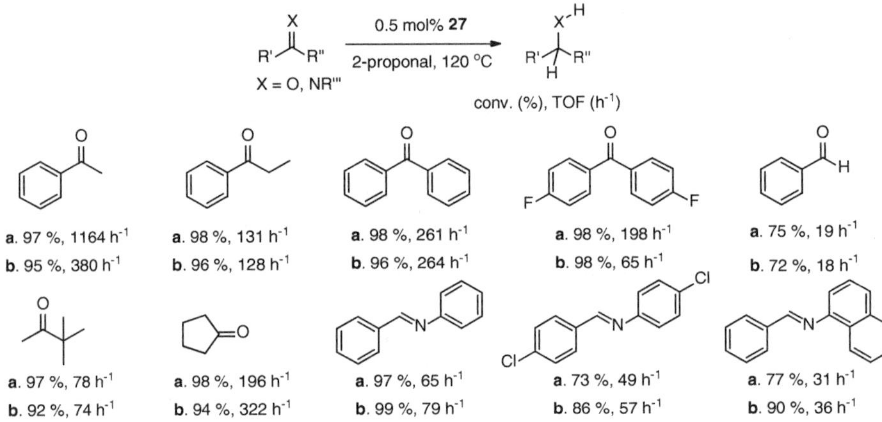

Scheme 30 Transfer hydrogenation of carbonyls catalyzed by the "Re–H/OH" system using 2-propanol as both the solvent and the H_2 donor

The ligand–metal bifunctional hydroxycyclopentadiene hydrido Re(I) complexes [Re(H)(NO)(PR_3)(C_5H_4OH)] (**27**, R = *i*Pr **a**, Cy **b**) were prepared from the reaction of the diphosphine hydrides **17a, b** with Li[$C_5H_4OSiMe_2$*t*Bu] followed by silyl deprotection with TBAF and subsequent acidification with NH_4Br. In nonpolar solvents compounds **27** are in equilibrium with the isomeric *trans*-dihydride cyclopentadienone species [Re$(H)_2$(NO)(PR_3)(C_5H_4O)] (**28**, R = *i*Pr **a**, Cy **b**), as yet an untypical process in the realm of bifunctional catalysis apparently expressing the extraordinary strength of Re–H bonds. Deuterium labeling studies of compounds **27** with D_2 and D_2O showed H/D exchange at the Re–H and O–H positions. The bifunctional complexes **27** are active catalysts in transfer hydrogenation reactions of ketones and imines with 2-propanol as both the solvent and the H_2 donor (Scheme 30). For instance, the transfer hydrogenation of acetophenone to 1-phenylethanol afforded at 120°C within 10 min a conversion of 97% corresponding to a TOF of 1,164 h^{-1}. Aryl ketones were hydrogenated more rapidly than alkyl ketones. The hydrogenations of imines were less efficient. Noteworthy is the fact that transfer hydrogenations with nonpolar olefinic double bonds could not be achieved. Quite unexpectedly, the transfer hydrogenation of benzaldehyde was found to be least efficient. Tracing the reaction course revealed that in this case severe decomposition of the catalyst had occurred, presumably due to a blocking effect of the formed benzyl alcohol at some catalytic stage. It should also be pointed out that in the case of

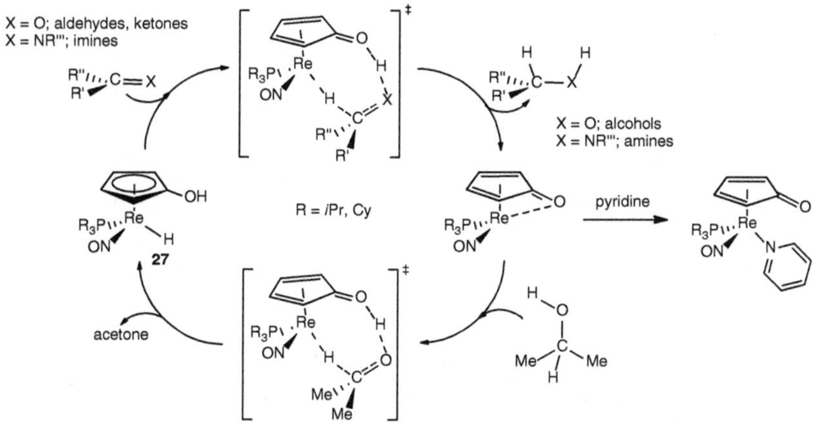

Scheme 31 Proposed mechanism for the "Re–H/OH" system catalyzed transfer hydrogenations of carbonyls

catalyst **27b** the presence of up to 40 equiv. of H_2O relative to the catalyst promoted the transfer hydrogenation performance.

DFT calculations suggested a concerted secondary coordination sphere double H transfer mechanism of the cascade type, as depicted in Scheme 31 [87, 88]. The acidic OH proton and the hydridic Re–H hydrogen **27** are transferred from the hydrogen loaded rhenium complex onto the double bonds of unsaturated substrates via a six-membered metallacyclic transition state. The concomitantly formed cyclopentadienone hydrogen acceptor gets recharged, accepting the hydrogen atoms from 2-propanol with regeneration of the catalyst **27**. Remarkably, the 16e cyclopentadienone intermediate acting as a hydrogen acceptor could be trapped by pyridine, further supporting the proposed catalytic cycle.

Re–H/NH Systems

Another bifunctional mononitrosyl system is the Noyori-type "Re/amido" and "Re–H/NH"-type rhenium system bearing a pincer-type PNP ligand reported by Gusev et al. [89, 90]. The amido hydride $[ReH(NO)(N(C_2H_4P\textit{i}Pr_2)_2)]$ (**29**) was prepared by the reaction of the Re(II) precursor $[NEt_4]_2[Re(NO)Br_5]$ with the pincer-type ligand $HN(C_2H_4P\textit{i}Pr_2)_2$ followed by dehydrobromination with *t*BuOK. Due to the presence of a π-acceptor NO ligand *trans* to the π-donor amido nitrogen, the five-coordinate complex **29** is stabilized by a push–pull interaction of the two ligands mediated by a d_π metal orbital. Under hydrogen atmosphere the dihydride *trans*-$[ReH_2(NO)(HN(C_2H_4P\textit{i}Pr_2)_2)]$ (**30**) was obtained from **29** via a slow H–H heterolytic splitting process. The proposed dihydrogen coordinated intermediate of such transformation could however not be observed, presumably due to the fact that the hydridic Re–H bond is relatively close to the protic H_N atom, complex **30** is not stable and slowly undergoes H_2 elimination regenerating **29**.

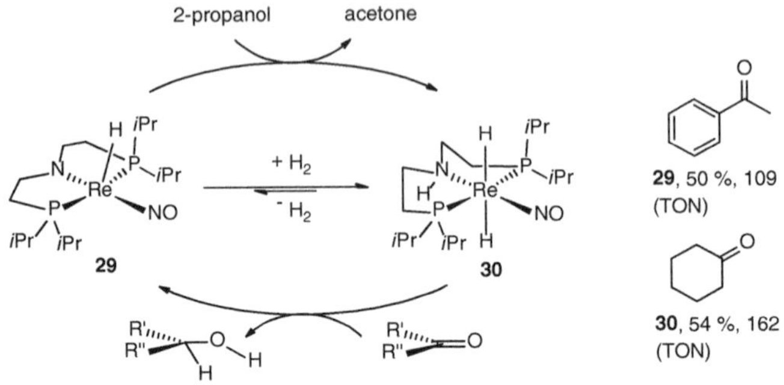

Scheme 32 Bifunctional mononitrosyl Re–H/NH systems

The bifunctional complexes **29** and **30** demonstrated low activities in transfer hydrogenations of acetophenone and cyclohexanone (Scheme 32). For instance, the reaction of acetophenone in 2-propanol catalyzed by 0.45 mol% of complex **29** afforded within 1 h a 50% conversion corresponding to a TON value of 109. A maximum conversion of 66% was eventually achieved reaching the equilibrium state between acetophenone and 1-phenylethanol. The transfer hydrogenation of cyclohexanone catalyzed by 0.33 mol% of complex **30** afforded within 65 min a conversion of 54% corresponding to a TON of 162. The low TON values were attributed to the decomposition of the catalyst due to the instability of **30** in 2-propanol.

3.3 Group 8 Iron Nitrosyl Complexes in Homogeneous Catalysis

Nitric oxide has been shown to interact with many enzymatic sites to reversibly form stable Fe–NO complexes. The binding in these complexes is similar to the coordination of dioxygen to metal centers, and therefore transition-metal iron nitrosyl complexes have been extensively investigated modeling the interactions of O_2 with Fe centers [15]. Despite the rich organometallic chemistry of iron nitrosyl species that has developed within the past decades, the application of Fe–NO complexes in homogeneous catalysis was rarely addressed. This scenario changed only recently, when the Plietker group found that the mononitrosyl Fe(−II) complex $[Bu_4N][Fe(CO)_3(NO)]$ (**31**), its π-allyl derivative $[Fe(\eta^3\text{–}CH_2\text{=}CH\text{–}CHMe_2)(CO)_2(NO)]$ (**32**), and dinitrosyl Fe(−II) complexes $[Bu_4N][Fe(NO)_2(SBn)]_2$ (**33**) (Scheme 33) are active catalysts in organic transformations, particularly in nucleophilic addition catalysis [91, 92]. Considering the nontoxicity, abundance, and biological relevance of iron, such ferrates-based catalyses are particularly significant with respect to sustainability aspects [93].

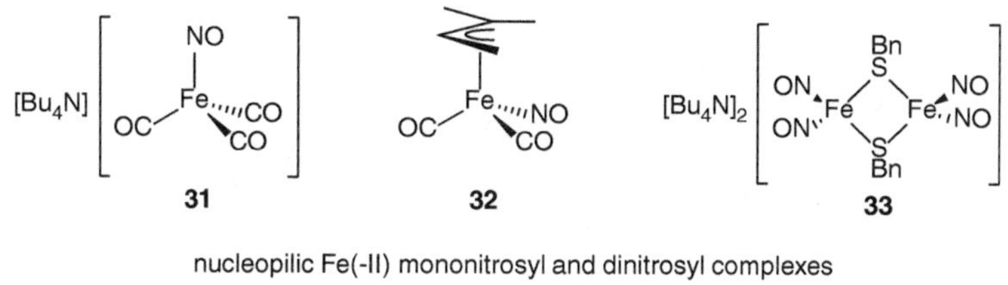

Scheme 33 Representatives of nucleophilic Fe(-II) nitrosyl complexes

3.3.1 Allylic Substitution Reactions

Allylic alkylations are among the most widely applied catalytic C–C bond formation reactions in organic chemistry. In the 1980s already mononitrosyl ferrate complexes of type **31** were reported to be active in regioselective allylic alkylations [94, 95]. However, this pioneering work suffered from low turnover numbers and the reaction had to be carried out under CO atmosphere.

Careful screening of the catalytic conditions by the Plietker group led to exploration of a very efficient allylic substitution system [96]. The essential improvements need to be noted here. (1) In order to avoid deactivation of the catalyst by excess of the external base, a carbonate function was employed in the allyl substrate acting as the leaving group and as an in situ generated base to deprotonate the pro-nucleophile. (2) The CO atmosphere could be replaced by catalytic amounts of PPh_3, which efficiently protected the active species in the catalytic course. (3) DMF was employed as a solvent, which tremendously enhanced the nucleophilicity of the C-nucleophiles. As a result, the allylic alkylation using 2.5 mol% of **31** afforded at 80°C within 24 h satisfactory TONs up to 40 and improved regioselectivities up to 98:2 in favor of the *ipso*-substitution product, as shown in Scheme 34. Using the same conditions, a variety of different allylic carbonates could be alkylated in good yields and with high regioselectivities. Different C-nucleophiles were also tested. Invariably the allylations occurred in high regioselectivities affording the products in good to excellent yields. The cyano group appeared to significantly increase the reactivity of the system, however, at the expense of the regioselectivity.

A reaction mechanism for the iron(II) nitrosyl **31** catalyzed allylic alkylation is depicted in Scheme 35. The nitrosyl group functions as an ancillary ligand tuning in addition the nucleophilicity of the metal center. Initially the iron center is coordinated to the allyl group via interaction with the allyl π^*-orbital. Upon liberation of the leaving group concomitant formation of the σ-enyliron complex takes place, which is anticipated to be stabilized by the added PPh_3 ligand acting as a co-catalyst. The liberated carbonate deprotonates the pro-nucleophile by releasing alcohol and CO_2 in an irreversible acid–base reaction. The generated nucleophile attacks the ally iron complex affording the allyl alkylated product and closes the catalytic cycle. Overall, the new bonds are formed in the products with conservation of the stereochemistry. In this cycle, the oxidation state of the iron center changes between –II and –I. The ancillary nitrosyl group, which is strongly electron-withdrawing, is expected to

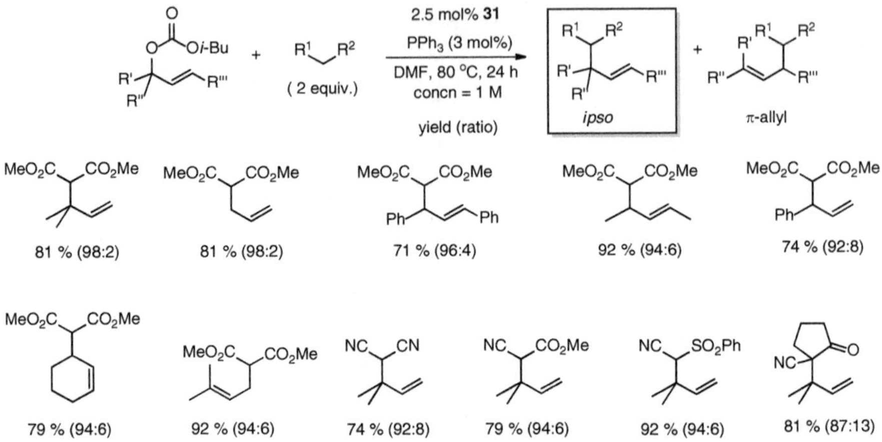

Scheme 34 Allylic substitution reaction catalyzed by Fe(-II) complex **31**

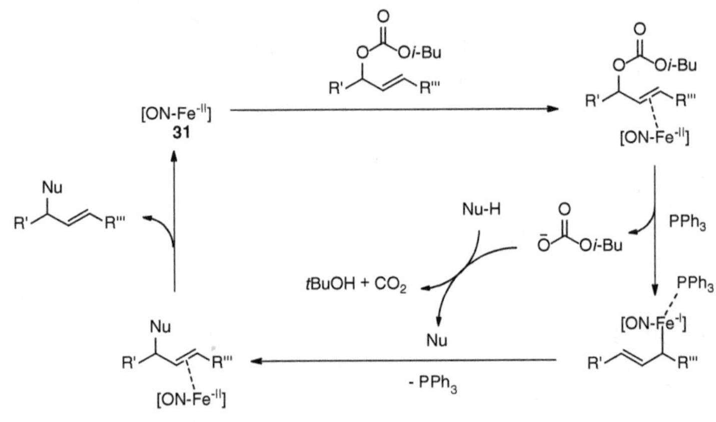

Scheme 35 Proposed mechanism for allylic substitution reaction catalyzed by Fe(-II) complex

facilitate the –I to –II transformation. It cannot be excluded that NO bending gets involved stabilizing the Fe(−I) species.

The scope of the reaction can be expanded to other allylic substitution processes employing heteronucleophiles. While the use of acidic O-pro-nucleophiles proved to be unsuccessful, the reaction with aromatic amines turned out to be quite promising [97]. In order to prevent the deactivation of the iron catalyst under basic conditions, an organic buffer, piperidine hydrochloride, was employed, which enhances the allylic amination reaction. As depicted in Scheme 36, the allylic amination of phenyl amine was achieved in DMF at 80°C in 87% yield in the presence of 5 mol% of **31** and PPh_3 together with 30 mol% of the buffer. Using the same conditions, various aromatic amines were allylated in good to excellent yields with high regioselectivities up to 98:2 in favor of *ipso*-products.

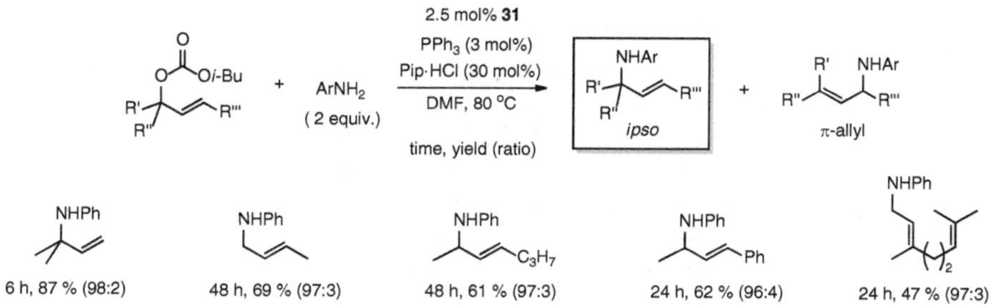

Scheme 36 Allylic substitution reaction using aromatic amines as nucleophiles

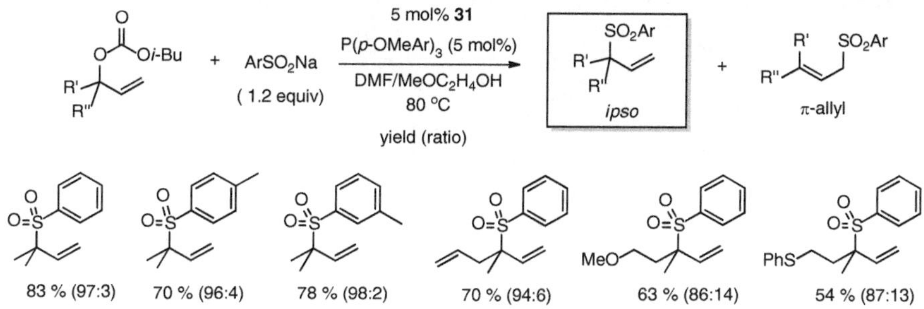

Scheme 37 Allylic sulfonylations catalyzed by Fe(-II) complex **31**

Allylic sulfonylations can also be accomplished by using the iron nitrosyl complex **31** as catalysts [98]. In these cases, 2-methoxylethanol was used as a co-solvent to DMF due to the poor solubility of the sodium sulfinate substrate to be applied as a nucleophile. The effect of the phosphine ligand was also investigated. Tuning efforts revealed an outstanding performance of $P(p\text{-}MeOAr)_3$ as co-catalyst. These efforts rendered a suitable protocol for efficient and regioselective allylic sulfonylation allowing C–S bond formations, as depicted in Scheme 37.

The π-allyl iron mononitrosyl **32** was also found to be highly active in allylic alkylation reactions [99]. Significantly, ligand-dependent regioselectivity was explored based on **32**, and a completely inverse regioselectivity was obtained when the SIMes ligand was used as a co-catalyst. As depicted in Scheme 38, complex **32** is more active than **31**, since a 1 mol% loading of the catalyst afforded good to excellent yields for the allylated product. The SIMes was in situ generated from the carbene salt in the presence of a base. The DMF, which was used as a solvent in the case of the catalyses with **31**, could be replaced by MTBE (methyl *tert*-butyl ether) due to its higher stability in nucleophilic reactions. Overall regioselectivities up to 98:2 were obtained in favor of the π-allyl product. This unusual regioselectivity was attributed to both electron-donating and steric bulk of the SIMes ligand, which boosts formation of the π-allyl iron SIMes mononitrosyl intermediate. The nitrosyl ligand functions in such transformation as an ancillary ligand adjusting the electron properties of the iron center.

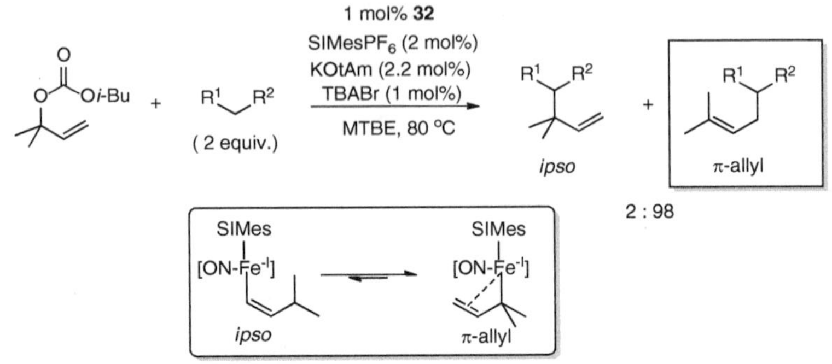

Scheme 38 Allylic substitution reaction showing ligand-dependent regioselectivity

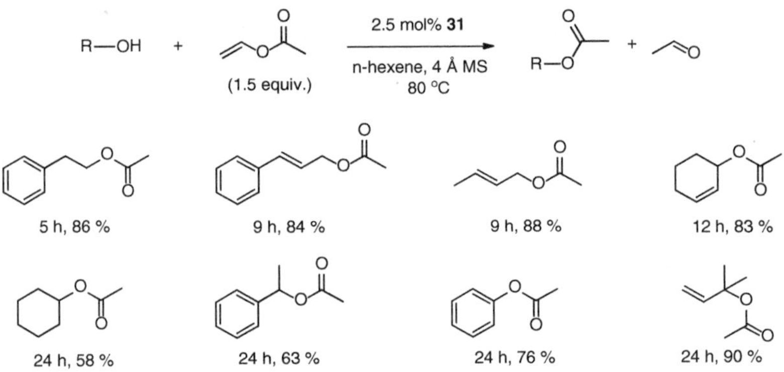

Scheme 39 Transesterification catalyzed by Fe(-II) complex **31**

3.3.2 Carbonyl Activation in Transesterifications and Deprotections of Allyloxycarbonyl Compounds

Similar to the reaction course of the allylic substitution, which involves formation of π-allyl moieties followed by subsequent nucleophilic addition across the π-bond, the mononitrosyl iron(−II) complex was expected to be active in transesterifications involving activation of carbonyl group and nucleophilic addition to the electrophilic carbon atom [100]. This assumption could be verified by experimental tests. Under neutral conditions without addition of a ligand co-catalyst, the iron complex **31** exhibited high activity in the transesterification of vinyl acetate. Good to excellent yields were obtained affording a new ester bond, as depicted in Scheme 39.

The mechanism is similar to that of the allylic substitution, as depicted in Scheme 40. Initially the iron(−II) complex coordinates to the C=O double bond, which is followed by elimination of the leaving group affording the acyl iron(−I) intermediate. The nucleophile in situ generated by the leaving group attacks the electrophilic carbonyl carbon atom reestablishing the iron(−II) species. Finally the

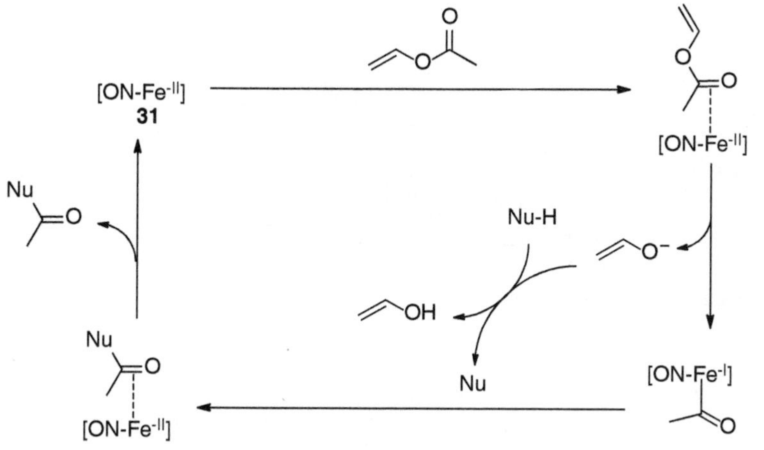

Scheme 40 Proposed mechanism for **31** catalyzed transesterification reaction

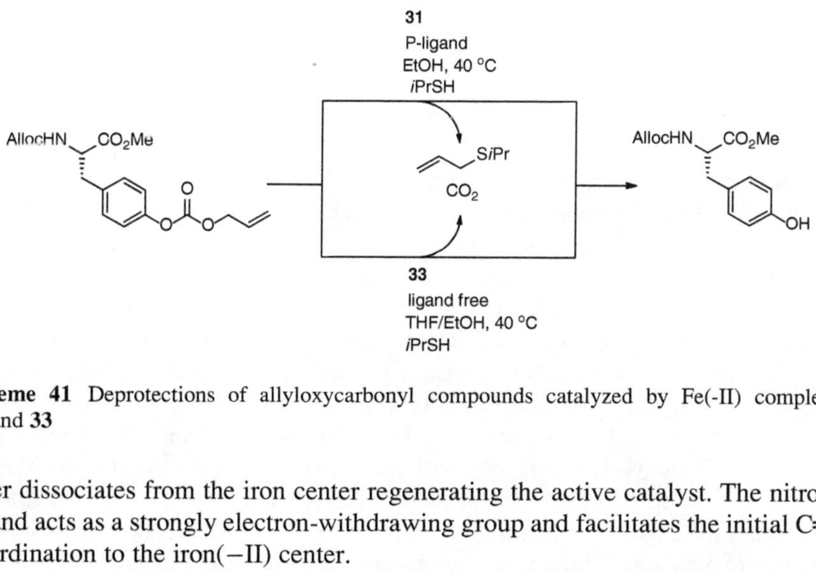

Scheme 41 Deprotections of allyloxycarbonyl compounds catalyzed by Fe(-II) complexes **31** and **33**

ester dissociates from the iron center regenerating the active catalyst. The nitrosyl ligand acts as a strongly electron-withdrawing group and facilitates the initial C=O coordination to the iron(−II) center.

Another versatile application of such carbonyl activation methodologies by iron (−II) complexes is the deprotection reactions of allyloxy carbonyl groups [101]. Different from most of such processes employing precious metal catalysts, the mononitrosyl iron(−II) complex exhibits high activities in the C–O cleavage course under neutral conditions showing a broad functional group tolerance. As depicted in Scheme 41, the cleavage of the allyl carbonates for liberation of the desired alcohol was accomplished by using catalytic amounts of **31** and a phosphine ligand in the presence of isopropyl thiol as an allyl scavenger at 40°C in ethanol. Significantly, use of the dinitrosyl iron(−II) complex **33** afforded comparable catalytic performance even in the absence of ligands.

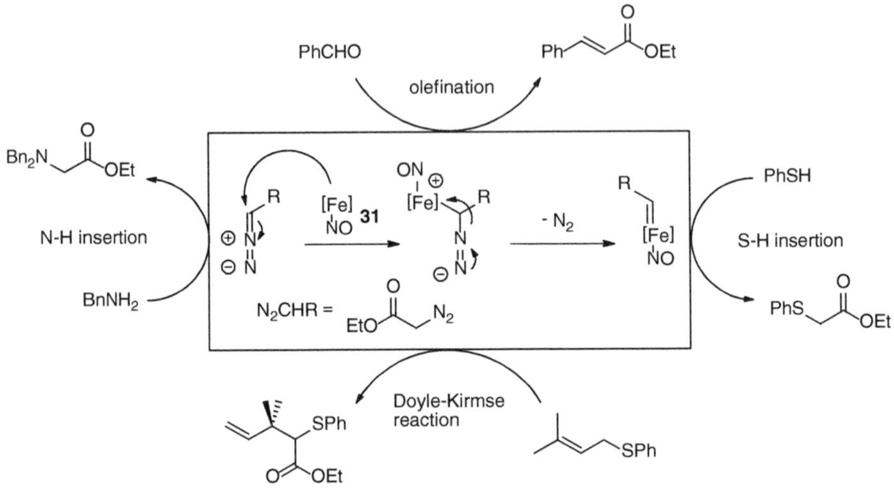

Scheme 42 Carbene-transfer reactions catalyzed by Fe(-II) complex **31**

3.3.3 Carbene-Transfer Reactions

Remarkably, mononitrosyl iron(−II) complexes displayed great potential in the activation of diazo compounds and carbene-transfer reactions [102]. Generally, the activation of diazo compound can be realized by electrophilic transition metal complexes. However, according to the concept of "Umpolung" [103], the electron-rich, nucleophilic iron(−II) compound **31** is expected to react with diazo compounds of electron-poor carbenes, such as ethyl diazoacetate (Scheme 42). At first the iron center would add the C=N bond of the diazo compound followed by release of N_2 and formation of the electrophilic iron carbene moiety. The nitrosyl group in such transformations is assumed to support as an ancillary ligand the N_2 release by pulling electron density to the iron center.

In the presence of catalytic amounts of **31**, an insertion into heteroatom-hydrogen S–H and N–H bonds, olefination of carbonyl groups, and Doyle–Kirmse reactions could be accomplished even in absence of ligands [102]. It should be noted that in the Doyle–Kirmse reaction a 2.5 mol% loading of the catalyst **31** is sufficient to afford in dichloromethane at 40°C good to excellent conversions. This transformation displays broad functional group tolerance.

3.3.4 Allylic C–C Bond Activations in Allylic Substitutions and [3+2]-Cycloadditions

Significantly, the limits in the activation of allylic C–C bond could be pursued further to include also electron-poor moieties by applying catalyses with nucleophilic iron complexes [104]. The combination of mononitrosyl iron(−II) complex

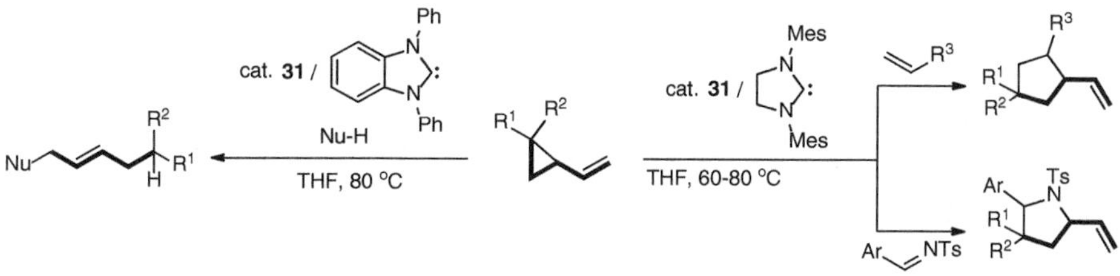

Scheme 43 Allylic C–C bond activation by Fe(-II) complex **31**

31 and the NHC ligand proved in THF to be successful in the formation of allylic C–C bonds via cleavage of vinylcyclopropanes. As depicted in Scheme 43, the catalytic reaction of a variety of acidic pro-nucleophiles using 10 mol% of **31** with addition of a NHC ligand afforded in THF at 80°C 1,5-dipolar addition products in good to excellent yields and regioselectivities in favor of the linear substitution products. A broad functional group tolerance was observed in this transformation. It was assumed that the allyl–Fe intermediate, which is formed upon action of **31** to cleave the C–C bond of the allyl group, functions as the true active species in the catalysis.

Furthermore, the "**31**/SIMes" catalytic system was explored to be active in [3+2]-cycloadditions of vinylcyclopropanes via a Michael-addition pathway. A variety of Michael acceptors bearing electron-withdrawing groups were used providing good to excellent yields of the five-membered ring products. Such coupling reactions were also applicable to the activation of electron-deficient π-bonds of imines. The reaction of vinylcyclopropanes with *N*-tosyl arylimines afforded the corresponding pyrrolidines in good yields. Similarly, the "**31**/SIMes" catalyzed [3+2]-cycloaddition reactions displayed broad functional group tolerance.

4 The (Reversible) $NO^+ \leftrightarrow NO^-$ Ligand Transformation in Homogeneous Catalysis

4.1 H–X (X = H, Si) Bond Activation by Cationic Re(−I) Dinitrosyl Complexes

A series of Re(−I) dinitrosyl hydride complexes **34–40** (Scheme 44) were studied in terms of the nitrosyl bending effect to be utilized in small molecule activation and catalysis [32, 33]. The five-coordinate dinitrosyl Re(−I) hydride complexes $[ReH(NO)_2(PR_3)_2]$ (**34**, R = *i*Pr **a**, Cy **b**) were prepared via stepwise introduction of NO ligands starting from rhenium metal [33]. Complexes of the type **34** are 18e five-coordinate species adopting pseudo-trigonal bipyramidal geometries with two essentially linear nitrosyl ligands and two phosphines bending toward the hydride.

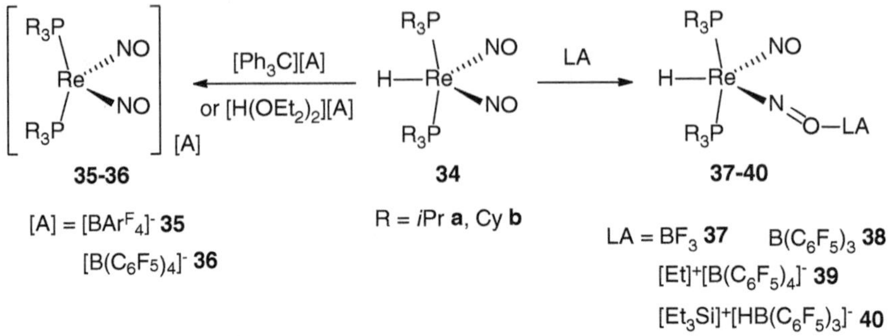

Scheme 44 Reactivity of Re(-I) dinitrosyl hydride complexes

Due to the two strong π-accepting NO ligands and the two σ-donating phosphines, a characteristic hydridic reactivity is expected for the metal bonded hydrogen. Indeed, temperature-dependent measurements of the DQCC revealed high bond ionicities of 72% for **31a** and 70% for **34b**, respectively.

The hydridic character of the rhenium-bonded hydrogen of **34a, b** facilitated the hydride abstraction upon treatment with either Brönsted acids or Lewis acids [32]. Depending on the nature of the phosphine substituent, the 16e Re(−I) dinitrosyl cations $[Re(NO)_2(PR_3)_2][BAr^F{}_4]$ (**35**, R = *i*Pr **a**, Cy **b**) could be accessed from the reaction of either **34a** with $[H(OEt_2)_2][BAr^F{}_4]$ or **34b** with $[Ph_3C][BAr^F{}_4]$. Further anion modification to access the corresponding $[B(C_6F_5)_4]^-$ salts $[Re(NO)_2(PR_3)_2][B(C_6F_5)_4]$ (**36**, R = *i*Pr **a**, Cy **b**) showed enhanced solubilities in nonpolar solvents [105].

Beside the hydride as a reactive site of **34a, b**, the nitrosyl group was considered as the other functional site. The O_{NO} atom is Lewis basic, which was found to react with external Lewis acids, such as BF_3, $B(C_6F_5)_3$, and $[Et_3O][B(C_6F_5)_4]$ to form Lewis acid/base adducts of the general composition $[ReH(NO)(NOLA)(PR_3)_2]$ (LA = BF_3 **37**, $B(C_6F_5)_3$ **38**, Et^+ **39**, R = *i*Pr **a**, Cy **b**) [105]. The reaction of **34** with a mixture of Et_3SiH and $B(C_6F_5)_3$ afforded the silylium coordinated adducts $[ReH(NO)(NOSiEt_3)(PR_3)_2][HB(C_6F_5)_3]$ (**40**, R = *i*Pr **a**, Cy **b**). A related NO-$SiEt_3$ derivatized rhenium hydride bearing different counter-ions can be accessed from the reaction of **35** and **36** with Et_3SiH and $B(C_6F_5)_3$ via H–Si heterolytic cleavage [32].

Taking into account both the coordinative unsaturation of **35** and **36**, and the Lewis basicity of the nitrosyl oxygen atoms, the 16e cationic fragment $[Re(NO)_2(PR_3)_2]^+$ can be considered as a bifunctional complex possessing the free acidic site provided by the metal and the basic site provided by the O_{NO} atom. This was anticipated to provoke bipolar or bifunctional reactivity in small molecule activation. Indeed, **35** and **36** demonstrated facile reactions activating H–X (X = H, Si) bonds [32].

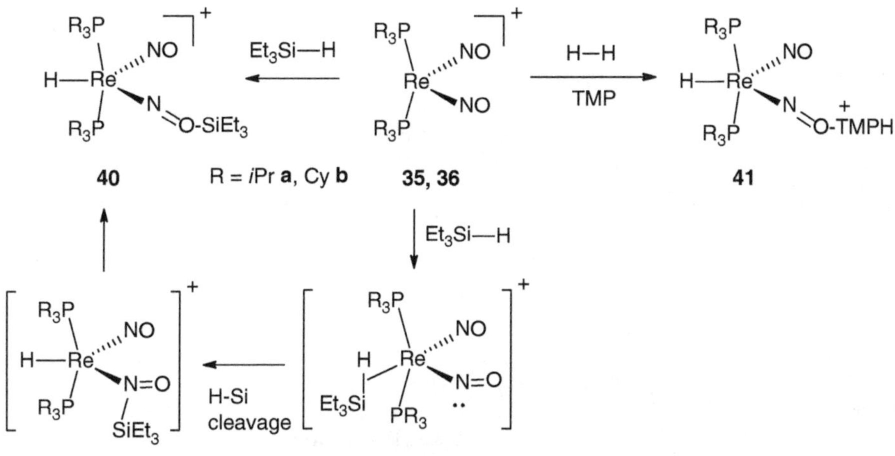

Scheme 45 Heterolytic Si–H and H–H bond cleavage by Re(-I) dinitrosyl complexes

For instance, the reaction of **35** with an excess of Et_3SiH afforded at room temperature clean formation of silylium cation coordinated Re(−I) hydride complexes of type **40**. These reactions are remarkable, since they are accomplished by reaction with an electron-rich d^8 Re(−I) center, which is in contrast to the heterolytic cleavage of a Si–H bond generally accomplished by electron-deficient, poor σ-donating systems enhancing the acidity of the Si–H bond upon coordination. Although the hydrosilane coordinated intermediate is not observable, it is postulated that coordination of the hydrosilane to the metal center is rate-limiting followed by a very fast Si–H bond heterolytic cleavage. A reaction pathway is proposed in Scheme 45. Coordination of Si–H bond to the rhenium center facilitates NO bending leading to a trigonal-pyramidal geometry, in which the hydrosilane is acidified by the linear *trans* NO ligand as a strong π-acceptor causing electron deficiency. Intramolecular heterolytic cleavage of the Si–H bond occurs by the neighboring basic N_{NO} atom, which further undergoes 1,2-shift of the silylium cation to afford the O_{NO} atom coordination.

In a similar way H–H bond activation could be achieved by the 16e cationic fragment $[Re(NO)_2(PR_3)_2]^+$. In this case, however, the presence of a sterically hindered base, such as tetramethylpiperidine (TMP), is a prerequisite to stabilize the parts of H–H splitting affording the NO ligand functionalized Re(−I) hydride complexes $[Re(H)(NO)(NOHTMP)(PR_3)_2][BAr^F{}_4]$ (**41**, R = *i*Pr **a**, Cy **b**). Most probably due to the nonpolar nature of the H–H bond, the presumed dihydrogen ligand on the metal center is not acidic enough to undergo spontaneous splitting. Therefore an external base must be present to assist the splitting course. The presence of a strong base might also serve to deactivate (or stabilize) the proton of the NO–H moiety, so as to prevent the protonation of the Re–H and regeneration of the 16e cationic reactant.

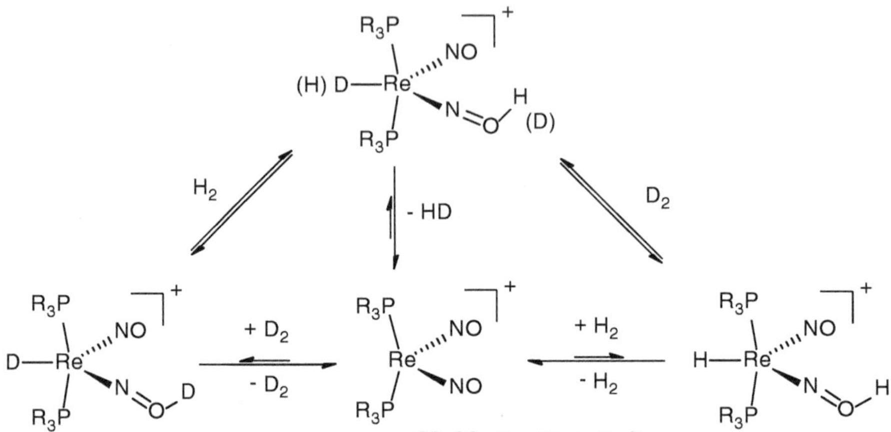

Scheme 46 H–H bond activation by Re(-I) dinitrosyl complexes

4.2 Catalytic H_2/D_2 Scrambling Triggered by Nitrosyl Bending

In the absence of an external base, complexes **35** and **36** are active catalysts in the scrambling of H_2/D_2 to afford HD under very mild conditions [32]. For instance, when a solution of **35a** was exposed to 1 bar of an equimolar mixture of H_2/D_2, complete H/D scrambling was observed within a few minutes at room temperature. The rate of the scrambling reaction is solvent dependent. Equilibration in toluene or chlorobenzene is finished within a few minutes, while in THF it requires several hours to be completed. This implies that the availability of the coordinative vacancy for H_2/D_2 binding is crucial to the H_2/D_2 scrambling. Such catalysis can be interpreted in terms of reversible H–H or D–D splitting affording the transient Re–H(D)/NOH(D) species, which then undergo H/D exchange leading to the formation of HD, as depicted in Scheme 46.

The catalytic H_2/D_2 scrambling can also be accomplished by the 18e Lewis acid coordinated Re(−I) dinitrosyl complexes **38–40** [32]. Taking into account the coordinative saturation of the metal center, the catalysis can only be interpreted in terms of the generation of a vacant site *cis* to the rhenium hydride bond. Therefore one might envisage bending of one nitrosyl ligand, which would be accompanied by a change in formal oxidation state of the metal center from −I to +I by creating a vacant site. Such NO bending is particularly plausible in the case of **38–40** type complexes bearing a coordinated Lewis acid, which increases the π-back donation from the metal to nitrosyl ligand so as to facilitate the bending course. A reaction mechanism for the **38–40** catalyzed H_2/D_2 scrambling is depicted in Scheme 47. The Lewis acid triggered nitrosyl bending affords a 16e Re(I) intermediate possessing a vacant site *trans* to the bent NO ligand, which further uptakes D_2. Then the rearrangement from the [Re–H(D–D)] to the [Re–D

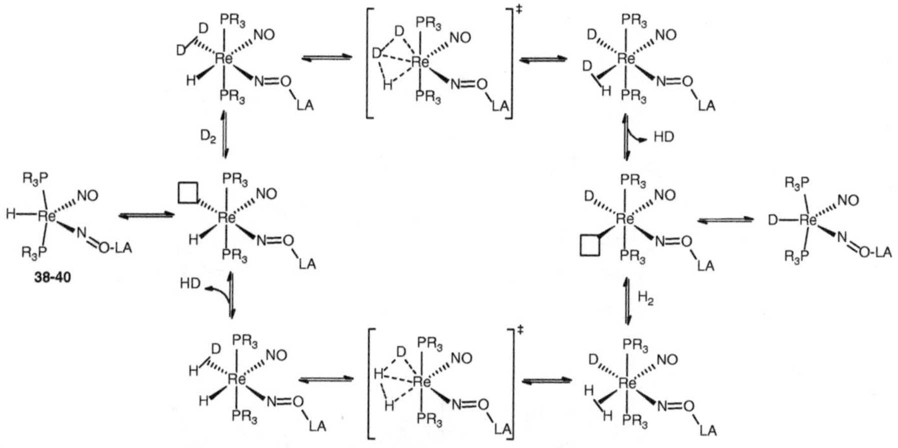

Scheme 47 Proposed pathway for H_2/D_2 scrambling catalyzed by Lewis acid functionalized Re(-I) dinitrosyl hydride complexes

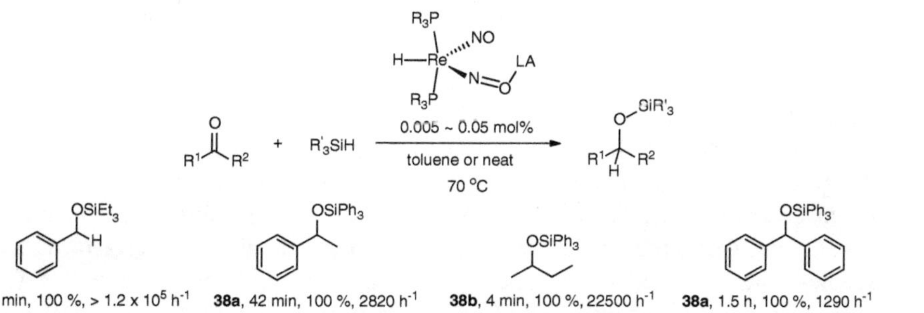

Scheme 48 Hydrosilylation of carbonyls catalyzed by Lewis acid functionalized Re(-I) dinitrosyl hydride complexes

(H–D)] occurs via a [Re–(H)(D)(D)] transition state, in which the H/D exchange can take place [24, 71]. The catalytic cycle could eventually be closed proceeding along the other rearrangement course from the [Re–D(H–H)] to the [Re–H(H–D)].

4.3 Hydrosilylation of Carbonyl Compounds Triggered by Nitrosyl Bending

Not surprising, the same protocol for H_2/D_2 scrambling can be employed in the hydrosilylation of carbonyl compounds [105]. Complexes **35** were found to be extremely active catalysts for the hydrosilylation of carbonyl compounds. As depicted in Scheme 48, the hydrosilylation catalyzed by 0.005–0.05 mol% of **38** was clean and efficient. Quantitative conversions can be achieved under mild conditions affording TONs up to 9,000 and TOFs up to $1.2 \times 10^5\ h^{-1}$. Remarkably,

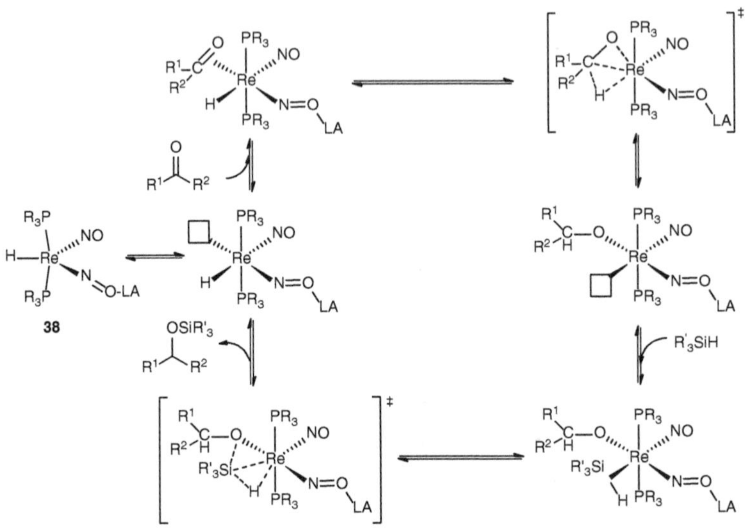

Scheme 49 Proposed mechanism for **38** catalyzed hydrosilylation of carbonyls

the hydrosilylation of ketones can be carried out under neat conditions without addition of solvents, which would be quite beneficial for industrial applications.

A mechanism for the highly efficient hydrosilylation of carbonyl compounds catalyzed by **38** is depicted in Scheme 49. Nitrosyl bending assisted by the coordinated Lewis acid leads to generation of the 16e Re(I) intermediate bearing one vacant site. Coordination of the C=O bond to the rhenium center affords an 18e species with *cis*-aligned C=O and Re–H units. Most probably via a "Re(H–C–O)" transition state, insertion of the C=O bond into the Re–H occurs to give a 16e rhenium–alkoxide intermediate possessing a vacant site. Subsequent binding of hydrosilane to the rhenium affords a six-coordinated intermediate, in which the Si–H bond is positioned *trans* to the linear nitrosyl. Due to the strong π-accepting *trans*-nitrosyl group, the Si–H bond becomes highly polarized toward the $Si^{+}–H^{-}$ form rendering an extremely oxophilic silylium cation. Eventually both the Re–O and Re–silyl bonds are eliminated via a Re(O–Si–H) transition state expelling silyl alcohols as the hydrosilylation product.

4.4 *Highly Efficient Hydrogenation of Alkenes Triggered by Nitrosyl Bending Applying Re(I) Diiodo Complexes and Lewis Acids*

Lately the Berke group has reported that enforced NO bending can boost the performance of alkene hydrogenations via co-catalysis of Re(I) diiodo mononitrosyl complexes and attached silylium Lewis acids [31] (Scheme 50). Combinations of the

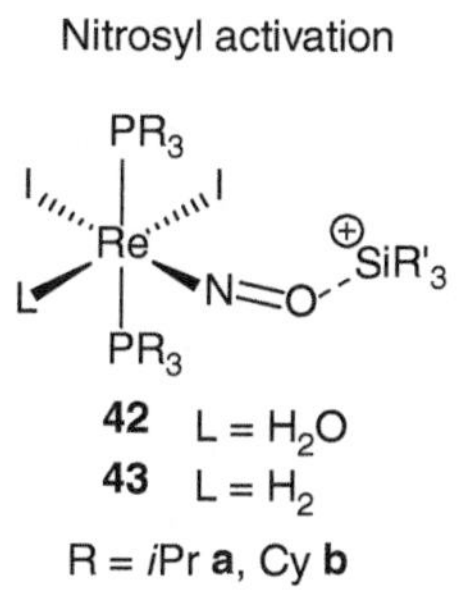

Scheme 50 Silylium functionalized Re(I) diiodo complexes

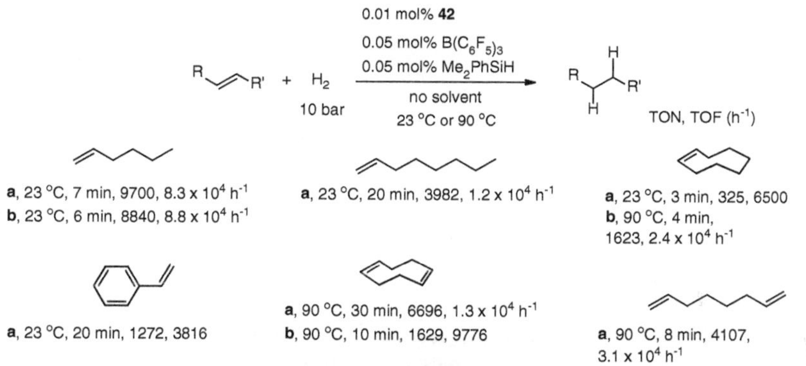

Scheme 51 Highly efficient hydrogenation of alkenes catalyzed by the "**42**/$B(C_6F_5)_3$/Me_2PhSiH" system

aqua complexes $[ReI_2(NO)(PR_3)_2(H_2O)]$ (**42**, R = *i*Pr **a**, Cy **b**) or the dihydrogen complexes $[ReI_2(NO)(PR_3)_2(\eta^2\text{-}H_2)]$ (**43**, R = *i*Pr **a**, Cy **b**) with hydrosilane and B $(C_6F_5)_3$ turned out to be effective catalytic systems. They were therefore termed as systems operating via the "catalytic nitrosyl effect". Two facts should be mentioned in this context. (1) The ligand field strength and donicity of halogen ligands decreases in the order of F > Cl > Br > I. The weakest ligand is iodide, which exhibits also least tendency toward attack by a hard external Lewis acid. (2) The silylium species R'_3Si^+ generated in situ from the $R_3'SiH/B(C_6F_5)_3$ mixture is one of the strongest oxophiles [106–109], which is expected to modify the nitrosyl properties through coordination to the O_{NO} atom.

The catalytic system of "**42(43)**/hydrosilane/$B(C_6F_5)_3$" indeed generates highly efficient catalytic systems showing excellent activities and longevities in the hydrogenation of terminal and internal alkenes, as depicted in Scheme 51. At 23°C under 10 bar of H_2 the "**43**/$Me_2PhSiH/B(C_6F_5)_3$" system afforded in the hydrogenation of 1-hexene a 97% conversion within 7 min corresponding to a TON of 9,700 and a TOF of $8.3 \times 10^4\ h^{-1}$. Using the "**43**/$Me_2PhSiH/B(C_6F_5)_3$" system a conversion of 89% was found within 6 min resulting in the high TOF of $8.8 \times 10^4\ h^{-1}$. Significantly, under 40 bar of H_2 at 23°C, almost full conversions

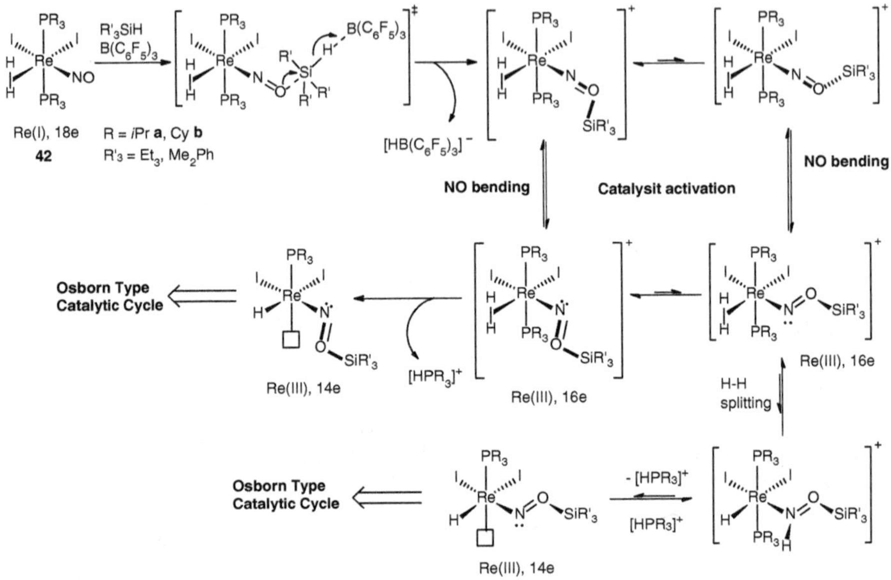

Scheme 52 Proposed mechanism for NO bending boosting high efficiency of Re(I) diiodo complexes in hydrogenation of alkenes

could be accomplished within 1 min resulting in the highest TOF of $6.0 \times 10^5\ h^{-1}$. In the case of internal alkenes, such as cyclohexene, cyclooctene (COE), and 1,5-cyclooctadiene (COD), the hydrogenations showed at 23°C poorer performance presumably due to a relative faster deactivation of the catalysts or due to the reduced binding ability of disubstituted olefins. But increasing the temperature to 90°C eventually resulted in improved TONs and TOFs. The hydride donating abilities of the hydrosilane components were crucial to the catalytic performance and were found to correlate with the catalytic activities in the order of $Ph_3SiH \approx$ $i$$Pr_3SiH \ll MePh_2SiH < Et_3SiH < Me_2PhSiH$ [110].

Comprehensive mechanistic studies showed an inverse kinetic isotope effect, fast H_2/D_2 scrambling, and slow alkene isomerizations pointing to an Osborn-type hydrogenation cycle with a rate-determining reductive elimination of the alkane. Both spectroscopic experiments and dispersion corrected DFT calculations revealed a reaction pathway as depicted in Scheme 52. In the initiation stage reversible coordination of the $B(C_6F_5)_3$ Lewis acid to the Si–H bond of the hydrosilane occurs affording an adduct intermediate with as yet not fully broken, but activated hyperconjugative B··H–Si and B–H··Si bonds. The R'_3Si^+ transfers to the Lewis basic O_{NO} atom occurs via a state with full hydride transfer from the hydrosilane to the $B(C_6F_5)_3$ reagent. As verified by DFT calculations, an intermediate with the Si atom oriented toward the H_2 ligand is favored over its isomer with the Si atom oriented toward the iodo ligand. Subsequently, nitrosyl bending occurs affording the $16e^-$ Re(III) species, in which the lone pair electron of the bent nitrosyl moiety is *anti* to the H_2 ligand. In this species the Re center is more electron-deficient turning the η^2-H_2 ligand acidic, or in other words, highly polarized toward $H^{\delta+}$–$H^{\delta-}$. The crypto-H^+ may be readily transferred to the neighboring in-plane *cis*-phosphorus atom. Via concomitant loss of a phosphine ligand a "superelectrophilic" 14e$^-$ Re(III) hydride species forms, which serves as the basic catalytic intermediate driving the hydrogenation cycle along an Osborn-type scheme with alkene before H_2 addition [46].

Parallel to this, another pathway is proposed. In a similar fashion to Noyori–Morris-type catalysts [37, 111, 112], H–H heterolytic cleavage of the polarized H_2 ligand across the Re–N bond could occur. This step demonstrates a novel cooperative function of the bent nitrosyl ligand especially suited for H_2 catalyses with polar reactivity characteristics [88]. The N–H moiety is then thought to be deprotonated by a labile *cis*-positioned phosphine ligand affording 14e$^-$ Re (III) hydrides, which seem to be more activated than the related 16e$^-$ complexes.

5 The L_nM–NO + ⟨O⟩ ↔ L_nM–NO_2 Equilibrium Functioning as Oxene ⟨O⟩ Source in Homogeneous Catalysis

More than 50% of the chemicals turned over in industry are prepared by oxidation of petrochemicals. Traditional oxidation courses are mostly based on metal peroxo complexes, which however often suffer from the co-oxidation of the ligands and nonspecific radical autooxidations. Due to the change in price and availability of the petrochemicals, novel and specific oxidation processes are needed. In the late 1970s and early 1980s, transition metal nitrosyl complexes were studied for oxygen atom transfer reactions based on a "nitrosyl/nitro" redox couple [113–122]. In comparison to the traditional methods, oxidation of organic substrates was developed operating via an oxygen atom transfer from a ligand (such as nitro) of the transition metal complex. These reactions are non-radical processes and therefore are expected to show higher chemoselectivities. After the oxene ⟨O⟩ transfer the reduced ligand (nitrosyl) would be re-oxidized by molecular oxygen completing the catalytic cycle. Since the redox process occurs in the secondary coordination sphere on the ligand, the oxidation state of the metal center would not change, and coordinative unsaturation of the metal complexes would not be a prerequisite for well-functioning catalyses.

It should be emphasized that tetradentate ligands, such as porphyrin and salen, are dominating in this chemistry, since they are nonoxidizable ligands and tend to form square-pyramidal complexes with bent nitrosyls at axial positions that facilitate the oxidation by molecular oxygen to form the desired nitro ligands.

In 1973, it was found that five-coordinate cobalt nitrosyl complexes possessing a bent NO ligand and a tetradentated ligand, such as salen, could react with molecular oxygen at room temperature in the presence of a base to form the six-coordinate cobalt nitro complexes (Scheme 53) [123]. In these formally Co(III) complexes the nitro group is monodentate and N bound. Taking into account the bent nitrosyl ligand, the oxidation takes place without changing in the oxidation state of cobalt.

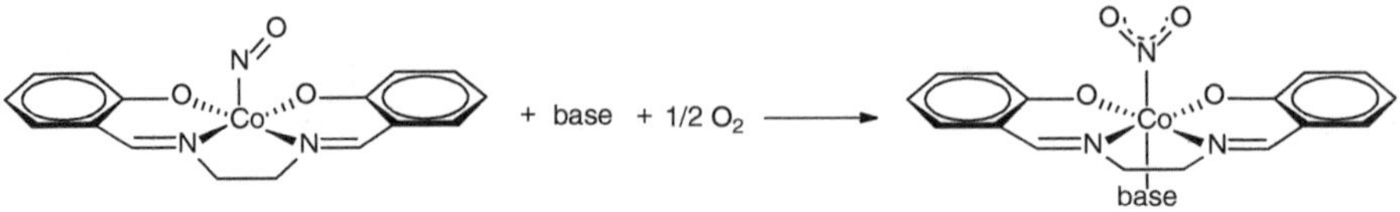

Scheme 53 Oxidation of Co–NO to Co–NO_2 by molecular O_2

5.1 Co–NO$_2$/NO Redox Couples

The following schematic reaction

$$[Co]{-}NO_2 \rightleftharpoons [Co]{-}NO + \langle O \rangle$$

became basis for studies of the oxygen atom transfer from the nitro ligand to a phosphine substrate in order to eventually establish full catalytic cycles [114]. Using the six-coordinate Co(III) nitro complex [Co(saloph)(py)(NO_2)] (**43**, saloph = *N,N′*-bisalicylidene-*o*-phenylenediamino) as a catalyst, oxidation of triphenylphosphine to phosphine oxide was accomplished in 1,2-dichloroethane with excess pyridine as a base. A maximum TON of 9 was achieved under 1 bar of O_2 at 60°C within 16 h.

A catalytic cycle involving the "nitrosyl/nitro" redox couple was proposed in Scheme 54. Oxene transfer from the nitro ligand of **43** to a phosphine was found to be accompanied by the formation of corresponding five-coordinate nitrosyl complex. The catalytic cycle is then completed by re-oxidation of the nitrosyl ligand by molecular oxygen in the presence of a base. Such non-radical catalytic process was supported by several facts. (1) Stoichiometric experiments demonstrate that **43** oxidizes PPh_3 without NO ligand dissociation from the metal center. (2) Tracing the IR spectrum of the cobalt complex during the reaction course confirmed that also dissociation of NO_2 ligand did not occur. (3) A radical oxidation pathway was ruled out, since inhibition of the reaction was not observed adding radical scavengers.

An important application of the "nitrosyl/nitro" redox couple was to accomplish catalytic oxidation of olefins [115]. The nitro ligand in the cobalt complex is formally regarded as a nitrogen-bound monoanionic ligand (Co^+–NO_2^-). It may function as a weak, oxygen-centered nucleophile. On the other hand, it is well known that transition metal coordinated alkenes experience to some extent "Umpolung" from a nucleophile to an electrophile via a shift of the olefin along its axis [103]. This leads to the formation of incipient carbon cations, which now are electrophilic in character. Therefore nucleophilic attack of a nitro ligand onto a metal coordinated alkene is expected to be facile.

The combination of the cobalt nitro complex **43** or [Co(TPP)(py)(NO_2)] (**44**, TPP = tetraphenylporphyrin) with palladium(II) complex [$(PhCN)_2PdCl_2$] indeed demonstrated catalytic activity toward oxidation of olefins to aldehydes or ketones. At 70°C in a gas flow of olefin and O_2, the "Co–NO_2/Pd(II)" catalytic system (1:2) afforded the formation of acetaldehyde from ethylene in TONs up to

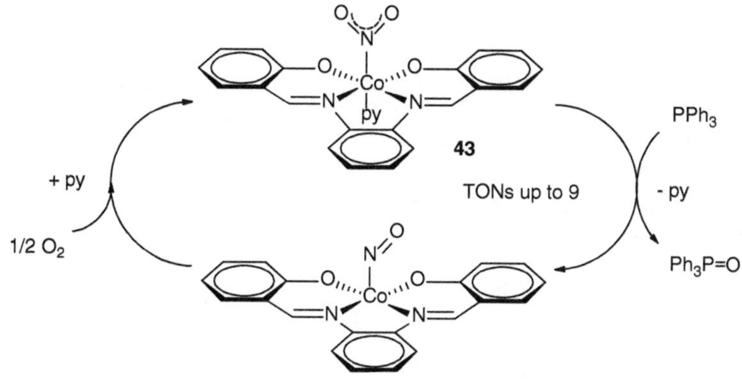

Scheme 54 Catalytic cycle for phosphine oxidation based on the Co–NO_2/NO redox couple

12 within 4 h, and the formation of acetone from propylene in TONs up to 2 within 8 h. The activities of **43** and **44** were comparable. The low TON values were attributed to the release of pyridine in the catalytic process that inhibits the ability of the palladium(II) to coordinate olefins and to assist the re-oxidation of the cobalt–nitrosyl complexes.

A mechanism for such novel models of catalytic olefin oxidation by molecular oxygen was depicted in Scheme 55. It involves oxygen atom transfer from the nitro ligand of cobalt–nitro complexes to palladium(II)-bound olefins followed by re-oxidation of the reduced nitrosyl ligand by molecular oxygen. The oxygen atom transfer proceeds most likely via an intermediate possessing [Pd–CH_2–CHR–(ONO)–Co] moiety, and the formation of acetaldehyde and ketone is accounted for by a subsequent β-hydride elimination process to form a Pd–H bond followed by a hydride transfer from Pd to the terminal CH_2 group affording the oxidization product and cobalt–nitrosyl complex related to the Wacker oxidation of substituted olefin [124, 125]. The palladium remains in the divalent state throughout the catalytic cycle and serves exclusively as a co-catalyst.

The above mechanism suggested that the use of olefin activators other than palladium(II), which are not capable of promoting the β-hydride elimination, may lead to other types of olefin oxidization products, such as epoxides. Since thallium(III) is a known oxidant for olefin epoxidation, it was therefore postulated that replacement of the palladium(II) activator by Tl(III) benzoate in the Co–NO_2/NO redox system would lead to the accomplishment of olefin epoxidation [118].

At 60°C in THF, the combination of **43** or **44** with thallium(III) afforded within 5 h the epoxide of 1-octene or propylene in over 50% yield. Formation of ketones could not be observed, which is most probably due to the fact that β-hydride elimination cannot be accomplished at a Tl(III) center, as depicted in Scheme 56. In principle, if the Tl(III) species could remain in the trivalent oxidation state throughout the reaction, a catalytic cycle with the use of molecular oxygen should be achievable. However, this goal could not be reached in the case of the Co–NO_2/Tl(III)-mediated olefin epoxidation, which was mainly due to the loss of the oxidant Tl(III) by a competing side reaction with reduction from Tl(III) to Tl(I).

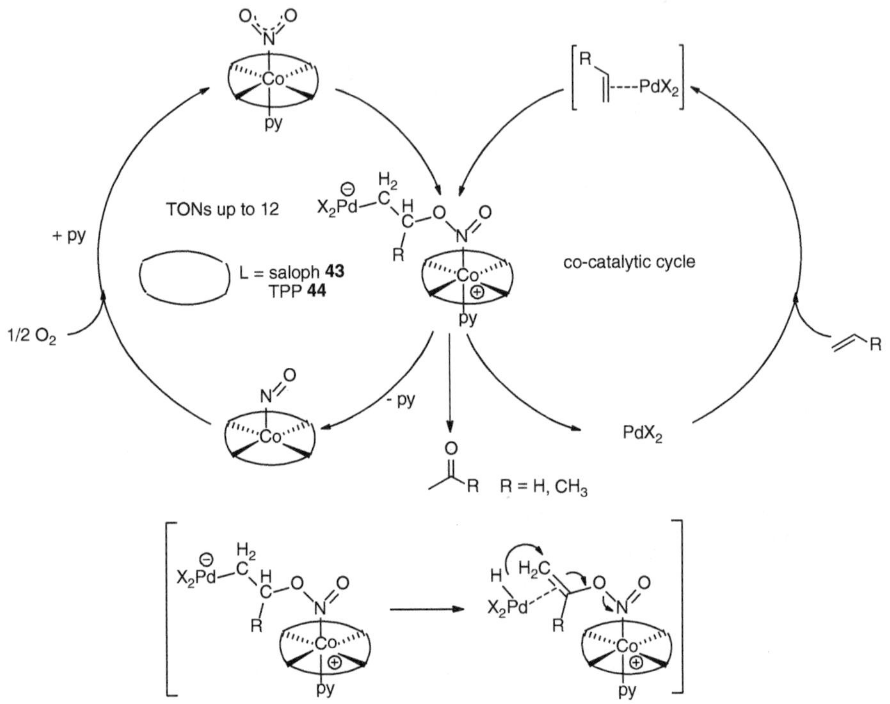

Scheme 55 Olefin oxidation by molecular O_2 based on the Co–NO_2/NO redox couple

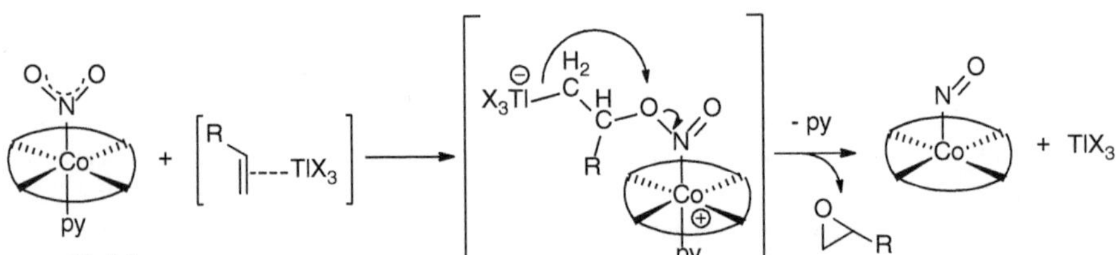

Scheme 56 Epoxidation of olefins catalyzed by Co–NO_2 complex using Tl(III) as co-catalyst

5.2 *Pd–NO_2/NO Redox Couples*

Olefin oxidations can also be accomplished by the palladium nitro complex [$(MeCN)_2PdCl(NO_2)$] (**45**), which has the capability to act at the same time as an oxygen atom transfer agent and olefin activator [116, 119–121]. Stoichiometric reaction of **45** and 1-decene under nitrogen atmosphere revealed the formation of palladium nitrosyl species identified as $[PdCl(NO)]_n$. At 60°C in air, the toluene solution of **45** catalyzed the oxidation of 1-decene to 2-decanone with a TON of 4. The same reaction carried out at room temperature afforded within 24 h a TON of 2.

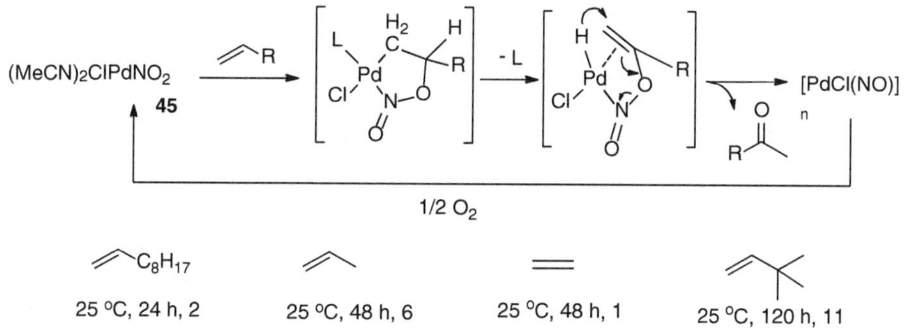

Scheme 57 Oxidation of olefins based on the Pd–NO_2/NO redox couple

The poor conversion was not only due to the inefficiency of the regeneration of the nitro complex from the nitrosyl one, but also was attributed to a side reaction forming 2-nitro-1-decene via nitro transfer instead of oxygen atom transfer to the olefin. Other terminal alkenes, such as ethylene, propylene, and *t*-butylethylene, have also been tested in **45** catalyzed oxidation reactions, as depicted in Scheme 57 [121]. The reaction with ethylene was the least efficient giving a TON of merely 1. The reaction with *t*-butylethylene exhibited the best performance, wherein the highest TON of 11 was reached. Regarding the mechanism of such a monometallic Pd–NO_2/NO redox couple-based olefin oxidation, an intramolecular nucleophilic oxygen atom transfer via a metallacyclic intermediate was proposed. Such heterometallacycles are in rapid equilibrium with the alkene coordinated nitro precursor. It undergoes β-hydride elimination most probably via the dissociation of an acetonitrile ligand followed by hydride transfer to afford the oxidized product concomitantly with the formation of the palladium nitrosyl complex [122].

Ketones are the principle products in the case of terminal alkenes. In contrast, with cyclic olefins, such as cyclopentene, cycloheptene and norbornene, epoxides were formed as the major products in the oxygen atom transfer reactions catalyzed by **45** [121]. This observation is significant, since it preludes the development of catalytic systems for direct epoxidation of olefins using both oxygen atoms of molecular O_2, which is contrast to the traditional methods converting only one oxygen atom of O_2 to generate the desired epoxide.

The formation of epoxide is markedly dependent on the ring size and stereochemistry. With cyclopentene, only a trace amount (0.15%) of the epoxide was produced by the palladium nitro complex **45** in 1,2-dichloroethane at 60°C within 3 h. In comparison, the reaction with cycloheptene afforded the epoxide in an improved TON of 1. In this case, a certain amount of α,β-unsaturated ketone was also generated. The selectivity could be enhanced by addition of catalytic amounts of a silver salt with non-coordinating anions. The highest TON of 7 was obtained in the oxygen atom transfer reaction with the activated bicyclic alkene norbornene at 60°C within 12 h. Significantly, the proposed metallacyclic intermediate could be isolated in the reaction of **45** with norbornene, which slowly decomposed yielding the corresponding epoxide, as depicted in Scheme 58 [121].

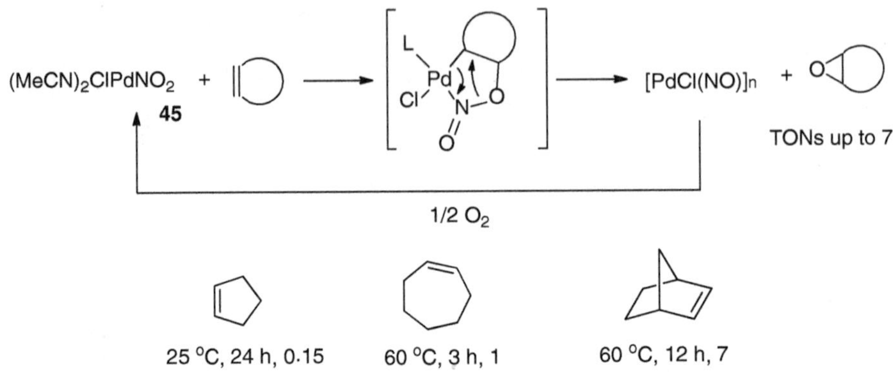

Scheme 58 Epoxidation of cyclic olefins with O_2 catalyzed by the Pd–NO_2 complex

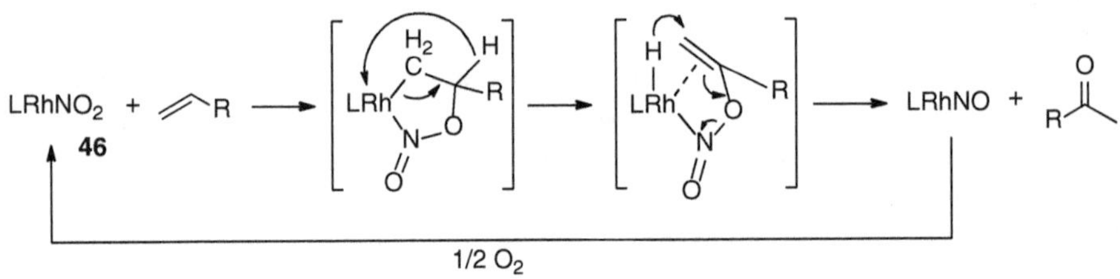

Scheme 59 Oxidation of olefins based on the Rh–NO_2/NO redox couple

5.3 *Rh–NO_2/NO Redox Couples*

Rhodium–nitro complexes have also been examined in the secondary oxygen atom transfer reaction based on Rh–NO_2/NO redox couples [118]. In this case, it was found that olefin oxidation can only be accomplished by cationic or dicationic rhodium nitro complexes in the general formula of $[LRhNO_2]^{n+}$ ($n = 1$ or 2), such as $[(MeCN)_4Rh(NO_2)][BF_4]_2$ (**46**). In acetonitrile, complex **46** undergoes oxygen atom transfer to 1-octene to afford 80% of 2-octanone at 60°C within 10 h. Using such a monometallic catalytic system the reaction most probably proceeds along an intramolecular pathway via initial formation of a rhodacyclic intermediate, which further undergoes β-hydride elimination and hydride transfer to afford the oxidized product accompanied by generation of the dicationic rhodium(I) nitrosyl complex $[(MeCN)_4Rh(NO)]$ (**47**). Oxidation of the nitrosyl species **47** by molecular oxygen occurs to regenerate the Rh(I) nitro precursor and closes the catalytic cycle, as depicted in Scheme 59. However, only poor TON values could be obtained based on such redox couples.

Interestingly, solvents of strong coordinating abilities turned out to be beneficial to the oxidization process from the nitrosyl to the nitro complex. While the dicationic rhodium(I) nitrosyl complex **47** undergoes oxidation with molecular oxygen, the neutral Rh(I) nitrosyl derivative $[(MeCN)Cl_2Rh(NO)]$ (**48**) appears to

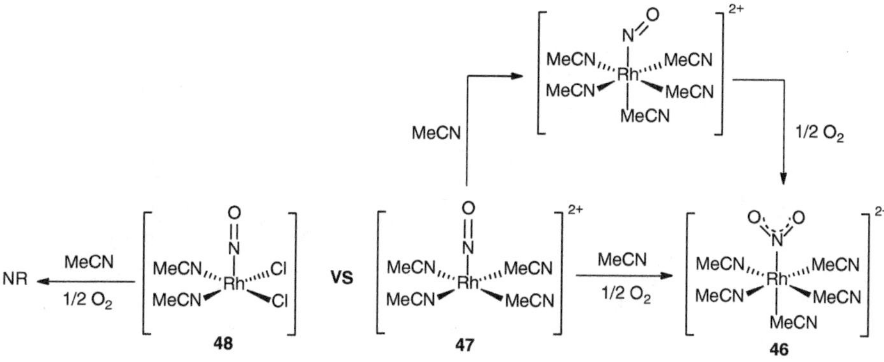

Scheme 60 Solvent triggered NO bending facilitating the oxidation of NO to NO_2

be inert toward oxygen and cannot regenerate the nitro species to render a catalytic cycle. This difference in reactivity was interpreted in terms of solvent triggered NO bending that facilitates oxidation of the nitrosyl ligand, as depicted in Scheme 60. In the case of the dicationic pentacoordinate complex **47**, a strongly coordinating solvent such as acetonitrile may convert the originally trigonal-pyramidal geometry containing an equatorial linear nitrosyl to a hexacoordinate, pseudo-octahedral complex possessing a bent nitrosyl in the axial position. Only the bent form of the nitrosyl ligand can be oxidized by molecular oxygen. In the case of the neutral pentacoordinate complex **48**, the central metal is apparently not electron-deficient enough exhibiting a very low tendency to coordinate even relatively basic ligands like acetonitrile. As a result, the complex retains a trigonal-pyramidal geometry with a linear and nonoxidizable nitrosyl ligand.

5.4 Ru–NO₂/NO Redox Couples

The Ru(III)–edta–NO/Ru(V)–edta–NO_2 redox couple (edta = ethylenediaminetetraacetate) was explored to act as a potential oxygen atom transfer agent in the oxidation of terminal olefins to ketones and cyclic olefins to epoxides [126, 127]. The nitrosyl complex [Ru(III)–edta–NO] (**49**) adopted a pseudo-octahedral geometry at the Ru center bearing a pentacoordinate edta ligand and the linear nitrosyl group located *trans* to one nitrogen atom of edta. This Ru-based redox couple turns out to be much more efficient than any of the previously developed oxygen atom transfer systems. In the presence of molecular oxygen, the nitrosyl Ru (III) complex **49** catalyzed the oxidation of 1-hexene to 2-hexan-2-one and cyclohexene to epoxide leading to TOFs of 55 and 44 h^{-1}, respectively. Such fast reactions provoked the kinetic studies, which indicated a first-order dependence of the rate on both the olefin and the catalyst concentration. Significantly, the rate of the reaction is half order with respect to molecular O_2 concentration. These kinetic observations pointed to a mechanism involving the formation of the nitro complex

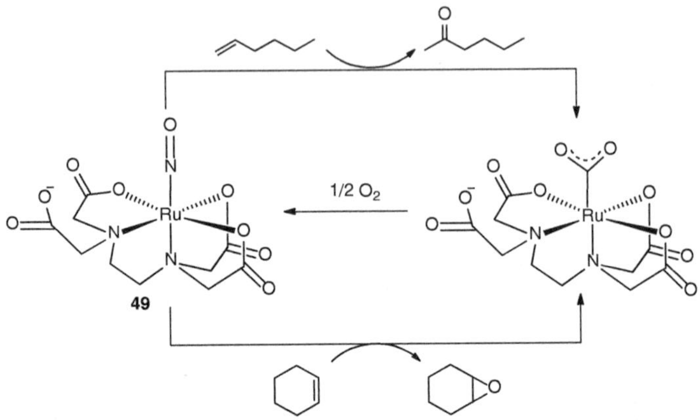

Scheme 61 Oxygen transfer reaction based on the Ru–NO_2/NO redox couple

by oxidization of the nitrosyl ligand followed by secondary coordination sphere oxygen atom transfer from the nitro group to the olefin. Similar to the monometallic systems mentioned before, a heterometallacycle was presumably involved, which led to different oxidation pathway depending on the type of the alkene (Scheme 61).

6 Conclusions

This comprehensive review described the present state of various catalytic applications of transition metal nitrosyl complexes. The nitrosyl ligand was found to play three major catalytically exploitable roles, namely as an ancillary ligand endowing *trans*-effect and *trans*-influence, as a non-innocent ligand undergoing reversible linear/bent transformations making vacant sites available, or as a redox-active ligand involved in oxygen atom transfer reactions with the "nitrosyl/nitro" redox couple. Many examples of this chapter demonstrated that the function as an ancillary ligand is dominant in homogeneous catalysis. The non-innocent function was recognized in coordination chemistry since long, but only recently it became successfully applied in catalysis. It can be predicted that the catalytic potential of non-innocent nitrosyl complexes is still huge and is waiting for appropriate exploitation. For the same token the "nitrosyl/nitro" redox couple applied in oxygen atom transfer reactions deserves re-investigation due to its predictably unique and beneficial reaction patterns. This chapter could indeed disclose a sound base for catalytically oriented nitrosyl chemistry providing great perspectives for revival and progress of this field.

Acknowledgments We are grateful to the financial support from the Swiss National Science Foundation, Lanxess AG, Leverkusen, Germany, the Funds of the University of Zurich, the DFG and SNF within the project "Forschergruppe FOR1175 – Unconventional Approaches to the Activation of Dihydrogen."

References

1. Cossee P (1964) Ziegler-Natta catalysis.1. Mechanism of polymerization of alpha-olefins with Ziegler-Natta catalysts. J Catal 3:80–88
2. Britovsek GJP, Gibson VC, Wass DF (1999) The search for New-generation olefin polymerization catalysts: life beyond metallocenes. Angew Chem Int Ed 38:428–447
3. Noyori R (2002) Asymmetric catalysis: science and opportunities (Nobel lecture). Angew Chem Int Ed 41:2008–2022
4. Noyori R, Hashiguchi S (1997) Asymmetric transfer hydrogenation catalyzed by chiral ruthenium complexes. Acc Chem Res 30:97–102
5. Noyori R, Kitamura M (1991) Enantioselective addition of organometallic reagents to carbonyl-compounds - chirality transfer, multiplication, and amplification. Angew Chem Int Ed 30:49–69
6. Noyori R, Yamakawa M, Hashiguchi S (2001) Metal-ligand bifunctional catalysis: a non-classical mechanism for asymmetric hydrogen transfer between alcohols and carbonyl compounds. J Org Chem 66:7931–7944
7. Noyori R, Ohkuma T (2001) Asymmetric catalysis by architectural and functional molecular engineering: practical chemo- and stereoselective hydrogenation of ketones. Angew Chem Int Ed 40:40–73
8. Trnka TM, Grubbs RH (2001) The development of L2X2Ru = CHR olefin metathesis catalysts: an organometallic success story. Acc Chem Res 34:18–29
9. Schrock RR (2006) Multiple metal-carbon bonds for catalytic metathesis reactions (Nobel lecture). Angew Chem Int Ed 45:3748–3759
10. Grubbs RH (2006) Olefin-metathesis catalysts for the preparation of molecules and materials (Nobel lecture). Angew Chem Int Ed 45:3760–3765
11. Miyaura N, Suzuki A (1995) Palladium-catalyzed cross-coupling reactions of organoboron compounds. Chem Rev 95:2457–2483
12. Beletskaya IP, Cheprakov AV (2000) The heck reaction as a sharpening stone of palladium catalysis. Chem Rev 100:3009–3066
13. Hassan J, Sevignon M, Gozzi C, Schulz E, Lemaire M (2002) Aryl-aryl bond formation one century after the discovery of the Ullmann reaction. Chem Rev 102:1359–1469
14. McCleverty JA (2004) Chemistry of nitric oxide relevant to biology. Chem Rev 104:403–418
15. Miranda KM (2005) The chemistry of nitroxyl (HNO) and implications in biology. Coord Chem Rev 249:433–455
16. Derry TTK, Williams TI (1993) A short history of technology: from the earliest time to A.D. 1900. Dover, Mineola
17. Enemark JH, Feltham RD (1974) Principles of structure, bonding, and reactivity for metal nitrosyl complexes. Coord Chem Rev 13:339–406
18. Hayton TW, Legzdins P, Sharp WB (2002) Coordination and organometallic chemistry of metal-NO complexes. Chem Rev 102:935–991
19. Richter-Addo GB, Legzdins P, Burstyn J (2002) Introduction: nitric oxide chemistry. Chem Rev 102:857–859
20. Ford PC, Lorkovic IM (2002) Mechanistic aspects of the reactions of nitric oxide with transition-metal complexes. Chem Rev 102:993–1017
21. Hoshino M, Laverman L, Ford PC (1999) Nitric oxide complexes of metalloporphyrins: an overview of some mechanistic studies. Coord Chem Rev 187:75–102
22. Grützmacher H (2008) Cooperating ligands in catalysis. Angew Chem Int Ed 47:1814–1818
23. Hindson K, de Bruin B (2012) Cooperative & redox non-innocent ligands in directing organometallic reactivity (Eur. J. Inorg. Chem. 3/2012). Eur J Inorg Chem 2012:340–342
24. Jiang Y, Hess J, Fox T, Berke H (2010) Rhenium hydride/boron lewis acid cocatalysis of alkene hydrogenations: activities comparable to those of precious metal systems. J Am Chem Soc 132:18233–18247

25. Kubas GJ (2007) Fundamentals of H-2 binding and reactivity on transition metals underlying hydrogenase function and H-2 production and storage. Chem Rev 107:4152–4205
26. Kubas GJ (2004) In: van Eldik R (ed) Advances in inorganic chemistry – including bioinorganic studies, vol 56. Advances in inorganic chemistry. Elsevier Academic Press, Amsterdam, pp 127–177
27. Heinekey DM, Lledos A, Lluch JM (2004) Elongated dihydrogen complexes: what remains of the H-H Bond? Chem Soc Rev 33:175–182
28. Heinekey DM, Oldham WJ (1993) Coordination chemistry of dihydrogen. Chem Rev 93:913–926
29. Bullock RM (2004) Catalytic ionic hydrogenations. Chem Eur J 10:2366–2374
30. Esteruelas MA, Oro LA (2001) The chemical and catalytic reactions of hydrido-chloro-carbonylbis (triisopropylphosphine)osmium(II) and its major derivatives. Adv Organomet Chem 47(47):1–59
31. Jiang Y, Schirmer B, Blacque O, Fox T, Grimme S, Berke H (2013) The "Catalytic Nitrosyl Effect": NO bending boosting the efficiency of rhenium based alkene hydrogenations. J Am Chem Soc 135(10):4088–4102
32. Llamazares A, Schmalle HW, Berke H (2001) Ligand-assisted heterolytic activation of hydrogen and silanes mediated by nitrosyl rhenium complexes. Organometallics 20:5277–5288
33. Gusev D, Llamazares A, Artus G, Jacobsen H, Berke H (1999) Classical and nonclassical nitrosyl hydride complexes of rhenium in various oxidation states. Organometallics 18:75–89
34. Goodwin J, Bailey R, Pennington W, Rasberry R, Green T, Shasho S, Yongsavanh M, Echevarria V, Tiedeken J, Brown C, Fromm G, Lyerly S, Watson N, Long A, De Nitto N (2001) Structural and oxo-transfer reactivity differences of hexacoordinate and pentacoordinate (nitro) (tetraphenylporphinato)cobalt(III) derivatives. Inorg Chem 40:4217–4225
35. Kurtikyan TS, Gulyan GM, Dalaloyan AM, Kidd BE, Goodwin JA (2010) Six-coordinate nitrosyl and nitro complexes of meso-tetratolylporphyrinatocobalt with trans sulfur-donor ligands. Inorg Chem 49:7793–7798
36. Afshar RK, Eroy-Reveles AA, Olmstead MM, Mascharak PK (2006) Stoichiometric and catalytic secondary O-atom transfer by Fe(III)-NO2 complexes derived from a planar tetradentate non-heme ligand: reminiscence of heme chemistry. Inorg Chem 45:10347–10354
37. Clapham SE, Hadzovic A, Morris RH (2004) Mechanisms of the H2-hydrogenation and transfer hydrogenation of polar bonds catalyzed by ruthenium hydride complexes. Coord Chem Rev 248:2201–2237
38. Berke H, Burger P (1994) Nitrosyl substituted hydride complexes - an activated class of compounds. Comments Inorg Chem 16:279–312
39. Jacobsen H, Berke H (2002) Tuning of the transition metal hydrogen bond: how do trans ligands influence bond strength and hydridicity? J Chem Soc Dalton Trans:3117–3122
40. Liang FP, Jacobsen H, Schmalle HW, Fox T, Berke H (2000) Carbonylhydridonitrosyltris (trimethylphosphine)molybdenum(0): an activated hydride complex. Organometallics 19:1950–1962
41. Liang FP, Schmalle HW, Fox T, Berke H (2003) Hydridic character and reactivity of Di 1,2-bis (dimethylphosphino)ethane hydridonitrosylmolybdenum(0). Organometallics 22:3382–3393
42. Chen Z, Schmalle HW, Fox T, Berke H (2005) Insertion reactions of hydridonitrosyltetrakis (trimethylphosphine) tungsten(0). Dalton Trans:580–587
43. Zhao Y, Schmalle HW, Fox T, Blacque O, Berke H (2006) Hydride transfer reactivity of tetrakis(trimethylphosphine) (hydrido)(nitrosyl)molybdenum(0). Dalton Trans:73–85
44. Dybov A, Blacque O, Berke H (2010) Molybdenum and tungsten nitrosyl complexes in hydrogen activation. Eur J Inorg Chem:3328–3337
45. Robinson SD, Wilkinson G (1966) New diene and carbonyl complexes of ruthenium. J Chem Soc A:300–301
46. Osborn JA, Jardine FH, Young JF, Wilkinson G (1966) Preparation and properties of tris (triphenylphosphine)halogenorhodium(1) and some reactions thereof including catalytic

homogeneous hydrogenation of olefins and acetylenes and their derivatives. J Chem Soc A:1711–1729
47. Bullock RM, Song JS (1994) Ionic hydrogenations of hindered olefins at low-temperature - hydride transfer-reactions of transition-metal hydrides. J Am Chem Soc 116:8602–8612
48. Luan L, Song JS, Bullock RM (1995) Ionic hydrogenation of alkynes by HOTF and CP(CO)(3)WH. J Org Chem 60:7170–7176
49. Song JS, Szalda DJ, Bullock RM, Lawrie CJC, Rodkin MA, Norton JR (1992) Hydride transfer by hydride transition-metal complexes - ionic hydrogenation of aldehydes and ketones, and structural characterization of an alcohol complex. Angew Chem Int Ed 31:1233–1235
50. Dybov A, Blacque O, Berke H (2011) Molybdenum nitrosyl complexes and their application in catalytic imine hydrogenation reactions. Eur J Inorg Chem:652–659
51. Schnider P, Koch G, Pretot R, Wang GZ, Bohnen FM, Kruger C, Pfaltz A (1997) Enantioselective hydrogenation of imines with chiral (phosphanodihydrooxazole)iridium catalysts. Chem Eur J 3:887–892
52. Jiang Y, Blacque O, Fox T, Frech CM, Berke H (2009) Highly selective dehydrogenative silylation of alkenes catalyzed by rhenium complexes. Chem Eur J 15:2121–2128
53. Jiang Y, Blacque O, Fox T, Frech CM, Berke H (2009) Development of rhenium catalysts for amine borane dehydrocoupling and transfer hydrogenation of olefins. Organometallics 28:5493–5504
54. Jiang Y, Berke H (2007) Dehydrocoupling of dimethylamine-borane catalysed by rhenium complexes and its application in olefin transfer-hydrogenations. Chem Commun:3571–3573
55. Hamilton CW, Baker RT, Staubitz A, Manners I (2009) B-N compounds for chemical hydrogen storage. Chem Soc Rev 38:279–293
56. Marder TB (2007) Will we soon be fueling our automobiles with ammonia-borane? Angew Chem Int Ed 46:8116–8118
57. Denney MC, Pons V, Hebden TJ, Heinekey DM, Goldberg KI (2006) Efficient catalysis of ammonia borane dehydrogenation. J Am Chem Soc 128:12048–12049
58. Pons V, Baker RT, Szymczak NK, Heldebrant DJ, Linehan JC, Matus MH, Grant DJ, Dixon DA (2008) Coordination of aminoborane, NH2BH2, dictates selectivity and extent of H-2 release in metal-catalysed ammonia borane dehydrogenation. Chem Commun:6597–6599
59. Jaska CA, Temple K, Lough AJ, Manners I (2003) Transition metal-catalyzed formation of boron-nitrogen bonds: catalytic dehydrocoupling of amine-borane adducts to form aminoboranes and borazines. J Am Chem Soc 125:9424–9434
60. Clark TJ, Russell CA, Manners I (2006) Homogeneous, titanocene-catalyzed dehydrocoupling of amine-borane adducts. J Am Chem Soc 128:9582–9583
61. Corey JY, Braddock-Wilking J (1999) Reactions of hydrosilanes with transition-metal complexes: formation of stable transition-metal silyl compounds. Chem Rev 99:175–292
62. Marciniec B (2005) Catalysis by transition metal complexes of alkene silylation - recent progress and mechanistic implications. Coord Chem Rev 249:2374–2390
63. Marciniec B, Pietraszuk C (1997) Silylation of styrene with vinylsilanes catalyzed by RuCl(SiR3)(CO)(PPh3)(2) and RuHCl(CO)(PPh3)(3). Organometallics 16:4320–4326
64. LaPointe AM, Rix FC, Brookhart M (1997) Mechanistic studies of palladium(II)-catalyzed hydrosilation and dehydrogenative silation reactions. J Am Chem Soc 119:906–917
65. Christ ML, Saboetienne S, Chaudret B (1995) Highly selective dehydrogenative silylation of ethylene using the bis(dihydrogen) complex RUH2(H-2)(2)(PCY(3))(2) as catalyst precursor. Organometallics 14:1082–1084
66. Gutmann V (1976) Solvent effects on reactivities of organometallic compounds. Coord Chem Rev 18:225–255
67. Beckett MA, Brassington DS, Coles SJ, Hursthouse MB (2000) Lewis acidity of tris(pentafluorophenyl) borane: crystal and molecular structure of B(C(6)F(5))(3)center dot OPEt(3). Inorg Chem Commun 3:530–533

68. van der Veen LA, Keeven PH, Schoemaker GC, Reek JNH, Kamer PCJ, van Leeuwen P, Lutz M, Spek AL (2000) Origin of the bite angle effect on rhodium diphosphine catalyzed hydroformylation. Organometallics 19:872–883
69. Kranenburg M, Vanderburgt YEM, Kamer PCJ, Vanleeuwen P, Goubitz K, Fraanje J (1995) New diphosphine ligands based on heterocyclic aromatics inducing very high regioselectivity in rhodium-catalyzed hydroformylation - effect of the bite angle. Organometallics 14:3081–3089
70. Dudle B, Rajesh K, Blacque O, Berke H (2011) Rhenium nitrosyl complexes bearing large-bite-angle diphosphines. Organometallics 30:2986–2992
71. Dudle B, Rajesh K, Blacque O, Berke H (2011) Rhenium in homogeneous catalysis: ReBrH (NO)(labile ligand) (large-bite-angle diphosphine) complexes as highly active catalysts in olefin hydrogenations. J Am Chem Soc 133:8168–8178
72. Grey RA, Pez GP, Wallo A (1981) Anionic metal hydride catalysts.2. Application to the hydrogenation of ketones, aldehydes, carboxylic-acid esters, and nitriles. J Am Chem Soc 103:7536–7542
73. Reguillo R, Grellier M, Vautravers N, Vendier L, Sabo-Etienne S (2010) Ruthenium-catalyzed hydrogenation of nitriles: insights into the mechanism. J Am Chem Soc 132 (23):7854–7855
74. Enthaler S, Addis D, Junge K, Erre G, Beller M (2008) A general and environmentally benign catalytic reduction of nitriles to primary amines. Chem Eur J 14:9491–9494
75. Rajesh K, Dudle B, Blacque O, Berke H (2011) Homogeneous hydrogenations of nitriles catalyzed by rhenium complexes. Adv Synth Catal 353:1479–1484
76. Morris RH (2009) Asymmetric hydrogenation, transfer hydrogenation and hydrosilylation of ketones catalyzed by iron complexes. Chem Soc Rev 38:2282–2291
77. Nishiyama H, Kondo M, Nakamura T, Itoh K (1991) Highly enantioselective hydrosilylation of ketones with chiral and c2-symmetrical bis(oxazolinyl)pyridine-rhodium catalysts. Organometallics 10:500–508
78. Shimada T, Mukaide K, Shinohara A, Han JW, Hayashi T (2002) Asymmetric synthesis of 1-aryl-1,2-ethanediols from arylacetylenes by palladium-catalyzed asymmetric hydrosilylation as a key step. J Am Chem Soc 124:1584–1585
79. Sprengers JW, de Greef M, Duin MA, Elsevier CJ (2003) Stable platinum(0) catalysts for catalytic hydrosilylation of styrene and synthesis of Pt(Ar-bian)(eta(2)-alkene) complexes. Eur J Inorg Chem:3811–3819
80. Choualeb A, Maccaroni E, Blacque O, Schmalle HW, Berke H (2008) Rhenium nitrosyl complexes for hydrogenations and hydrosilylations. Organometallics 27:3474–3481
81. Dong H, Berke H (2009) A convenient and efficient rhenium-catalyzed hydrosilylation of ketones and aldehydes. Adv Synth Catal 351:1783–1788
82. Jiang Y, Blacque O, Berke H (2011) Probing the catalytic potential of chloro nitrosyl rhenium (I) complexes. Dalton Trans 40:2578–2587
83. Casey CP, Singer SW, Powell DR, Hayashi RK, Kavana M (2001) Hydrogen transfer to carbonyls and imines from a hydroxycyclopentadienyl ruthenium hydride: evidence for concerted hydride and proton transfer. J Am Chem Soc 123:1090–1100
84. Shvo Y, Czarkie D, Rahamim Y, Chodosh DF (1986) A new group of ruthenium complexes - structure and catalysis. J Am Chem Soc 108:7400–7402
85. Blum Y, Czarkie D, Rahamim Y, Shvo Y (1985) (Cyclopentadienone)ruthenium carbonyl-complexes - a new class of homogeneous hydrogenation catalysts. Organometallics 4:1459–1461
86. Landwehr A, Dudle B, Fox T, Blacque O, Berke H (2012) Bifunctional rhenium complexes for the catalytic transfer-hydrogenation reactions of ketones and imines. Chem Eur J 18:5701–5714
87. Berke H (2010) Conceptual approach to the reactivity of dihydrogen. Chemphyschem 11:1837–1849

88. Berke H, Jiang Y, Yang X, Jiang C, Chakraborty S, Landwehr A (2013) Coexistence of Lewis acid and base functions: a generalized view of the frustrated lewis pair concept with novel implications for reactivity. Top Curr Chem 334. doi:10.1007/128_2012_400
89. Choualeb A, Lough AJ, Gusev DG (2007) Hydridic rhenium nitrosyl complexes with pincer-type PNP ligands. Organometallics 26:3509–3515
90. Choualeb A, Lough AJ, Gusev DG (2007) Hemilabile pincer-type hydride complexes of iridium. Organometallics 26:5224–5229
91. Plietker B, Dieskaul A (2009) The reincarnation of the Hieber anion Fe(CO)(3)(NO) (−) - a new venue in nucleophilic metal catalysis. Eur J Org Chem:775–787
92. Jegelka M, Plietker B (2011) In: Plietker B (ed) Iron catalysis: fundamentals and applications, vol 33. Topics in Organometallic Chemistry. Springer, Heidelberg, pp 177–213
93. Plietker B (2010) Sustainability in catalysis - concept or contradiction? Synlett:2049–2058
94. Roustan JL, Abedini M, Baer HH (1989) Catalytic alkylations of allylic carbonates under argon in the presence of nitrosylcarbonyliron complexes. J Organomet Chem 376:C20–C22
95. Zhou B, Xu YY (1988) Studies on the enantioselectivity in BU4N FE(CO)3NO -catalyzed nucleophilic-substitution of optically-active allylic carbonates with malonate. J Org Chem 53:4419–4421
96. Plietker B (2006) A highly regioselective salt-free iron-catalyzed allylic alkylation. Angew Chem Int Ed 45:1469–1473
97. Plietker B (2006) Regioselective iron-catalyzed allylic amination. Angew Chem Int Ed 45:6053–6056
98. Jegelka M, Plietker B (2009) Selective C-S bond formation via Fe-catalyzed allylic substitution. Org Lett 11:3462–3465
99. Plietker B, Dieskau A, Moews K, Jatsch A (2008) Ligand-dependent mechanistic dichotomy in iron-catalyzed allylic substitutions: sigma-allyl versus pi-allyl mechanism. Angew Chem Int Ed 47:198–201
100. Magens S, Plietker B (2010) Nucleophilic iron catalysis in transesterifications: scope and limitations. J Org Chem 75:3715–3721
101. Dieskau AP, Plietker B (2011) A mild ligand-free iron-catalyzed liberation of alcohols from allylcarbonates. Org Lett 13:5544–5547
102. Holzwarth MS, Alt I, Plietker B (2012) Catalytic activation of diazo compounds using electron-rich, defined iron complexes for carbene-transfer reactions. Angew Chem Int Ed 51:5351–5354
103. Seebach D (1979) Methods of reactivity umpolung. Angew Chem Int Ed 18:239–258
104. Dieskau AP, Holzwarth MS, Plietker B (2012) Fe-catalyzed allylic C-C-bond activation: vinylcyclopropanes as versatile a1, a3, d5-synthons in traceless allylic substitutions and 3+2 -cycloadditions. J Am Chem Soc 134:5048–5051
105. Huang WJ, Berke H (2005) Rhenium complexes as highly active catalysts for the hydrosilylation of carbonyl compounds. Chimia 59:113–115
106. Mewald M, Froehlich R, Oestreich M (2011) An axially chiral, electron-deficient borane: synthesis, coordination chemistry, lewis acidity, and reactivity. Chem Eur J 17:9406–9414
107. Rendler S, Oestreich M (2008) Conclusive evidence for an S(N)2-Si mechanism in the B(C (6)F(5))(3)-catalyzed hydrosilylation of carbonyl compounds: implications for the related hydrogenation. Angew Chem Int Ed 47:5997–6000
108. Parks DJ, Blackwell JM, Piers WE (2000) Studies on the mechanism of B(C6F5)(3)-catalyzed hydrosilation of carbonyl functions. J Org Chem 65:3090–3098
109. Blackwell JM, Sonmor ER, Scoccitti T, Piers WE (2000) B(C6F5)(3)-catalyzed hydrosilation of imines via silyliminium intermediates. Org Lett 2:3921–3923
110. Mayr H, Basso N, Hagen G (1992) Kinetics of hydride transfer-reactions from hydrosilanes to carbenium ions - substituent effects in silicenium ions. J Am Chem Soc 114:3060–3066
111. Egbert JD, Bullock RM, Heinekey DM (2007) Cationic dihydrogen/dihydride complexes of osmium: structure and dynamics. Organometallics 26:2291–2295

112. Jessop PG, Morris RH (1992) Reactions of transition-metal dihydrogen complexes. Coord Chem Rev 121:155–284
113. Tovrog BS, Diamond SE, Mares F (1979) Oxidation of ruthenium coordinated alcohols by molecular-oxygen to ketones and hydrogen-peroxide. J Am Chem Soc 101:5067–5069
114. Tovrog BS, Diamond SE, Mares F (1979) Oxygen-transfer from ligands - cobalt nitro complexes as oxygenation catalysts. J Am Chem Soc 101:270–272
115. Tovrog BS, Mares F, Diamond SE (1980) Cobalt-nitro complexes as oxygen-transfer agents - oxidation of olefins. J Am Chem Soc 102:6616–6618
116. Andrews MA, Kelly KP (1981) The transition-metal nitro-nitrosyl redox couple - catalytic-oxidation of olefins to ketones. J Am Chem Soc 103:2894–2896
117. Tovrog BS, Diamond SE, Mares F, Szalkiewicz A (1981) Activation of cobalt-nitro complexes by lewis-acids - catalytic-oxidation of alcohols by molecular-oxygen. J Am Chem Soc 103:3522–3526
118. Diamond SE, Mares F, Szalkiewicz A, Muccigrosso DA, Solar JP (1982) Cobalt nitro complexes as oxygen-transfer agents. 4. Epoxidation of olefins. J Am Chem Soc 104:4266–4268
119. Andrews MA, Chang TCT, Cheng CWF, Emge TJ, Kelly KP, Koetzle TF (1984) Synthesis, characterization, and equilibria of palladium(ii) nitrile, alkene, and heterometallacy-clopentane complexes involved in metal nitro catalyzed alkene oxidation reactions. J Am Chem Soc 106:5913–5920
120. Andrews MA, Chang TCT, Cheng CWF, Kapustay LV, Kelly KP, Zweifel MJ (1984) Nitration of alkenes by palladium nitro complexes. Organometallics 3:1479–1484
121. Andrews MA, Chang TCT, Cheng CWF, Kelly KP (1984) A new approach to the air oxidation of alkenes employing metal nitro complexes as catalysts. Organometallics 3:1777–1785
122. Andrews MA, Chang TCT, Cheng CWF (1985) Observations regarding the mechanisms of o atom transfer from metal nitro ligands to oxidizable substrates. Organometallics 4:268–274
123. Clarkson SG, Basolo F (1973) Study of reaction of some cobalt nitrosyl complexes with oxygen. Inorg Chem 12:1528–1534
124. Takacs JM, Jiang XT (2003) The Wacker reaction and related alkene oxidation reactions. Curr Org Chem 7:369–396
125. Jira R (2009) Acetaldehyde from ethylene—a retrospective on the discovery of the Wacker process. Angew Chem Int Ed 48:9034–9037
126. Khan MMT, Venkatasubramanian K, Shirin Z, Bhadbhade MM (1992) (Ethylenediaminete-traacetato)nitrosylruthenium, an efficient oxygen-atom transfer agent. J Chem Soc Dalton Trans:1031–1035
127. Khan MMT, Venkatasubramanian K, Shirin Z, Bhadbhade MM (1992) Mixed-ligand complexes of ruthenium(III) and ruthenium(II) with ethylenediaminetetraacetate and bidentate phosphines and arsines. J Chem Soc Dalton Trans:885–890

Index

FSC
www.fsc.org
MIX
Papier | Fördert gute Waldnutzung
FSC® C083411

Kolibri 360 GmbH
Ferdinand-Jühlke-Straße 7
99095 Erfurt, Deutschland
produktsicherheit@kolibri360.de